Microwave NDT

DEVELOPMENTS IN
ELECTROMAGNETIC THEORY AND APPLICATIONS

VOLUME 10

Editor

G. F. Roach, *Strathclyde University, Department of Mathematics, UK*

The titles published in this series are listed at the end of this volume.

Microwave NDT

by

N. Ida
Department of Electrical Engineering,
University of Akron, Ohio, U.S.A.

KLUWER ACADEMIC PUBLISHERS
DORDRECHT / BOSTON / LONDON

Library of Congress Cataloging-in-Publication Data

Ida, Nathan.
Microwave NDT / by N. Ida.
p. cm. -- (Developments in electromagnetic theory and applications ; 10)
Includes index.
ISBN 0-7923-2007-7 (alk. paper)
1. Non-destructive testing. 2. Microwave measurements.
I. Title. II. Series.
TA417.2.I33 1992
620.1'127--dc20 92-33611

ISBN 0-7923-2007-7

Published by Kluwer Academic Publishers,
P.O. Box 17, 3300 AA Dordrecht, The Netherlands.

Kluwer Academic Publishers incorporates
the publishing programmes of
D. Reidel, Martinus Nijhoff, Dr W. Junk and MTP Press.

Sold and distributed in the U.S.A. and Canada
by Kluwer Academic Publishers,
101 Philip Drive, Norwell, MA 02061, U.S.A.

In all other countries, sold and distributed
by Kluwer Academic Publishers Group,
P.O. Box 322, 3300 AH Dordrecht, The Netherlands.

Printed on acid-free paper

Printed in the Netherlands

Contents

Part II. Microwave Techniques and Devices

Chapter 4. Microwave Measurement Techniques

Chapter 5. Microwave Sources, Sensors, and Devices

Part III. Testing

Chapter 8.
Testing by Monitoring Material Properties

Part IV. Modeling of the Testing Environment

Chapter 9.
Methods of Modeling 274

Appendix A. Vector Relations

Preface

It is perhaps curious that a technique as developed as microwave testing has been given only scant attention as a method for nondestructive testing of materials. This is particularly so since much of the existing work with microwaves is directly applicable to nondestructive testing. In fact, microwave techniques have been used in a large number of applications that can be classified as nondestructive testing applications, ranging from large scale remote sensing to detection of tumors in the body. The instrumentation is also easily available and, unlike for other methods of *NDT*, require little modification. Perhaps the main reason for this state is the fact that the resolution one can expect of microwave testing is of the order of about half a wavelength which, with standard microwave equipment, is of the order of at most *1mm* (at *100 GHz* the wavelength is *3 mm*). Other reasons are that microwave measurements are viewed as noisy and "difficult" to perform. While some of these reasons were relevant in the past, they have been largely overcome. Current instrumentation is excellent, computer controlled, and with excellent noise figures. Microwave sources are more stable than have ever been and equipment in the *mm*-wave region is now available, further reducing the resolution to a fraction of a *mm*. This is fortunate since microwave testing offers some attractive features, especially for testing of composite and other nonmetallic materials.

Applications of microwaves to nondestructive testing of materials date back to at least the early 1950's. There were a number of attempts at testing of plastics, paper products, and composite materials. This activity has intensified with increased use of microwaves in processing of materials and availability of better, more accurate equipment. However, surprisingly, little has been written on microwave nondestructive testing. A notable exception is a small book written in 1982 by A.J. Bahr ("Microwave Nondestructive Testing Methods", Gordon and Breach Science Publishers, 1982).

The present volume attempts a unified approach to microwave nondestructive testing by presenting the three essential components of testing: theory, practice, and modeling. While recognizing that each of these subjects is wide enough to justify a volume of its own, the presentation of the three topics together is intended to show that these are interrelated and should be practiced together. While few will argue against a good theoretical background, modeling and simulation of the testing environment is seldom part of the *NDT* training in any method, but particularly so in microwave testing.

The text is divided into four parts. The first part presents the field theory background necessary for understanding the microwave domain. This includes chapters 1, 2, and 3. The second part, treats microwave measurements as well as devices and sources and includes chapters 4 and 5. Chapters 6, 7, and 8 discuss practical tests applicable to a variety of materials and geometries. The fourth part discusses modeling of microwave testing and consists of chapters 9, 10, and 11.

Chapter 1 presents Maxwell's equations in differential and integral forms, as well as the various material properties, potential functions, and derived equations. Aspects of field theory required for work in the microwave domain such as propagation of waves in materials, complex permittivity and energy relations are emphasized.

Aspects of electromagnetic field theory relating to transmission lines, waveguides and electromagnetic cavity resonators are given in chapter 2. Transmission line theory is presented first followed by TE and TM modes in rectangular and cylindrical waveguides. The properties of cavity resonators are discussed mainly from an energy point of view with emphasis on perturbation techniques.

Chapter 3 discusses reflection, transmission and scattering of waves in lossless, low loss, and lossy dielectrics. A variety of topics, including polarization, propagation between materials, reflection and transmission, coefficients, and propagation through layered materials are discussed as used in testing.

Microwave measurement techniques is the subject of chapter 4. While the topic of microwave measurements is extremely wide, the discussion here emphasizes those measurements that are commonly used for testing, from a very basic point of view. Only the techniques are given without attempting to introduce instrumentation. Measurements are described in simple block diagrams which allow understanding of the measuring technique without the need to discuss complicated equipment such as spectrum or network analyzers. While these instruments can, and often are used to accomplish the measurements described here, the diagrams are given in terms of the basic functions of measurement of power, amplitude, phase, frequency and the like.

Chapter 5 is a short chapter in which a number of microwave devices are described. As with previous chapters, only those aspects of microwave power generation, devices and instrumentation that are likely to be used in a nondestructive testing environment are given. Thus, magnetrons and klystrons are discussed because these are so common while solid state devices are given because they are the choice for simple, low cost, low power microwave sources for NDT. A number of passive devices used for measurement setup are described as well.

Chapter 6 is the first of three chapters on testing, discussing methods of dimensional testing. The various methods and test parameters are presented but, while all methods of testing are given, the emphasis in this chapter is on reflectometry and related topics (such as standing wave methods) since this is a universal method for testing of dimensional parameters. Other methods such as transmission, attenuation, scattering and resonant methods are also given but these are emphasized in following chapters where they are more useful. The chapter also contains a general discussion on testing which serves as an introduction to chapters 6, 7 and 8.

Chapter 7 discusses testing for discontinuities and emphasizes scattering methods. Detection of flaws and location of discontinuities are discussed from a general scattering point of view. In doing so, various techniques and notation used in radar work are borrowed. This, in effect, shows testing for discontinuity to be any means of detecting and measuring the scattered field.

In chapter 8, we emphasize resonant methods of testing. Resonant methods rely on detection of variation of material properties and are well adapted for testing of bulk properties such as moisture content or degree of curing in materials. However, the subject of the chapter is testing for material properties and a variety of other methods can be used for this purpose, including attenuation, transmission and reflection. These methods are also covered.

Chapter 9 introduces modeling for nondestructive testing purposes. The presentation is divided into finite difference, finite element and method of moments to emphasize three different, general methods for modeling. The basic methods are introduced here and some of these, will be then expanded upon in chapters 10 and 11. Some very simple examples of modeling are given to show that modeling need not be complicated.

The time-dependent wave equation and modeling of testing environments related to time-dependent wave propagation are introduced in chapter 10. The finite difference method is emphasized as the method of choice for time-dependent applications. Two-dimensional and axi-symmetric formulations are given first with examples from testing of composite materials. These are then extended to three dimensional modeling.

Chapter 11 gives a number of formulations for time-harmonic applications, using the finite element method. 2-D, 3-D and axisymmetric methods and applications are given including calculation of shift in resonant frequencies, and modes in cavities and waveguides.

In addition, a short chapter (chapter 12) gives information on material properties as well as describing hyperbolic and exponential functions and the Euler equation, in the form of short appendices.

Each chapter contains a bibliography intended to expand on the material given and, in particular, to point to subjects which could not be covered either as not appropriate or for lack of space.

Introduction

1. The Microwave Domain

The electromagnetic spectrum in Table 1 identifies the microwave domain roughly between *300 MHz* and *300 GHz*. This definition is somewhat arbitrary but has been followed for descriptive purposes and, to a large extent, for commercial purposes. At the lower end of the band, microwaves overlaps the *UHF* band and at the higher end, they border the far infrared region. In addition to the general domain, microwaves are often identified with subdomains or bands. These are designated by letters to allow simple identification of equipment and operating frequencies. The standard designations are shown in Table 2 and are known as microwave frequency bands. There is a separate military band designation shown in Table 3 and an IEEE band designation shown in Table 4. The relative location of the microwave domain in the electromagnetic spectrum is shown schematically in Figure 1.

Table 1. The electromagnetic spectrum.

Frequency		Designation	Usage
.003 - .03	*(MHz)*	*VLF*	
.03 - .3	*(MHz)*	*LF*	LF Radio
0.3 - 3	*(MHz)*	*MF*	AM radio
3 - 30	*(MHz)*	*HF*	SW radio
30 - 300	*(MHz)*	*VHF*	TV, FM radio
300 - 3,000	*(MHz)*	*UHF*	TV, Radar
3 - 30	*(GHz)*	*SHF*	Satellites, Radar
30 - 300	*(GHz)*	*EHF*	Radar
300 - 3,000	*(GHz)*	*mm and sub-mm*	
3 - 420	*(THz)*	*Infrared*	
420 - 790	*(THz)*	*Visible Light*	
790 THz - 30 PHz		*Ultraviolet*	
30 PHz - 3,000 EHz		*X rays*	

$MHz=10^6$ Hz, $GHz=10^9$ Hz, $THz=10^{12}$ Hz, $PHz=10^{15}$ Hz, $EHz=10^{18}$ Hz

Table 2. Band designation in the microwave region

Designated Letter	Frequency in *GHz*	Wavelength in *cm.* (in vacuum)
P	*.225 - .390*	*133.3-76.9*
L	*.390 - 1.55*	*76.9 - 19.3*
S	*1.55 - 5.2*	*19.3 - 5.77*
X	*5.2 - 10.9*	*5.77 - 2.75*
K	*10.9 - 36.0*	*2.75 - 0.834*
Q	*36.0 - 46.0*	*0.834 - 0.652*
V	*46.0 - 56.0*	*0.652 - 0.536*
W	*56.0 - 100.0*	*0.536 - 0.3*

Table 3. New military microwave band designation

Designated Letter	Frequency in *GHz*	Wavelength in *cm.* (in vacuum)
A	*0.1-0.25*	*300.0 - 120.0*
B	*0.25-0.5*	*120.0 - 60.0*
C	*0.5-1.0*	*60.0 - 30.0*
D	*1.0-2.0*	*30.0 - 15.0*
E	*2.0-3.0*	*15.0 - 10.0*
F	*3.0-4.0*	*10.0 - 7.5*
G	*4.0-6.0*	*7.5 - 5.0*
H	*6.0-8.0*	*5.0 - 3.75*
I	*8.0-10.0*	*3.75 - 3.0*
J	*10.0-20.0*	*3.0 - 1.5*
K	*20.0-40.0*	*1.5 - 0.75*
L	*40.0-60.0*	*0.75 - 0.5*
M	*60.0-100.0*	*0.5 - 0.3*

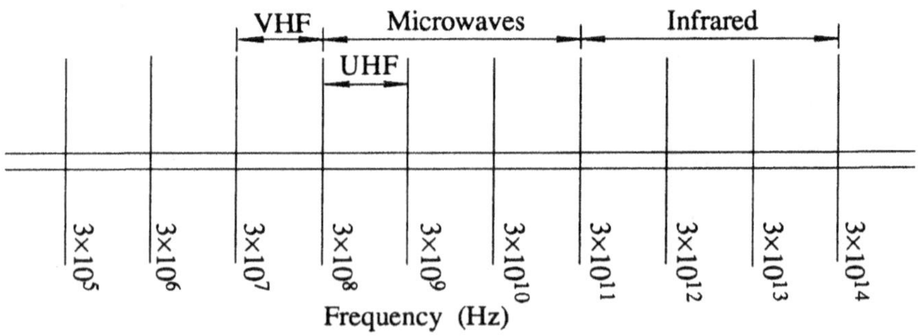

Figure 1. Relative location of the microwave domain in the electromagnetic spectrum.

Table 4. IEEE microwave band designation.

Designated Letters	Frequency in *GHz*	Wavelength in *m*. (in vacuum)
HF	*0.003-0.030*	*100.0 - 10.0*
VHF	*0.030-0.300*	*10.0 - 1.0*
UHF	*0.300-1.0*	*1.0 - 0.3*
L	*1.0-2.0*	*0.3 - 0.15*
S	*2.0-4.0*	*0.15 - 0.075*
C	*4.0-8.0*	*0.075 - 0.0375*
X	*8.0-12.0*	*0.0375 - 0.025*
Ku	*12.0-18.0*	*0.025 - 0.0167*
K	*18.0-27.0*	*0.0167 - 0.011*
Ka	*27.0-40.0*	*0.011 - 0.0075*
Millimeter	*40.0-300.0*	*0.0075 - 0.001*
Submillimeter	*300.0-*	*0.001 -*

As examples of various uses, many microwave ovens operate in the *L* and *S* bands of Table 1 (a typical frequency is *2.450 GHz*). Radar detectors used by police operate in the *X* and *K* bands. Communication with satellites is normally done in the *C* and *K* bands (*4* to *30 GHz*).

2. Historical

It is of some interest to note that the first electromagnetic experiment to show the existence of waves as predicted by Maxwell is, in a way, a microwave experiment. When Heinrich Hertz used a spark transmitter to transmit electromagnetic waves, the wavelengths he used were smaller than *1m* (about *60 cm*), well within the microwave range. Propagation of waves in waveguides was shown theoretically as early as 1897 by Lord Rayleigh [1], even though they could not be realized experimentally at the time for lack of appropriate generators. However, three years earlier, Sir Oliver Lodge, observed wave guiding when he surrounded a spark generator of the type used by Hertz by a conducting tube. Waveguides were not actually used until about 1936. The first reliable microwave sources were developed in the 1930's for radar applications. The first device was the magnetron, a microwave source that, in one form or another, still finds applications in high power microwave systems and, in vast quantities, in microwave ovens. The development of the klystron in the 1930's [2] was another important step, because it could be used as a microwave amplifier. The importance of these two microwave sources is amply proven by their continuous use ever since their invention. Solid state microwave devices are coming into wide use, especially for low power applications, but their earliest use can be traced to the use of the crystal diode in the 1930's. Active devices became available in the 1960's with the development of negative resistance devices and field effect transistors (*FET's*). The use of gallium arsenide (*GaAs*) and indium phosphide (*InP*) have also benefited this aspect of solid state device development. Currently there are a number of simple, reliable and inexpensive microwave devices including monolithic microwave integrated circuits (*MMIC's*) that the designer can use.

Passive devices were also developed in parallel with the development of microwave sources. These include a variety of antennas, couplers, filters, attenuators, waveguides, cavity resonators, absorbers, terminators, rotators and others.

Microwave circuit development followed steps similar to other circuits at lower frequencies. As examples, the idea of heterodyne receivers was used starting with the earliest radar equipment, while microwave integrated circuits find applications in many communication systems.

Measurement equipment was, for a long time, one of the stumbling blocks of microwave work. This however has been largely solved by the availability of computer controlled network and spectrum analyzers. Modern microwave equipment is as reliable and as accurate as any other equipment although, in most cases, it tends to be more expensive.

The use of microwaves for nondestructive testing of materials started in the 1950's, after relatively stable, inexpensive microwave sources became widely available.

3. Advantages and Disadvantages of Microwaves for Testing

Testing with microwaves is dominated by the basic properties of microwaves. Since their penetration in good conducting materials is minimal, they are mainly used to test nonconducting materials. This includes dielectric and lossy dielectric materials. Testing and measurements on conducting materials are limited to dimensional testing such as thickness gaging and surface measurements such as testing for surface breaking flaws.

On the other hand, microwaves are affected by a large number of material properties. In lossless or lossy dielectrics, such diverse properties as porosity, material composition, uniformity of the material, delamination of layers, moisture and contamination content, are only some of the properties that can be measured. The range of nonmetallic materials in which this is possible is extensive and growing steadily. This includes ceramics, plastics, polymers and composites, as well as organic materials such as wood products, foods or biological materials. Measurements and testing in all dielectrics and lossy dielectrics are possible.

The spatial resolution that can be expected of microwave tests depends on the wavelength of the wave. For microwaves and millimeter waves this is of the order of $1mm$. This resolution indicates the ability of the test to discern closely spaced discontinuities in the materials. However, more sensitive measurements are possible by correlating them with changes in amplitude or phase. This is the case with dimensional measurements where variations in thickness of a few microns are measurable. High resolution testing can be achieved by scanning of the microwave field and by moving the source. This is similar to synthetic aperture radar *(SAR)* methods and can be used for imaging. If this is done in the near field of the antenna, it may even be called "micrometry".

Another particular property of microwave testing is the means by which energy is coupled into the testing environment. This can be as simple as a horn antenna or an open microwave guide. In some other cases, an aperture is used for this purpose or, as with microwave cavities, a simple probe or loop serves the purpose. In all cases, the coupling can be done through air, free space or a convenient dielectric. Impedance matching can also be employed if necessary. While most testing is done in what may be termed the "near field" environment (close to the antenna or source), measurements in the far field are also possible where the waves propagate through a medium such as air. This is particularly applicable to scattering methods, including radar testing applications.

Because of the influence of so many effects and properties on the wave, the testing environment can be, and often is, noisy. Reflections from near and far surfaces, edges and other artifacts in the material are often encountered, resulting in noise and loss of resolution.

Microwave radiation is highly directive, and, because of the short wavelengths involved, the devices used are often very compact. While many of the applications are in high power communication and radar system, low power applications are just as common. In this range, the choice of microwave sources is relatively wide and includes low power solid state devices.

Of primary interest in this work is the interaction of microwaves with materials. This takes the form of absorption in materials, scattering, attenuation and transmission. These effects are exploited in various testing arrangements to allow for quantitative measurements in materials.

The absorption of microwaves in water is well known and widely used in microwave ovens. These ovens depend on this effect because it is typical of the whole microwave range. Thus, while most microwave ovens operate at specific frequency bands because of regulation, they can also operate at other frequencies. The absorption in water can be used either directly or indirectly for testing of moisture and related effects (e.g. curing, drying). Other materials absorb radiation at specific frequencies. Often, these are narrow bands that allow chemical analysis of materials. These resonant frequencies are extremely useful for material characterization and identification. Sometimes even traces of materials can alter the resonant frequency of a microwave cavity, leading to detection of materials in trace amounts. This again may be used for testing or detection. Typical applications of this type are contamination tests and detection of explosives.

Because microwaves border on the one hand the high frequency radio range and on the other the low infrared range, they have properties of both. More than any other frequency range, the microwave range is sometimes analyzed using circuit theory and sometimes using wave theory. Transmission lines are almost always analyzed as distributed parameter circuits while the aspects of refraction, transmission and propagation in waveguides are analyzed using wave theory.

Another aspect of microwaves is the special nature of the components used. The familiar conductors are now replaced with guiding structures. Energy is transmitted not through the flow of current but through propagation of fields. The fields are guided in the required direction by guiding structures. These can take the form of hollow conductors, parallel plates or dielectric slabs. The impression one gets of a certain "plumbing" character involved in microwaves is not without basis. A waveguide cannot be bent in sharp corners as one would do with a wire. The guides must be properly terminated and matched to loads and sources, and modification of propagation properties is often accomplished by physical structures in the waveguides.

Microwave radiation has other properties that are less important in testing than, for example, in communication. One of these is the large bandwidth possible, because of the high frequency of the waves. Microwaves also penetrate easily through the ionosphere, with obvious applications to satellite communication. In addition, microwaves penetrate into the body, causing a variety of effects, including heating of tissue. The safety aspects of testing with microwaves cannot be neglected.

4. Energy Associated with Microwaves

The radiation energy associated with microwaves can be estimated considering the quantum equivalent photon. The energy of a photon is equal to hf where h is the Plank constant ($h=4.14\times10^{-15}ev$). Thus the maximum energy of a photon in the microwave range is roughly $1.2\times10^{-3}ev$ (minimum is about $1.0\times10^{-6}ev$ at the lower frequency range). This energy is relatively low and is much lower than the energy needed for ionization. The energy is much lower than the energy in molecular links. Thus, because it cannot break these links it is considered a nonionizing form of radiation.

The danger from microwave radiation is considered to be primarily due to absorption rather than due to its intrinsic energy. Exposure to microwaves is not unlike absorption in a microwave oven, except for the levels encountered. For this reason, the safety levels of radiation are defined on the surface, in terms of power per unit area $(Watts/m^2)$. There is also some evidence that non-thermal effects of microwave radiation also play a role but this issue is still controversial and not very well researched.

The exposure levels allowed should serve as guidelines only. While there may be no harmful effects due to allowable levels of microwave radiation, the levels in the USA are much higher than in other countries. As a rule, one should avoid all exposure to microwave radiation if only because of its absorption effects.

The allowable energy density exposure in industry in the USA is $10\ mW/cm2$. As a means of understanding the thermal effects of this radiation level (nonthermal effects are not as well defined and are still being debated) it is useful to compare this radiation level with thermal radiation from the sun. The sun radiation is about $1,400\ W/m^2$ or 140 $mW/cm2$. To compare the fields associated with the two types of radiation, these two energy densities are viewed as the result of a Poynting vector and the equivalent electric and magnetic field intensities are calculated. For the allowable microwave exposure in the USA, the electric and magnetic field intensities are calculated as follows

$$P_{av} = \frac{E^2}{2\eta_0} = 10^{-2}\ [W/cm^2]$$

where η_0 is the intrinsic impedance of free space and is equal to $377\ \Omega$. From these, the electric field intensity is

$$E = \sqrt{2\times377\times10^{-2}} = 2.75\ [V/cm] = 275\ [V/m]$$

The magnetic field intensity is

$$H = \frac{E}{\eta_0} = \frac{275V/m}{377W} = 0.73\ [A/m]$$

For comparison, the electric and magnetic field intensities associated with radiation from the sun are

$$P_{av} = \frac{E^2}{2\eta_0} = 1400\ \ [W/m^2]$$

$$E = \sqrt{1400 \times 2 \times 377} = 1{,}027\ [V/m]$$

$$H = \frac{E}{\eta_0} = \frac{1027V/m}{377W} = 2.72\ [A/m]$$

This however, does not necessarily mean that microwaves are less "dangerous" than the sun's radiation at the levels given. There are two aspects of microwave energy that make them different than energy from the sun: one is the high absorption in water, and therefore the heating effects of microwave power. The second is its deeper penetration in tissue as well as penetration through clothing. While normal clothing protects against the sun, it does not protect against microwave radiation.

5. Properties of Fields at High Frequencies

While the properties of electromagnetic fields can be deduced from Maxwell's equations and material properties, the high frequency range has some common properties that will be emphasized throughout this work. These properties form the cornerstone of testing at high frequencies. For this reason they will be discussed qualitatively, and briefly here.

The first and foremost point is that the fields are always part of a wave. The electric field intensity and magnetic field intensity are used to describe a wave. Their relations with each other and with materials define a variety of properties such as mode of propagation, reflection, refraction, transmission and impedance. The wave nature of the fields also forces us to think in terms of speeds of propagation in different materials and delays in signals because of these speeds. The notion that an electromagnetic disturbance propagates instantaneously is convenient in the static or quasi-static domains but not in the dynamic case where propagation times are extremely important and, in fact, often serve for measurements and testing. Simple aspects of low frequency fields such as impedance matching take a whole new meaning at high frequencies. At low frequency, mismatched impedances mean mainly inefficiency but, at high frequencies, this also means reflection of waves back into the transmitting medium and standing waves.

Flow of currents is also different than at low frequencies. First, we must deal with displacement currents. Thus the whole space can support currents, not only conductors. In many cases, displacement currents are dominant or the only currents in existence. Without these displacement currents, many phenomena cannot be explained, including the existence of the wave itself. Just consider the simple dipole antenna to see that the low frequency, or quasistatic model is inadequate (in the quasistatic model, currents must close through conductors, but the dipole antenna is open circuited at both ends). Conduction currents are dominated by the skin effect and flow only close to the surface of conductors. At microwave frequencies, the skin depth is of the order of a few microns. A regular solid conductor is, therefore useless at these frequencies. If large currents exist, the surface of the conductor must be large. Reduction of impedance now means reduction of the "surface impedance" of the conductor. Thus, a thin layer of highly conductive material on a substrate (which can be nonconducting) is more efficient than a thick, heavy conducting wire. The idea of conductivity itself, while the same as low frequency conductivity takes a different meaning. Because of the skin depth, many of the conductors we use are often considered to be perfect conductors. Low conductivity material like sea water, in the microwave range are considered to be highly conducting (or, more accurately - high loss materials, since conductivity is most often associated with losses).

Material properties in the microwave range are considerably more complex and, to a certain extent, more prominent than at low frequencies. Many of the properties which are normally considered frequency independent must now be altered to include frequency dependence. The best example of this is the electric permittivity. Both the permittivity and permeability are complex numbers where the imaginary part is associated with losses. In addition to conduction losses, we can talk of dielectric losses and of attenuation of waves due to material properties.

This rich and complex system of properties complicates matters considerably but, also allows a very wide range of measurements and tests that are not possible at low frequencies. These tests are only limited by our instruments and our imagination. By direct or indirect measurement of material properties and associated effects, virtually any effect can be measured. From the testing point of view this is fortunate since a basically simple method allows measurements as diverse as simple dimensional measurements of a sheet of metal or the curing condition of a polymer. Techniques for measurements are also diverse ranging from simple time-of-flight measurements to spectroscopy to holography.

Many of the test techniques and measurements are borrowed from other disciplines such as radar measurements, antenna measurements and, basic microwave measurements. Many of these have been adapted for microwave testing and many remain to be used.

6. A Note on References and Bibliography

While microwave nondestructive testing is a rather specialized aspect of testing, it can be viewed as part of the vast domain of microwave measurements techniques. Since it is not possible to include even a small portion of this material, all aspects of measurements and testing that are not directly related to *NDT* are represented by references at the end of each chapter.

Reference are rather extensive and, whenever appropriate, additional notes are given. This is done to allow the reader to supplement the material given, and to give proper credit to sources used in the text, while avoiding references in the text itself. When appropriate, references are arranged by subjects and within each subject, by author's preference. Sometimes this reflects the author's opinion as to the value of the reference, sometimes its availability, and the reader should weigh this when referring to this material. Particular effort was invested in quoting only widely available references avoiding almost completely the use of technical reports. It has been the Author's experience that, with few exceptions, these are difficult to obtain and in most cases, the material has been published elsewhere. Whenever possible books are preferred over articles primarily because books often expand on a subject and the notation tends to be more uniform.

7. References

Lord Rayleigh's paper [1] is considered by many to be the first publication describing guided waves. This has led later on in the 1930's to experiments at Bell Laboratories on guiding waves and eventually to the waveguides the way we know them.

An interesting account on the invention of the Klystron is given by Ginzton [2].

[1] Lord Rayleigh, "On the passage of electric waves through tubes", Phil. Mag, Vol. 43, 1897, pp. 125-132.

[2] E.L. Ginzton, "The $100 Idea, " IEEE Spectrum, Vol. 12, Feb. 1975, pp.30-39.

Chapter 1. The Electromagnetic Field Equations and Theoretical Aspects

1.1. Introduction: The Electromagnetic Field Equations

The electromagnetic field equations as we know and use them are succinctly summarized in Maxwell's equations. This set of four equations relates the four vector field quantities **E**, **H**, **D** and **B** with each other and with the sources of the magnetic field. These equations, attributed to James Clerk Maxwell (1831-1879) are, as Maxwell himself made clear, a summary of the work of Faraday, Ampere, Gauss, Coulomb, Helmholtz and others. However, perhaps the main reason they are attributed to him is that he introduced the idea of displacement currents (or, alternatively, of the continuity of charge) and by doing so he was able to predict the propagation of electromagnetic waves as well as unifying the theory of electromagnetic fields. While much of the work with electromagnetic fields before and after Maxwell does not require the definition of displacement currents, propagation, and all microwave aspects of fields are based on it. To understand how important this aspect is it is sufficient to look at the current in a capacitor. The flow of current through a capacitor cannot be explained by the pre-Maxwell field equations. With all that, the equation as we know them today were actually written by Oliver Heaviside who introduced the notation we are familiar with.

The approach used here is to introduce the complete set of Maxwell's equations as a set of postulates. This inductive approach is based on the assumption that the reader is familiar with the basic laws of electromagnetism and its physical manifestations. However, beyond this, the approach is consistent with other methods of study. The deductive approach, often associated with a first course in electromagnetics, builds up to Maxwell's equations by introducing the same postulates, one at a time and expanding on them as new aspects of the field are introduced. One may, for example, start with Coulomb's law and derive the static electric field from it. Then the static electric field postulates can be expanded to include time dependent fields and so on. It is important to realize that either approach assumes the correctness of the various postulates since these are experimental laws that can only be validated but not proven.

We will first present the general form of Maxwell's equations and expand on its various aspects. Then these equations will be modified into forms that are useful for analysis. Particular attention is given to interaction with materials and to energy concepts of the electromagnetic field, as these are of vital importance in testing, interpretation of test results and in modeling of testing configurations. For consistency, the SI system of units is followed throughout.

1.2. Maxwell's Equations in Differential Form

The general, time dependent, Maxwell's equations are written in differential form (or in point or local form) as:

$$\nabla\times\mathbf{E} = -\frac{\partial\mathbf{B}}{\partial t} \tag{1.1}$$

$$\nabla\times\mathbf{H} = \mathbf{J} + \frac{\partial\mathbf{D}}{\partial t} \tag{1.2}$$

$$\nabla\cdot\mathbf{D} = \rho \tag{1.3}$$

$$\nabla\cdot\mathbf{B} = 0 \tag{1.4}$$

The four vector field quantities are defined as:

E - the electric field intensity, given in *Volts/meter* [V/m]
H - the magnetic field intensity, given in *Amperes/meter* [A/m]
D - the electric flux density, given in *Coulombs/m*^{2} [C/m^2]
B - the magnetic flux density, given in *Webers/m*^{2} [Wb/m^2]

The electric and magnetic sources are defined as:

J - the current density, given in *Amperes/m*^{2} [A/m^2]
ρ - the electric charge density, given in *Coulombs/m*^{3} [C/m^3]

Eq. (1.1) is known as Faraday's law of induction. Eq. (1.2) is Ampere's law, or, as is sometimes referred to the "modified Ampere's law". This distinction simply indicates that the law, as presented by Ampere did not include the term $\partial\mathbf{D}/\partial t$. Eq. (1.3) is Gauss' law. Eq. (1.4) is not associated with a particular law and simply states the nonexistence of isolated magnetic poles. Because of the form of the expression it is sometimes referred to as the magnetic form of Gauss' law.

In this form, the equations are linear but are not sufficient for practical solution of electromagnetic field problems. To understand this, we first note that each of the vector quantities, in general, has three components and therefore, the four Maxwell's equations are equivalent to twelve scalar equations in twelve unknowns. However, the four equations above are not independent. We can obtain the last two from the first two and therefore we can only consider the first two equations as independent. To show that this is the case, we take the divergence of Eq. (1.1)

$$\nabla\cdot(\nabla\times\mathbf{E}) = \nabla\cdot\left(-\frac{\partial\mathbf{B}}{\partial t}\right)$$

Since the divergence is independent of time, the divergence and time derivative can be interchanged. The left hand side is identically equal to zero for any vector (see appendix A) and we get

$$\frac{\partial(\nabla\cdot\mathbf{B})}{\partial t} = 0$$

This implies that $\nabla\cdot\mathbf{B}$ is constant in time and we can write

$$\nabla\cdot\mathbf{B} = 0$$

Similarly, by taking the divergence on both sides of Eq. (1.2) we have:

$$\nabla\cdot(\nabla\times\mathbf{H}) = \nabla\cdot\mathbf{J} + \frac{\partial(\nabla\cdot\mathbf{D})}{\partial t}$$

where the same consideration as above was used to interchange between the time derivative and the divergence. Since the left hand side is identically zero we have

$$\nabla\cdot\mathbf{J} + \frac{\partial(\nabla\cdot\mathbf{D})}{\partial t} = 0$$

Using the continuity equation

$$\nabla\cdot\mathbf{J} = -\frac{\partial\rho}{\partial t} \tag{1.5}$$

we get

$$\nabla\cdot\mathbf{D} = \rho$$

This latter consideration was in fact what led Maxwell to introduce the displacement current into Ampere's law, and to predict the propagation of electromagnetic waves.

If the four Maxwell's equations are not independent, what are then the additional relations we must consider? The answer to this is in the interaction of the electromagnetic fields with materials (including vacuum). The two relation we must use are the material constitutive relations:

$$\mathbf{B} = \mu\mathbf{H} \tag{1.6}$$

$$\mathbf{D} = \varepsilon\mathbf{E} \tag{1.7}$$

These two equations are equivalent to six scalar equations. We shall not comment on these at this stage other than to say that they provide relationships between **B** and **H** and between **D** and **E**. The nature of these relationships will be discussed in detail in subsequent sections, but, at least in their linear, isotropic form, they are familiar relation with μ representing permeability of materials and ε representing permittivity. The system of equation we normally use to determine the field quantities are Eqs. (1.1), (1.2), (1.6) and (1.7).

In addition to these, the Lorentz force equation is often associated with the complete set of field relations. This is written as:

$$\mathbf{F} = q\mathbf{E} + q\mathbf{v}\times\mathbf{B} \tag{1.8}$$

where q is a charge and **v** is the velocity of the charge. The first term is the electric force on charges and exists whenever an electric field exists. The second is the force in the magnetic field and only exists if the charge moves.

1.2.1. The Time Harmonic Equations

If the electromagnetic field vectors are characterized by sinusoidal variation in time, we can place their phasor equivalents in Eqs. (1.1) through (1.4). As an example, the electric field intensity is written as

$$\mathbf{E} = \mathbf{E}_0 e^{j\omega t} \tag{1.9}$$

where the vector notation has been retained but now **E** is a phasor. ω is the frequency of the field and j is the complex number. The derivative with respect to time is therefore

$$\frac{\partial\mathbf{E}}{\partial t} = j\omega\mathbf{E}_0 e^{j\omega t} = j\omega\mathbf{E} \tag{1.10}$$

By substituting $j\omega\mathbf{E}$ for $\partial\mathbf{E}/\partial t$ and $j\omega\mathbf{B}$ for $\partial\mathbf{B}/\partial t$, and using the notation above we get the time harmonic form of Maxwell's equations:

$$\nabla\times\mathbf{E} = -j\omega\mathbf{B} \tag{1.11}$$

$$\nabla\times\mathbf{H} = \mathbf{J} + j\omega\mathbf{D} \tag{1.12}$$

$$\nabla\cdot\mathbf{D} = \rho \tag{1.13}$$

$$\nabla\cdot\mathbf{B} = 0 \tag{1.14}$$

1.2.2. The Source Free Equations

A simpler form of Maxwell's equations, one that is often convenient, and used extensively in the microwave domain, is the source free equations. If both the current density **J** and the charge density ρ do not exist, the equations take the following form for the time harmonic case

$$\nabla\times\mathbf{E} = -j\omega\mathbf{B} \tag{1.15}$$

$$\nabla\times\mathbf{H} = j\omega\mathbf{D} \tag{1.16}$$

$$\nabla\cdot\mathbf{D} = \rho \tag{1.17}$$

$$\nabla\cdot\mathbf{B} = 0 \tag{1.18}$$

The same can be done for the general time dependent equations in Eqs. (1.1) through (1.4).

1.3. Maxwell's Equations in Integral Form

The differential representation of Maxwell's equations is very useful in many instances, including for some of the numerical methods used for computation. Their integral representation is often more convenient for analytic calculations and for physical interpretation of phenomena. Eqs (1.1) through (1.4) can be converted into integral expressions by using the integral theorems of vector calculus, and vector identities. In particular, the use of the divergence theorem and Stokes' theorems are necessary for this purpose. We shall explicitly show the process for the time dependent form of Maxwell's equations.

By integrating both sides of Eq. (1.1) over a surface S, we get

$$\int_S (\nabla\times\mathbf{E})\cdot d\mathbf{s} = -\int_S \frac{\partial\mathbf{B}}{\partial t}\cdot d\mathbf{s}$$

Using Stokes' theorem (see Appendix A), the left hand side becomes

$$\int_S (\nabla\times\mathbf{E})\cdot d\mathbf{s} = \oint_C \mathbf{E}\cdot d\mathbf{l}$$

where the contour C encloses the surface S. The integral form of Eq. (1.1) is therefore

$$\oint_C \mathbf{E}\cdot d\mathbf{l} = -\int_S \frac{\partial\mathbf{B}}{\partial t}\cdot d\mathbf{s} \tag{1.19}$$

This may be written in a more familiar form by observing that the surface integrals of the flux density **B** over the surface S is the total flux Φ through the surface

$$\oint_C \mathbf{E}\cdot d\mathbf{l} = -\frac{\partial\Phi}{\partial t}$$

This form of Faraday's law is often used in low frequency applications, especially in electric machines. For our purposes, the form in Eq. (1.19) is more convenient.

Ampere's law in Eq. (1.2) is treated in an identical fashion, by integrating both sides over a surface S. The result is

$$\oint_C \mathbf{H}\cdot d\mathbf{l} = \int_S \left(\mathbf{J} + \frac{\partial \mathbf{D}}{\partial t}\right)\cdot d\mathbf{s}$$

We may wish to integrate the current density **J** over the surface S and obtain the total current through S

$$\oint_C \mathbf{H}\cdot d\mathbf{l} = I + \int_S \frac{\partial \mathbf{D}}{\partial t}\cdot d\mathbf{s} \qquad (1.20)$$

where I is the total current enclosed by the contour C. The other two equations are integrated over a volume V and use is made of the divergence theorem. For Eq. (1.3), we have

$$\int_V \nabla\cdot\mathbf{D}dv = \oint_S \mathbf{D}\cdot d\mathbf{s} = \int_V \rho dv = Q$$

where S is the closed surface enclosing the volume V and Q is the total charge density in volume V. Similarly, for Eq. (1.4)

$$\int_V \nabla\cdot\mathbf{B} = \oint_S \mathbf{B}\cdot d\mathbf{s} = 0$$

Maxwell's equations in differential and integral forms are summarized in Tables 1.1 and 1.2 for the time dependent and time-harmonic fields.

Table 1.1. The electromagnetic field equations.

In differential form	In integral form
$\nabla\times\mathbf{E} = -\frac{\partial\mathbf{B}}{\partial t}$	$\oint_c \mathbf{E}\cdot d\mathbf{l} = -\int_s \frac{\partial\mathbf{B}}{\partial t}\cdot d\mathbf{s}$
$\nabla\times\mathbf{H} = \mathbf{J} + \frac{\partial\mathbf{D}}{\partial t}$	$\oint_c \mathbf{H}\cdot d\mathbf{l} = \int_s \left(\mathbf{J} + \frac{\partial\mathbf{D}}{\partial t}\right)\cdot d\mathbf{s}$
$\nabla\cdot\mathbf{B} = 0$	$\oint_s \mathbf{B}\cdot d\mathbf{s} = 0$
$\nabla\cdot\mathbf{D} = \rho$	$\oint_s \mathbf{D}\cdot d\mathbf{s} = \int_v \rho dv$

Table 1.2. The time-harmonic electromagnetic field equations.

In differential form	In integral form
$\nabla\times\mathbf{E} = -j\omega\mathbf{B}$	$\oint_c \mathbf{E}\cdot d\mathbf{l} = -\int_s \frac{\partial\mathbf{B}}{\partial t}\cdot d\mathbf{s}$
$\nabla\times\mathbf{H} = \mathbf{J} + j\omega\mathbf{D}$	$\oint_c \mathbf{H}\cdot d\mathbf{l} = \int_s (\mathbf{J} + j\omega\mathbf{D})\cdot d\mathbf{s}$
$\nabla\cdot\mathbf{B} = 0$	$\oint_s \mathbf{B}\cdot d\mathbf{s} = 0$
$\nabla\cdot\mathbf{D} = \rho$	$\oint_s \mathbf{D}\cdot d\mathbf{s} = \int_v \rho dv$

1.4. Material Properties and Constitutive Relations

As mentioned above, the interaction of the electromagnetic field with materials (including vacuum) provides the additional relations necessary for the solutions of Maxwell's equations. These relations are also the essence of all nondestructive testing methods in the microwave region, as well as in any other electromagnetic domain. The study and understanding of materials is therefore essential to our goal.

1.4.1. Conductivity

Conductivity of a material can be broadly defined as its ability to conduct electric current. Since electric current is the movement of electrons, conductivity is closely related to the ability of electrons to move. Under the influence of an electric field, free electrons move at various velocities. If we assume a volume v, with N free electrons a current density $\mathbf{J}$ exists in a cross section of the conductor

$$\mathbf{J} = Nq\mathbf{u}$$

where q is the charge of the electron and $\mathbf{u}$ its velocity. The velocity $\mathbf{u}$ can be directed in any direction but, under an external electric field, we may assume that the velocity is directed in the direction of the electric field and, that all electrons have the same velocity. This is certainly the case in metals and other "good" conductors. The velocity is directly proportional to the electric field. While electrons will have drift velocities regardless of external fields these are random and are low compared with the velocities imparted by external fields. Under these conditions, the current density $\mathbf{J}$ in a conductor can be directly related to the applied electric field intensity $\mathbf{E}$ as

$$\mathbf{J} = \sigma\mathbf{E} \tag{1.21}$$

where σ is a material parameter called conductivity. Although we have written this relation as linear (assuming for the moment that σ is constant), this need not be the case. Conductivity can be field dependent, it is almost always temperature dependent and, in the microwave range it can be frequency dependent. We shall have ample opportunity to discuss these aspects later. The relation in Eq. (1.21) is one of the three constitutive relations we use and, although it is strictly speaking defined for conductors (i.e. high conductivity materials), the arguments that led to its definition are general. We often use this as a general relation in any material. The form in Eq. (1.21) is known as Ohm's law, written here in its vector form.

1.4.2. Complex Permittivity

In dielectric materials, an external electric field polarizes the molecules of the material to a certain degree. Since charges in these materials are bound, conduction is negligible but, polarizations of charges (i.e. alignment of dipoles with the external electric field) may be significant. This polarization, shown schematically in Figure 1.1 increases the electric flux density. For this reason we can write

$$\mathbf{D} = \varepsilon_0\mathbf{E} + \mathbf{P} \tag{1.22}$$

where $\mathbf{E}$ is the external electric field intensity, $\mathbf{P}$ is the polarization vector (characteristic of the material) and ε_0 is the permittivity of free space. The polarization vector is proportional to the electric field. This relation may be quite complex but, for simplicity, we will assume here that it is linear:

$$\mathbf{P} = \varepsilon_0 \chi \mathbf{E}$$

where χ is the electric susceptibility of the material. This material property is, in general, a complex quantity. Eq. (1.22) now becomes

$$\mathbf{D} = \varepsilon_0 \mathbf{E} + \varepsilon_0 \chi \mathbf{E} = \varepsilon_0 (1+\chi) \mathbf{E}$$

We define the quantity

$$\varepsilon = \varepsilon_0 (1+\chi)$$

as the complex permittivity. This is often written as

$$\varepsilon = \varepsilon' + j\varepsilon'' \tag{1.23}$$

to signify the fact that permittivity is a complex quantity. By substituting the complex permittivity in Eq. (1.22) we get

$$\mathbf{D} = \varepsilon \mathbf{E} \tag{1.24}$$

Eq. (1.24) is the second constitutive relation. Again, as with conductivity, ε does not have to be linear or isotropic. The imaginary part of ε represents dielectric losses. These losses are due to damping of the dipole moments but, as we shall see shortly, we can view these losses as conduction losses in the dielectric since dielectric losses generate heat, and are indistinguishable from conduction losses. Lossless materials have a real permittivity.

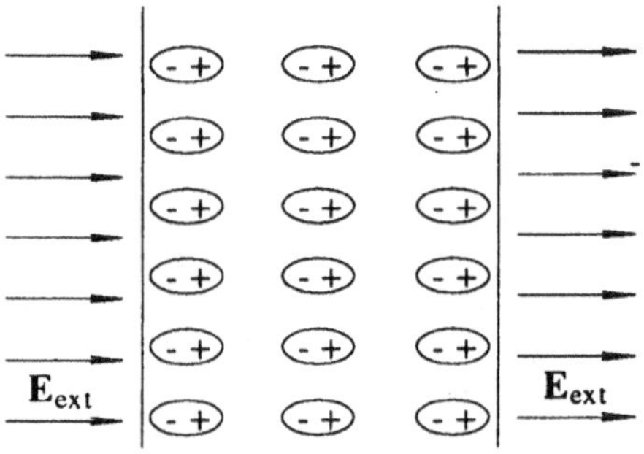

Figure 1.1. Polarization of dielectric materials in the presence of an external electric field.

1.4.3. Complex Permeability

The basic model of materials indicates that the orbiting electrons in atoms generate magnetic dipoles with dipole moments **m**. Normally, these dipoles are randomly oriented such that their net external effect is zero as shown in Figure 1.2a. In the presence of an external magnetic field, some of the dipoles orient themselves with the external field as in Figure 1.2b. Without discussing the details of magnetic materials here (which include many interesting phenomena) we indicate that a total magnetization vector **M** exists in the material, in the presence of an external magnetic field. Using this magnetization vector as analogous to the polarization vector in the electric field, the magnetic flux density **B** may be written as

$$\mathbf{B} = \mu_0\mathbf{H} + \mu_0\mathbf{M} \tag{1.25}$$

where **M** is the magnetization vector and μ_0 is the permeability of free space. Magnetization can be written in terms of the magnetic field intensity as

$$\mathbf{M} = \mu_0\chi_m\mathbf{H}$$

where χ_m is the magnetic susceptibility of the material and, is generally, complex. From Eq. (1.24) we have an expression for the complex permeability μ

$$\mu = \mu_0(1+\chi_m) = \mu' + j\mu''$$

By substituting this into Eq. (1.24) we get the third constitutive relation

$$\mathbf{B} = \mu\mathbf{H} \tag{1.26}$$

The imaginary part of the complex permeability represents losses due to damping forces in the magnetic material. Since there are no magnetic currents, we cannot view these losses as conducting losses. The real part represents materials without magnetic losses.

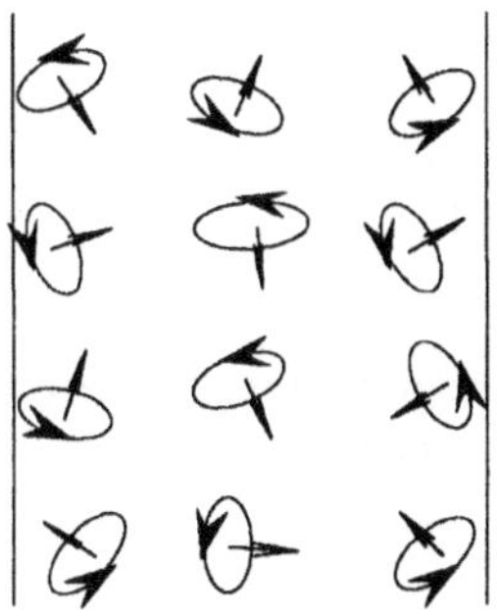

Figure 1.2a. Magnetic dipoles, randomly oriented.

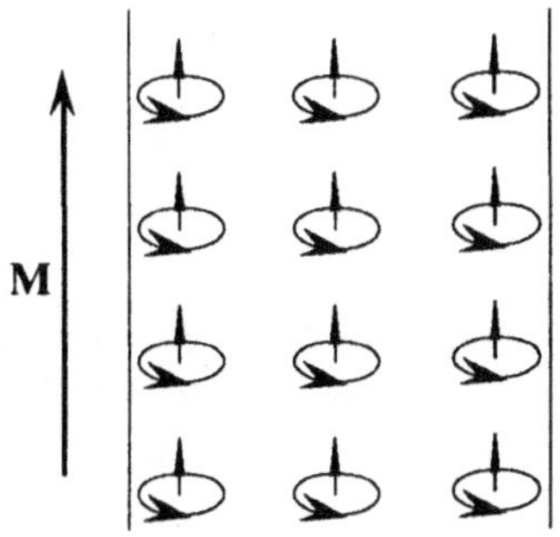

Figure 1.2b. Magnetic dipoles oriented in the direction of the external field

1.4.4. Anisotropic Materials

All material properties may be isotropic or anisotropic. While many materials are isotropic (i.e. metals are often isotropic in conductivity and permeability, and most dielectrics have an isotropic permittivity) there are important engineering materials that are not. For example, composite materials are often anisotropic. Graphite reinforced composites are anisotropic in their conductivity with a relatively high conductivity in the direction along the fibers and low conductivity in directions transverse to fibers. Anisotropy can also show up in dielectrics (i.e. layered dielectrics, crystals) and in terms of permeability in ferrites.

The constitutive equations can be modified to take into account these relations. To do so, they are written as 3×3 tensors

$$\mathbf{J} = \|\sigma\|\mathbf{E} \tag{1.27}$$

$$\mathbf{D} = \|\varepsilon\|\mathbf{E} \tag{1.28}$$

$$\mathbf{B} = \|\mu\|\mathbf{H} \tag{1.29}$$

or, in explicit form

$$\begin{bmatrix} \mathbf{J}_x \\ \mathbf{J}_y \\ \mathbf{J}_z \end{bmatrix} = \begin{bmatrix} \sigma_{xx} & \sigma_{xy} & \sigma_{xz} \\ \sigma_{yx} & \sigma_{yy} & \sigma_{yz} \\ \sigma_{zx} & \sigma_{zy} & \sigma_{zz} \end{bmatrix} \begin{bmatrix} \mathbf{E}_x \\ \mathbf{E}_y \\ \mathbf{E}_z \end{bmatrix}$$

$$\begin{bmatrix} \mathbf{B}_x \\ \mathbf{B}_y \\ \mathbf{B}_z \end{bmatrix} = \begin{bmatrix} \mu_{xx} & \mu_{xy} & \mu_{xz} \\ \mu_{yx} & \mu_{yy} & \mu_{yz} \\ \mu_{zx} & \mu_{zy} & \mu_{zz} \end{bmatrix} \begin{bmatrix} \mathbf{H}_x \\ \mathbf{H}_y \\ \mathbf{H}_z \end{bmatrix}$$

$$\begin{bmatrix} \mathbf{D}_x \\ \mathbf{D}_y \\ \mathbf{D}_z \end{bmatrix} = \begin{bmatrix} \varepsilon_{xx} & \varepsilon_{xy} & \varepsilon_{xz} \\ \varepsilon_{yx} & \varepsilon_{yy} & \varepsilon_{yz} \\ \varepsilon_{zx} & \varepsilon_{zy} & \varepsilon_{zz} \end{bmatrix} \begin{bmatrix} \mathbf{E}_x \\ \mathbf{E}_y \\ \mathbf{E}_z \end{bmatrix}$$

In all these relations, linear material properties were assumed (i.e. independent of field intensities). Nonlinear relations may also be taken into account.

1.5. The Poynting Theorem and Energy

To examine the energy in the magnetic field it is convenient to look first at the general time dependent expression for energy that includes stored magnetic energy, stored electric energy and dissipated energy. To do so, the following vector identity is used

$$\nabla\cdot(\mathbf{E}\times\mathbf{H}) = \mathbf{H}\cdot(\nabla\times\mathbf{E}) - \mathbf{E}\cdot(\nabla\times\mathbf{H})$$

Into this expression we substitute the expression for the curl of **E** and **H** from the first and second of Maxwell's equations:

$$\nabla\cdot(\mathbf{E}\times\mathbf{H}) = -\mathbf{H}\cdot\frac{\partial\mathbf{B}}{\partial t} - \mathbf{E}\cdot\left(\mathbf{J} + \frac{\partial\mathbf{D}}{\partial t}\right) = -\mathbf{H}\cdot\frac{\partial\mathbf{B}}{\partial t} - \mathbf{E}\cdot\frac{\partial\mathbf{D}}{\partial t} - \mathbf{E}\cdot\mathbf{J}$$

Assuming that we look at the energy flow in a volume V, bounded by an area S, we can integrate this expression over the volume V to obtain the total energy flow in the volume.

$$\int_v \nabla\cdot(\mathbf{E}\times\mathbf{H})dv = -\int_v\left(\mathbf{H}\cdot\frac{\partial\mathbf{B}}{\partial t} + \mathbf{E}\cdot\frac{\partial\mathbf{D}}{\partial t}\right)dv - \int_v \mathbf{E}\cdot\mathbf{J}dv \tag{1.30}$$

The left hand side is transformed from a volume integral to a surface integral using the divergence theorem. We also use the following identities:

$$\mathbf{E}\cdot\frac{\partial\mathbf{D}}{\partial t} = \frac{\partial}{\partial t}\left(\frac{\mathbf{D}\cdot\mathbf{E}}{2}\right) \qquad \mathbf{H}\cdot\frac{\partial\mathbf{B}}{\partial t} = \frac{\partial}{\partial t}\left(\frac{\mathbf{H}\cdot\mathbf{B}}{2}\right) \tag{1.31}$$

With these, Eq. (1.30) becomes

$$\oint_s (\mathbf{E}\times\mathbf{H}).\mathbf{ds} = -\frac{\partial}{\partial t}\int_v\left(\frac{\mathbf{H}\cdot\mathbf{B}}{2} + \frac{\mathbf{E}\cdot\mathbf{D}}{2}\right)dv - \int_v \mathbf{E}\cdot\mathbf{J}dv \tag{1.32}$$

Strictly speaking, the identities in Eq. (1.31) are only valid for constant ε and μ. We will nevertheless use these here for simplicity.

The left side of the expression represents the total flow of energy through the area bounding the volume. The expression **E**×**H** has units of *Watts/m*^{2} and is therefore a power density. This power density is called a Poynting vector **P**:

$$\mathbf{P} = \mathbf{E}\times\mathbf{H} \tag{1.33}$$

The advantage of this expression is that it also indicates the direction of the energy flow, information that is important for wave propagation calculations.

The first term on the right side of Eq. (1.32) represents the time rate of increase in the potential or stored energy in the system. It has two components: one is the stored electric energy and the other is the stored magnetic energy.

The second term on the right hand side of Eq. (1.32) is the power dissipated as well as the power due to any sources that may exist in the volume V. If no such sources exist, this term represents losses.

Thus, the Poynting theorem describes all energy relations in a system be they electrostatic, magnetostatic or time dependent. Because the vector product between the electric field intensity and the magnetic field intensity is taken, these two quantities must be related (i.e. they must be generated by the same sources) otherwise the results obtained will have no meaning.

The expression in Eq. (1.32) is an instantaneous quantity. For practical purposes, a time averaged quantity is more useful. For periodic time variation of fields, this can be done by averaging over a time T (usually a cycle of the field), giving the time average Poynting vector:

$$\mathbf{P}_{av} = \frac{1}{T}\int_0^T \mathbf{P}(t)dt \tag{1.34}$$

1.5.1. The Complex Poynting Vector

In most microwave applications, fields and energy are handled in the frequency domain, assuming sinusoidal excitation. Thus it is useful to define the Poynting vector in the frequency domain. The complex Poynting vector is defined as

$$\mathbf{P} = \frac{1}{2}(\mathbf{E}\times\mathbf{H}^*) \tag{1.35}$$

where * indicates a complex conjugate and **P**, **E** and **H** are phasors. From this we can also define a time average complex Poynting vector as

$$\mathbf{P}_{av} = \frac{1}{2}Re(\mathbf{E}\times\mathbf{H}^*) \tag{1.36}$$

where Re indicates the real part of the complex Poynting vector.

To define the complex Poynting vector we start with Maxwell's first two equations (Eqs. (1.11) and (1.12).

$$\nabla\times\mathbf{E} = -j\omega\mu\mathbf{H} \tag{1.37}$$

$$\nabla\times\mathbf{H} = \mathbf{J} + j\omega\varepsilon\mathbf{E} \tag{1.38}$$

The conjugates of Eqs. (1.37) and (1.38) are

$$\nabla\times\mathbf{E}^* = -j\omega\mu\mathbf{H}^* \tag{1.39}$$

$$\nabla\times\mathbf{H}^* = \mathbf{J}^* - j\omega\varepsilon\mathbf{E}^* \tag{1.40}$$

The scalar product between $\mathbf{H}^*$ and Eq. (1.37) is

$$\mathbf{H}^*\cdot(\nabla\times\mathbf{E}) = -j\omega\mu\mathbf{H}\cdot\mathbf{H}^* \tag{1.41}$$

The scalar product between **E** and Eq. (1.40) is

$$\mathbf{E}\cdot(\nabla\times\mathbf{H}^*) = (\mathbf{J}^* + j\omega\varepsilon\mathbf{E}^*)\cdot\mathbf{E} \tag{1.42}$$

Subtracting Eq. (1.41) from Eq. (1.42) we get

$$\mathbf{E}\cdot(\nabla\times\mathbf{H}^*) - \mathbf{H}^*\cdot(\nabla\times\mathbf{E}) = (\mathbf{J}^* - j\omega\varepsilon\mathbf{E}^*)\cdot\mathbf{E} + j\omega\mu\mathbf{H}\cdot\mathbf{H}^*$$

Using the fact that $\mathbf{E}\cdot\mathbf{E}^*=|E|^2$ and $\mathbf{H}\cdot\mathbf{H}^*=|H|^2$ and rearranging terms

$$\mathbf{E}\cdot(\nabla\times\mathbf{H}^*) - \mathbf{H}^*\cdot(\nabla\times\mathbf{E}) = \mathbf{E}\cdot\mathbf{J}^* - j\omega(\varepsilon|E|^2 - \mu|H|^2)$$

From the vector identity (see Appendix A):

$$\mathbf{H}^*\cdot(\nabla\times\mathbf{E}) - \mathbf{E}\cdot(\nabla\times\mathbf{H}^*) = \nabla\cdot(\mathbf{E}\times\mathbf{H}^*)$$

and rearranging terms we get

$$\nabla\cdot(\mathbf{E}\times\mathbf{H}^*) = j\omega(\varepsilon|E|^2 - \mu|H|^2) - \mathbf{E}\cdot\mathbf{J}^*$$

From the definition of the complex Poynting vector in Eq. (1.35) we write

$$\nabla\cdot\mathbf{P} = j\omega\left(\frac{1}{2}\varepsilon|E|^2 - \frac{1}{2}\mu|H|^2\right) - \frac{1}{2}\mathbf{E}\cdot\mathbf{J}^*$$

The first two terms on the right hand side represent the electric and magnetic energy densities. The third term represents the input and dissipated energy densities. To show this we separate **J** into an applied current density and an induced current density since **J** in Maxwell's equations includes all current densities except the displacement current density

$$\mathbf{J}^* = \mathbf{J}_0^* + \mathbf{J}_e^* = \mathbf{J}_0^* + \sigma\mathbf{E}^*$$

where the eddy currents were written in terms of the conductivity and electric field. Thus

$$\nabla\cdot\mathbf{P} = j\omega\left(\frac{1}{2}\varepsilon|E|^2 - \frac{1}{2}\mu|H|^2\right) - \frac{1}{2}\mathbf{E}\cdot\mathbf{J}_0^* - \frac{1}{2}\sigma\mathbf{E}\cdot\mathbf{E}^*$$

To write this in terms of power rather than power density, we integrate this expression over an arbitrary volume V

$$\int_v \nabla\cdot\mathbf{P}dv = j\omega\int_v\left(\frac{1}{2}\varepsilon|E|^2 - \frac{1}{2}\mu|H|^2\right)dv - \int_v\frac{1}{2}\mathbf{E}\cdot\mathbf{J}_0^*dv - \int_v\frac{1}{2}\sigma|\mathrm{E}|^2\mathrm{dv}$$

It is customary to write the first term on the right hand side in terms of the magnetic energy density

$$w_m = \frac{1}{2}\mu|H|^2{}_{av} = \frac{1}{4}\mu\mathbf{H}\cdot\mathbf{H}^*$$

and the electric energy density

$$w_e = \frac{1}{2}\varepsilon|E|^2{}_{av} = \frac{1}{4}\varepsilon\mathbf{E}\cdot\mathbf{E}^*$$

Thus, using the divergence theorem we have

$$\int_s \mathbf{P}\cdot\mathbf{ds} = j2\omega\int_v \left(w_e - w_m\right)dv - \int_v \frac{1}{2}\mathbf{E}\cdot\mathbf{J}_0^* dv - \int_v \frac{1}{2}\sigma|\mathrm{E}|^2 \mathrm{dv} \qquad (1.43)$$

The term on the left hand side is the total power flowing out of the volume V. The first term on the right side is the total electric and magnetic power while w_e and w_m are the time averaged electric and magnetic power densities. The second term on the right hand side is the total (complex) power of the source, if such a source exists. The last term is the dissipated power in the volume.

1.6. Potential Functions

Potential functions are often used either for computation or for representation of electromagnetic fields. A potential function is derived from the field relation to achieve a stated purpose. Often, the reason to define a potential function is to simplify a field relation or to simplify its solution.

1.6.1. The Electric Scalar Potential

The electric scalar potential is perhaps the best known potential function. Its definition is based on the irrotational nature of the static electric field. Since for the static electric field, the curl of the electric field intensity is zero (the static electric field is irrotational)

$$\nabla\times\mathbf{E} = 0$$

we can define the electric field intensity as the gradient of a scalar function

$$\mathbf{E} = -\nabla V \qquad (1.44)$$

The static potential, is of little use in the microwave region. However it clearly shows the reason for defining a potential function. In this case, a single scalar function is sufficient to describe the vector behavior of the electric field. If the calculation, or measurement of the potential function can be performed, then finding the electric field is a simple matter of differentiating this function.

1.6.2. The Magnetic Scalar Potential

A magnetic scalar potential can also be defined provided that the same conditions are satisfied: that is, if the magnetic field is irrotational. From Maxwell's equations it is obvious that this can only happen if there are no sources in the region of interest. Then, if

$$\nabla\times\mathbf{H} = 0$$

the magnetic field intensity can be written in terms of the magnetic scalar potential as

$$\mathbf{H} = -\nabla\psi \qquad (1.45)$$

The magnetic scalar potential is often used for calculation of fields in magnetic materials, especially for static fields and for dynamic fields in nonconducting materials because under these conditions there are no currents in materials and the condition above is satisfied.

1.6.3. The Magnetic Vector Potential

More interesting for our purposes is the magnetic vector potential. Because the magnetic flux density **B** is solenoidal in nature (i.e. $\nabla\cdot\mathbf{B}=0$), it can be derived as the curl of another vector (see appendix A)

$$\mathbf{B} = \nabla\times\mathbf{A} \qquad (1.46)$$

We note here that this can be done even if the potential function has no particular meaning. In the case of the electric scalar potential the meaning is clear in terms of potential energy while the magnetic scalar potential has units of *Ampere* (or *Ampere.turns*) The magnetic vector potential is much more difficult to explain in terms of physical quantities. We will simply take Eq. (1.46) as its definition and the solenoidal nature of the field as the condition for its existence.

Based on the same property, namely, that the field must be irrotational in order to define it in terms of the curl of another vector, we can define a second order vector potential **W** if $\nabla\cdot\mathbf{A}=0$. Thus,

$$\mathbf{A} = \nabla\times\mathbf{W} \qquad (1.47)$$

and the magnetic flux density can be written in terms of the second order magnetic vector potential as

$$\mathbf{B} = \nabla\times(\nabla\times\mathbf{W})$$

The question wether we should use this or any other potential function is a separate issue. At this point we merely comment on the fact that this can be done. Also to be noted is that any vector must be defined in terms of its curl and divergence (see Helmholtz's theorem in appendix A). The curl and divergence of **E** and **H** are defined in Eqs. (1.1) through (1.4). For the magnetic vector potential, this must be done separately. One choice is to make the magnetic vector potential solenoidal ($\nabla\cdot\mathbf{A}=0$) as above. However, we will see cases when this is not necessarily the best choice.

To see why we often resort to potential functions, we substitute $\nabla\times\mathbf{A}$ for **B** into Eq. (1.1):

$$\nabla\times\mathbf{E} = -\frac{\partial}{\partial t}(\nabla\times\mathbf{A})$$

Interchanging the time derivative with the curl and rearranging:

$$\nabla\times\left(\mathbf{E} + \frac{\partial\mathbf{A}}{\partial t}\right) = 0$$

The expression in brackets is an irrotational field and it is also obvious that $\partial\mathbf{A}/\partial t$ is an electric field. Based on the definition of the electric scalar potential as the gradient of the electric field intensity we can write the following:

$$\mathbf{E} + \frac{\partial\mathbf{A}}{\partial t} = -\nabla V$$

Now, the electric field can be written as

$$\mathbf{E} = -\nabla V - \frac{\partial\mathbf{A}}{\partial t} \qquad (1.48)$$

This representation of the electric field intensity has a physical meaning: it shows that an electric field intensity can be generated by accumulation of charges (through ∇V), by time variation of the magnetic field (through $\partial\mathbf{A}/\partial t$), or both. A second advantage of introducing the magnetic vector potential is clear from the relation $\mathbf{B}=\nabla\times\mathbf{A}$. Consider a two dimensional case where the magnetic field intensity varies in the x and y direction but does not vary in the z direction. By writing this expression for the curl explicitly we have

$$\mathbf{B} = \hat{\mathbf{x}}\left(\frac{\partial A_z}{\partial y} - \frac{\partial A_y}{\partial z}\right) + \hat{\mathbf{y}}\left(\frac{\partial A_x}{\partial z} - \frac{\partial A_z}{\partial x}\right) + \hat{\mathbf{z}}\left(\frac{\partial A_y}{\partial x} - \frac{\partial A_x}{\partial y}\right)$$

If **B** is two dimensional (in the x-y plane), the last term must be zero. This means either that A_y and A_x are constant in the x and y directions respectively, or that they are zero. Taking the latter, we have

$$\mathbf{B} = \hat{\mathbf{x}}\frac{\partial A_z}{\partial y} - \hat{\mathbf{y}}\frac{\partial A_z}{\partial x}$$

Thus, instead of requiring a two component vector (**B**), a single component vector (**A**) is sufficient for solution in the two-dimensional case. Solution of a field problem in two dimensions in terms of the magnetic vector potential can be done as a scalar solution rather than a vector solution, considerably simplifying the process.

1.7. The Field Equations in Terms of Potential Functions

1.7.1. Vector Potentials

By direct substitution of the definitions of potential functions into Maxwell's equations we obtain a potential function representation of the equations. As pointed out earlier this is done in order to take advantage of the particular properties of the potential functions. This may be as simple as requiring a scalar calculation or being continuous across an interface. It should be pointed out that unless there is an advantage to using a potential function we should use the field variables directly.

In the case of the magnetic vector potential, substituting Eq. (1.46) into the second of Maxwell's equations and using the electric field **E** as represented in Eq. (1.48) we get

$$\nabla\times\left(\frac{1}{\mu}\nabla\times\mathbf{A}\right) = \mathbf{J} + \varepsilon\frac{\partial}{\partial t}\left(-\nabla V - \frac{\partial \mathbf{A}}{\partial t}\right) \qquad (1.49)$$

where the relations $\mathbf{B}=\mu\mathbf{H}$ and $\mathbf{D}=\varepsilon\mathbf{E}$ were used in their scalar form. Assuming linear media, and using the vector identity

$$\nabla\times(\nabla\times\mathbf{A}) = \nabla(\nabla\cdot\mathbf{A}) - \nabla^2\mathbf{A}$$

Eq. (1.49) becomes

$$\nabla(\nabla\cdot\mathbf{A}) - \nabla^2\mathbf{A} = \mu\mathbf{J} - \mu\varepsilon\nabla\left(\frac{\partial V}{\partial t}\right) - \mu\varepsilon\left(\frac{\partial^2\mathbf{A}}{\partial t^2}\right) \qquad (1.50)$$

This equation represents Maxwell's equations in terms of the magnetic vector potential **A** and the electric scalar potential V. If we require an equation in terms of the vector potential only, the scalar potential must be removed. This is done observing that the divergence of **A** has not been specified. We may choose this divergence in any way we wish, as long as this choice is consistent with the field equations. This aspect of potential functions will be discussed in a subsequent section on the gage conditions but for now we simply set

$$\nabla\cdot\mathbf{A} + \mu\varepsilon\frac{\partial V}{\partial t} = 0 \qquad (1.51)$$

Substitution of this expression in Eq. (1.50) yields the following:

$$\nabla^2\mathbf{A} + \mu\mathbf{J} - \mu\varepsilon\left(\frac{\partial^2\mathbf{A}}{\partial t^2}\right) = 0 \tag{1.52}$$

This is the nonhomogeneous wave equation for the magnetic vector potential but, for the purpose of this discussion, the most important point is that we have obtained an equation in terms of a single potential function, representing Maxwell's equations.

Solution of this equation, and subsequent use of Eqs (1.46), (1.48) and (1.51) provides the required field quantities.

1.7.2. Scalar Potentials

If we wish to obtain a representation in terms of the electric scalar potential, we start with Maxwell's third equation:

$$\nabla\cdot\varepsilon\mathbf{E} = \rho$$

where the scalar form of $\mathbf{D}=\varepsilon\mathbf{E}$ was again used. If we substitute the value for $\mathbf{E}$ from Eq. (1.48) we get

$$\nabla\cdot\varepsilon\left(-\nabla V - \frac{\partial\mathbf{A}}{\partial t}\right) = \rho$$

Assuming ε is constant and can be taken outside the divergence gives

$$\nabla\cdot\nabla V + \frac{\partial}{\partial t}(\nabla\cdot\mathbf{A}) = -\frac{\rho}{\varepsilon}$$

Since $\nabla\cdot\nabla V \equiv \nabla^2 V$, and using again Eq. (1.51) for the divergence of $\mathbf{A}$ we get

$$\nabla^2 V - \mu\varepsilon\frac{\partial^2 V}{\partial t^2} = -\frac{\rho}{\varepsilon} \tag{1.53}$$

This is again a nonhomogeneous wave equation in terms of the electric scalar potential V. Note however that there is a fundamental difference between Eq. (1.52) and Eq. (1.53). The first contains current densities. The current density is a vector and therefore, the potential function required for solution must also be a vector. In the second case, we have only scalar sources (charge density) and a scalar potential is sufficient. In this sense, our choice of potential function is restricted by sources. We are free to choose any potential function as long as it represents the sources properly.

1.7.3. Gage Conditions

The use of vector potential functions requires the specification of the curl and the divergence of the potential functions for unique representation of field quantities. The curl is normally specified based on the properties of the field (e.g. in the case of the magnetic vector potential the solenoidal nature of **B** allows the definition of **A** from $\mathbf{B}=\nabla\times\mathbf{A}$). The divergence must be then specified to be consistent with the field equations. To arrive at Eq. (1.52) we used the relation

$$\nabla\cdot\mathbf{A} + \mu\varepsilon\frac{\partial V}{\partial t} = 0$$

This is known as the Lorentz gage condition. We say that this relation is consistent with the field equations because it leads to the continuity equation. To show that this is correct, consider Figure 1.3 where we assume that both a volume current density **J** and a volume charge density ρ exist in the volume. The magnetic vector potential can be obtained directly from the current density and the electric scalar potential from the charge density as

$$\mathbf{A} = \int_{v'} \frac{\mu\mathbf{J}}{4\pi R}dv'$$

$$V = \int_{v'} \frac{\rho}{4\pi\varepsilon R}dv'$$

These relations, which, are quite familiar from static solutions also apply to general time dependent fields. The current density **J** is the total current density and may include conduction, convection, displacement and polarization currents. The charge density ρ represents both free and bound (polarization) charges. Substituting these into the Lorentz gage in Eq. (1.51) gives

$$\frac{\mu}{4\pi}\int_{v'}\left(\nabla\cdot\frac{\mathbf{J}}{\mathrm{R}}V + \frac{1}{R}\frac{\partial\rho}{\partial t}\right)dv' = 0$$

The primed coordinates indicate that the integration is performed over the source region while the potentials **A** and V are anywhere in space. We can therefore write

$$\nabla\cdot\frac{\mathbf{J}}{R} = \frac{1}{R}\nabla\cdot\mathbf{J} + \mathbf{J}\cdot\nabla\left(\frac{1}{R}\right) = \mathbf{J}\cdot\nabla\left(\frac{1}{R}\right) = -\mathbf{J}\cdot\left(\frac{1}{R}\right)$$

where the $\nabla\cdot\mathbf{J}$ is always zero since **J** represents all current densities. The prime coordinates again represent the source region. To represent the quantity $\nabla'(1/R)$ we write

$$\nabla\cdot\left(\frac{\mathbf{J}}{R}\right)=\frac{1}{R}\nabla^{'}\mathbf{J}-\nabla^{'}\left(\frac{\mathbf{J}}{R}\right)$$

We can now write

$$\int_{v'}\left(\nabla\cdot\frac{\mathbf{J}}{\mathrm{R}}V+\frac{1}{R}\frac{\partial\rho}{\partial t}\right)\mathrm{d}v^{'}=\int_{v'}\frac{1}{R}\left(\nabla^{'}\cdot\mathbf{J}+\frac{\partial\rho}{\partial t}\right)dv^{'}=\int_{v'}\nabla^{'}\cdot\left(\frac{\mathbf{J}}{\mathrm{P}}\right)dv^{'}$$

Using the divergence theorem on the last term gives

$$\int_{v'}\nabla^{'}\cdot\left(\frac{\mathbf{J}}{R}\right)dv^{'}=\oint_{s'}\left(\frac{\mathbf{J.ds}}{R}\right)$$

by assuming a volume theat tends to infinity (i.e. $R\rightarrow\infty$ and therefore $S'\rightarrow\infty$) we get

$$\nabla^{'}\cdot\mathbf{J}+\frac{\partial\rho}{\partial t}=0 \tag{1.54}$$

which is the continuity equation. Thus, the use of the Lorentz gage is consistent with the field equations.

The Lorentz gage condition reduces to $\nabla\cdot\mathbf{A}=0$ in the static case. We will have no occasion to use this condition in the context of microwaves but, in the static case this is known as the Coulomb gage.

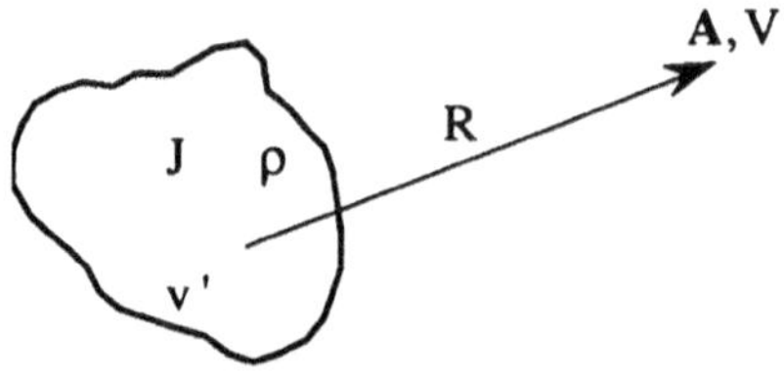

Figure 1.3. A volume containing both charge and current densities.

1.8. The General, Time Dependent Wave Equation

While Maxwell's equations are in themselves all that is necessary for field solutions, they are coupled equations in **E** and **H**. In most cases these equations cannot be solved other than by numerical means. Also, for the purpose of understanding the phenomena involved it is much easier to deal with a single variable rather than coupled variables. By uncoupling the equations we obtain partial differential equations in **E** or **H**, or any other function derivable from **E** and **H** (as, for example the magnetic vector potential **A**). We shall call the result of this uncoupling, wave equations since they represent the general, time dependent equations, whose solutions are waves. In the quasistatic case (that is, if the displacement current in Eq. (1.2) can be neglected), the same process results in diffusion equations while in the static case, they reduce to either a Poisson or a Laplace equation.

To obtain a wave equation in **E**, we first take the curl of both sides of Eq. (1.1)

$$\nabla\times(\nabla\times\mathbf{E}) = -\frac{\partial}{\partial t}(\nabla\times\mu\mathbf{H}) \tag{1.55}$$

where the constitutive relation $\mathbf{B}=\mu\mathbf{H}$ was used). From here on we will assume linear, isotropic materials. To remove **H** from the equations we use the second of Maxwell's equations and substitute it for $\nabla\times\mathbf{H}$ in Eq. (1.55)

$$\nabla\times(\nabla\times\mathbf{E}) = -\mu\frac{\partial\mathbf{J}}{\partial t} - \mu\varepsilon\frac{\partial^2\mathbf{E}}{\partial t^2}$$

Using the vector identity

$$\nabla\times(\nabla\times\mathbf{E}) = \nabla(\nabla\cdot\mathbf{E}) - \nabla^2\mathbf{E}$$

and substituting $\nabla\cdot\mathbf{E}=\rho/\varepsilon$ from Maxwell's third equation, we get

$$\nabla^2\mathbf{E} - \mu\varepsilon\frac{\partial^2\mathbf{E}}{\partial t^2} = \mu\frac{\partial\mathbf{J}}{\partial t} + \nabla\left(\frac{\rho}{\varepsilon}\right) \tag{1.56}$$

A similar procedure is followed for a wave equation in **H**. Starting with Maxwell's second equation, and taking the curl on both sides gives

$$\nabla\times(\nabla\times\mathbf{H}) = \nabla\times\mathbf{J} + \varepsilon\frac{\partial}{\partial t}(\nabla\times\mathbf{E})$$

We now use the same vector identity for $\nabla\times(\nabla\times\mathbf{H})$ as above, with $\nabla\cdot\mathbf{H}=0$ (from Maxwell's fourth equation) and substituting for $\nabla\times\mathbf{E}$ from Maxwell's first equation to get

$$\nabla^2\mathbf{H} - \mu\varepsilon\frac{\partial^2\mathbf{H}}{\partial t^2} = \nabla\times\mathbf{J} \tag{1.57}$$

Eq. (1.52) is also a wave equation in terms of the magnetic vector potential **A**:

$$\nabla^2\mathbf{A} - \mu\varepsilon\left(\frac{\partial^2\mathbf{A}}{\partial t^2}\right) = -\mu\mathbf{J} \tag{1.58}$$

Similarly, a wave equation in terms of the electric scalar potential V is written as (see Eq. (1.53) above).

$$\nabla^2 V - \mu\varepsilon \frac{\partial^2 V}{\partial t^2} = -\frac{\rho}{\varepsilon} \tag{1.59}$$

In these equations, and others like these, the right hand side represents the sources of the fields. The solutions to these equations is a propagating wave. The interaction of waves and materials is what we are normally interested in microwave nondestructive testing.

Note again that Eqs. (1.56) (1.57) (1.58) and (1.59) are second order, partial differential equations. Therefore, they have standard solutions. This does not imply that the solution to a particular physical situation is simple. This solution requires imposition of boundary conditions which may not be possible to do analytically. However, the equations can be solved much more readily than the original Maxwell's equations and, they can certainly be solved numerically.

1.8.1. The Time-Harmonic Wave Equation

In the time-harmonic domain, the phasor notation is again adopted (i.e. $\partial/\partial t = j\omega$) The four wave equations are now written as:

$$\nabla^2 \mathbf{E} + \omega^2 \mu\varepsilon \mathbf{E} = j\omega\mu \mathbf{J} + \nabla\left(\frac{\rho}{\varepsilon}\right) \tag{1.60}$$

$$\nabla^2 \mathbf{H} + \omega^2 \mu\varepsilon \mathbf{H} = \nabla \times \mathbf{J} \tag{1.61}$$

$$\nabla^2 \mathbf{A} + \omega^2 \mu\varepsilon \mathbf{A} = -\mu \mathbf{J} \tag{1.62}$$

$$\nabla^2 \mathrm{V} + \omega^2 \mu\varepsilon \mathrm{V} = -\frac{\rho}{\varepsilon} \tag{1.63}$$

In these equations the sources were left essentially as in Eqs. (1,56), (1.57), (1.58) and (1.59), although, sometimes it will become convenient to write the sources in different forms. As an example, in Eq. (1.60), the term $j\omega\mu\mathbf{J}$ is often written as $j\omega\mu\sigma\mathbf{E}$ by using the constitutive relation for **J**. The main advantage in doing so is to show that this term is a loss term by incorporation of the conductivity explicitly. Also, any of these sources may not exist in a particular situation.

1.8.2. The Helmholtz Equations

A particularly useful form of the wave equations are the source free (or homogeneous) wave equations. By removing all source terms in Eqs. (1.60) through (1.63), we also assume implicitly, that there are no losses in the space in which the equations apply. This is certainly true in free space, but, the results obtainable from the source free equations apply, at least to a good approximation, in more realistic situations. For example, in a microwave cavity, the source free equations are a good approximation since the losses are small and, normally, there are no applied sources within the cavity.
The source free equations are:

$$\nabla^2 \mathbf{E} + \omega^2 \mu\varepsilon \mathbf{E} = 0 \tag{1.64}$$

$$\nabla^2 \mathbf{H} + \omega^2 \mu\varepsilon \mathbf{H} = 0 \tag{1.65}$$

$$\nabla^2 \mathbf{A} + \omega^2 \mu\varepsilon \mathbf{A} = 0 \tag{1.66}$$

$$\nabla^2 \mathrm{V} + \omega^2 \mu\varepsilon \mathrm{V} = 0 \tag{1.67}$$

These are known as the homogeneous or Helmholtz type equations. Note that the general permittivity and permeability instead of free space permittivity and permeability were used to indicate that, under the conditions of no sources and no losses, these apply in general media. We note, however, that the equations are linear. Their solution is simple relative to the general wave equations and we shall make considerable use of these in subsequent chapters. It is also possible to write similar equations in terms of **D** or **B**.

1.9. Propagation of Waves: Plane Waves in Lossless Dielectrics

The solution to any of the wave equations above is fundamental to understanding of interactions with materials. A solution can be obtained to any of these equations but, for simplicity, we choose the source free wave equation for the electric field (Eq. (1.64)). This will allow us to point to the important aspects of the wave solution properties without having to resort to the more complex solution of the nonhomogeneous equation. The choice of a time-harmonic wave equation for this purpose reflects the importance of time-harmonic waves and, at the same time allows transformation back to the time domain. In addition, many of the more important aspects of wave propagation are easier to define and understand in the frequency domain. Some terms, such as phase, only have a meaning in the context of sinusoidal excitation. We will limit ourselves here to discussion of uniform plane waves. A uniform plane wave is a wave (that is: a solution to a wave equation), in which the electric and the magnetic field intensities are constant in magnitude and phase and each is in a constant direction on planes perpendicular to the direction of propagation. The assumption of plane waves simplifies discussion but does not, in any way, restrict it.

All results obtained are general enough especially since in many cases (such as at large distances from a source) we actually have waves which are very close to plane waves.

The general equation we need to solve is

$$\hat{\mathbf{x}}\nabla^2 E_x + \hat{\mathbf{y}}\nabla^2 E_y + \hat{\mathbf{z}}\nabla^2 E_z + \omega^2\mu\varepsilon(\hat{\mathbf{x}}E_x + \hat{\mathbf{y}}E_y + \hat{\mathbf{z}}E_z) = 0 \tag{1.68}$$

where the vector Laplacian ($\nabla^2\mathbf{E}$) is written in terms of the components of **E**. To simplify discussion, we assume that the electric field has only a component in the x direction and varies only in the z direction. These assumptions imply the following conditions:

$$E_y = E_z = 0 \qquad \text{and} \qquad \frac{\partial E_*}{\partial x} = \frac{\partial E_*}{\partial y} = 0$$

where * denotes any component of **E**. Substitution of these into Eq. (1.68) results in

$$\frac{\partial^2 E_x}{\partial z^2} + \omega^2\mu\varepsilon E_x = 0$$

Since E_x depends on z only, the equation can be written as an ordinary differential equation

$$\frac{d^2 E_x}{dz^2} + \omega^2\mu\varepsilon E_x = 0 \tag{1.69}$$

Before writing the solution we observe the following:

1. The term $\omega^2\mu\varepsilon$ has units of $(rad/m)^2$. For this reason, we normally define a quantity

$$k = \omega\sqrt{\mu\varepsilon} \tag{1.70}$$

and call it the "wave number" or a "phase constant". The term wave number will become obvious shortly but, for now, we note that, as a phase constant, it defines how the wave propagates. Analogously, in free space, we define the "free space wave number" as

$$k_0 = \omega\sqrt{\mu_0\varepsilon_0}$$

where μ_0 and ε_0 are the permeability and permittivity of vacuum. k_0 is a real number in vacuum, but, for a general material k is normally a complex number.

2. The term $\mu\varepsilon$ has units of $(sec/m)^2$. Therefore, a velocity can be defined as

$$v = \frac{1}{\sqrt{\mu\varepsilon}} \tag{1.71}$$

This velocity, is known as the velocity of propagation of the wave or phase velocity. In free space this velocity is the speed of light

$$c = \frac{1}{\sqrt{\mu_0 \varepsilon_0}}$$

The wave number can now be written as:

$$k = \omega\sqrt{\mu \varepsilon} = \frac{\omega}{v} \qquad or \qquad k_0 = \omega\sqrt{\mu_0 \varepsilon_0} = \frac{\omega}{c}$$

3. The square root of the ratio between the product of permeability and permittivity of any material and that of the product of permeability and permittivity of free space is the index of refraction of the material.

$$n = \sqrt{\frac{\mu \varepsilon}{\mu_0 \varepsilon_0}} = \sqrt{\mu_r \varepsilon_r} \tag{1.72}$$

where μ_r and ε_r are the relative permeability and relative permittivity respectively ($\mu = \mu_0 \mu_r$, $\varepsilon = \varepsilon_0 \varepsilon_r$).

Eq. (1.69) describes simple harmonic motion, therefore has the following solution:

$$E_x(z) = E_0^+ e^{-jkz} + E_0^- e^{jkz} \tag{1.73}$$

where E_0^+ and E_0^- are constants to be determined from the boundary conditions of the problem. The notation *(+)* and *(−)* indicates that the first term is a propagating wave in the positive *z* direction and the second a propagating wave in the negative *z* direction. The constants can be complex and are arbitrary.

Using the phasor transformation, we can write this in the time domain as

$$E_x(z,t) = E_0^+ cos(\omega t - kz + \phi) + E_0^- cos(\omega t + kz + \phi) \tag{1.74}$$

where the arbitrary phase angle ϕ was added for completeness.

If we assume a radiation boundary condition, E_0^- is zero (that is, the wave radiates out of the volume of interest and there is no backward propagating wave). On the other hand, if the forward propagating wave is reflected without losses (e.g. from a perfect conductor), the amplitude of the two waves is equal. Assuming only a forward propagating wave, the solution is

$$E_x(z) = E_0^+ e^{-jkz} \tag{1.75}$$

or, in the time domain

$$E_x(z,t) = E_0^+ cos(\omega t - kz + \phi) \qquad (1.76)$$

Looking at these expressions, it is obvious that what changes with time is the phase of the wave. In other words, the phase of the wave "travels" at a certain velocity. To see what this velocity is we use Figure 1.4. We follow a fixed point on the wave, for which $\omega t - kz = const$. Then

$$z = \frac{\omega t}{k} - const$$

To find the phase velocity, we write

$$\frac{dz}{dt} = v = \frac{\omega}{k} = \frac{1}{\sqrt{\mu\varepsilon}} \qquad [m/sec]$$

This velocity is the same velocity calculated in Eq. (1.71). Thus the phase velocity in free space is equal to the speed of light. The phase velocity is the speed of propagation of the electromagnetic wave in any medium. We note also, that the speed of propagation in any material is lower than c since ε_r and μ_r are both equal or larger than 1 for all materials (permeability of diamagnetic materials is slightly lower than permeability of free space but their permittivity is larger than that of free space).

Now we can also define the wavelength λ (in meters) which is that distance the wave travels in one cycle

$$\lambda = \frac{v}{f} = \frac{2\pi}{k} \qquad [m] \qquad (1.77)$$

The name "wave number" now becomes obvious. If we write

$$k = \frac{2\pi}{\lambda} \qquad (1.78)$$

k is the number of wavelengths in one cycle.

Up to this point we treated the electric field only, disregarding the magnetic field intensity. In propagation of electromagnetic waves the magnetic field intensity must be treated together with the electric field. Indeed, without this coupling the propagation would not exist in the first place. To see the relation between the electric and magnetic field we use Maxwell's second equation. The source free equation is

$$\nabla \times \mathbf{E} = -j\omega\mu\mathbf{H}$$

In terms of the components of the equation this is

$$\hat{\mathbf{x}}\left(\frac{\partial E_z}{\partial y}-\frac{\partial E_y}{\partial z}\right)+\hat{\mathbf{y}}\left(\frac{\partial E_x}{\partial z}-\frac{\partial E_z}{\partial x}\right)+\hat{\mathbf{z}}\left(\frac{\partial E_y}{\partial x}-\frac{\partial E_x}{\partial y}\right)=-j\omega\mu(\hat{\mathbf{x}}H_x+\hat{\mathbf{y}}H_y+\hat{\mathbf{z}}H_z)$$

From the assumption that only the term $\partial E_x/\partial z$ exists, we have

$$H_x = H_y = 0$$

and

$$\frac{\partial E_x}{\partial z} = -j\omega\mu H_y$$

or, writing this for the forward propagating wave, we can write for H_y

$$H_y^+ = \frac{j}{\omega\mu}\frac{\partial E_x^+}{\partial z}$$

Substituting for E_x^+ from Eq. (1.73), we get

$$\frac{\partial E_x^+}{\partial z} = \frac{\partial}{\partial z}(E_0^+ e^{-jkz}) = -jk(E_0^+ e^{-jkz}) = -jkE_x^+(z)$$

This gives

$$H_y^+(z) = \frac{k}{\omega\mu}E_x^+(z) \tag{1.79}$$

The ratio between $H(z)$ and $E(z)$, has units of $(1/\Omega)$. We define the quantity $\omega\mu/k$ as

$$\eta = \frac{\omega\mu}{k} = \sqrt{\frac{\mu}{\varepsilon}} \quad [\Omega] \tag{1.80}$$

This quantity is an impedance and is called the intrinsic impedance of the material since it is only dependent on material properties. The intrinsic impedance of free space is

$$\eta_0 = \sqrt{\frac{\mu_0}{\varepsilon_0}} = 377\ \Omega \tag{1.81}$$

The intrinsic impedance is a complex number in general media. Eq. (1.79) is now

$$H_y^+(z) = \frac{1}{\eta} E_x^+(z) \tag{1.82}$$

We note here that **H** and **E** propagate in the same direction. They are orthogonal to each other and to the direction of propagation. This property makes **E** and **H** transverse electromagnetic (*TEM*) waves. The ratio between **E** and **H** is equal to the intrinsic impedance of the material in which they propagate which, in general, is complex (the intrinsic impedance has both a resistive and a reactive part).

The discussion above was restricted to a single component of the electric and magnetic field intensities. However, we can do the same with any other component of the electric or magnetic field, and use any other direction of propagation. The restriction to lossless media, is also minor in that all that is necessary is replacement of the permittivity with the complex permittivity. This and the effect of losses are discussed next. Perhaps the most significant restriction in this discussion was the assumption of linear, isotropic materials. However, in the context of wave propagation, linearity is a good approximation.

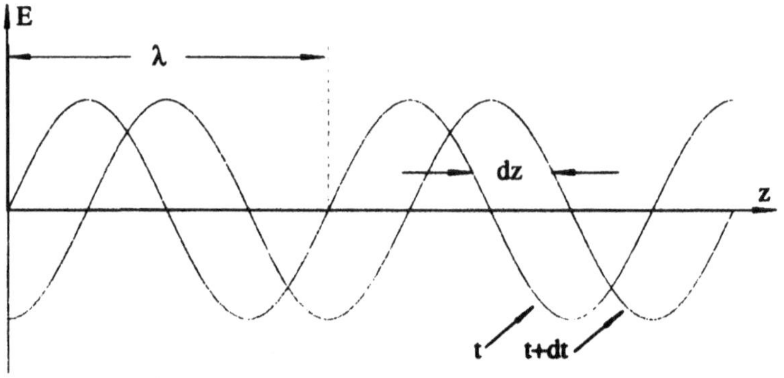

Figure 1.4. Definition of phase constant and wavelength.

1.10. Propagation of Plane Waves in Lossy Media

1.10.1. Losses in Materials

The loss mechanism in materials has two components: one is due to conduction currents, the other due to dielectric, or polarization losses. The complex permittivity was described in Eq. (1.23) as

$$\varepsilon = \varepsilon' - j\varepsilon''$$

The imaginary part of the permittivity represents the dielectric or polarization losses. By substituting this into Maxwell's second equation, we have

$$\nabla\times\mathbf{H} = j\omega\mathbf{D} + \mathbf{J} = j\omega\varepsilon\mathbf{E} + \sigma\mathbf{E}$$

Substitution of the complex permittivity for ε gives

$$\nabla\times\mathbf{H} = j\omega(\varepsilon' - j\varepsilon'')\mathbf{E} + \sigma\mathbf{E} = j\omega(\varepsilon' - j\varepsilon'' - \frac{j\sigma}{\omega})\mathbf{E}$$

Now, the losses are separated into two parts. The first, due to ε'' is the dielectric loss while the second due to σ is the conduction or Ohmic loss. For the purpose of analysis, the permittivity can now be generalized as

$$\varepsilon = \varepsilon' - j\left(\frac{\omega\varepsilon'' + \sigma}{\omega}\right) \tag{1.83}$$

This form shows that there is no difference in the effect of dielectric and Ohmic losses although their origins are quite different. Both manifest themselves as heat. We can, in fact consider the term $\omega\varepsilon''+\sigma$ to be an "effective conductivity" of the lossy dielectric.

The degree to which a dielectric is lossy is quantified by the ratio of the imaginary (lossy part) to the real (lossless part) parts of Eq. (1.83). This is written as

$$\tan\delta = \frac{\omega\varepsilon'' + \sigma}{\omega\varepsilon'} \tag{1.84}$$

This is called the loss tangent or the loss factor of the material. The loss tangent can also be seen as the ratio of the conduction current density to the displacement current density in the dielectric. It allows us to specify the behavior of the material by its real permittivity and a loss tangent. Since dielectric losses are frequency dependent, it is more convenient to specify the loss tangent rather than conductivity and dielectric losses. From a practical point of view, the loss tangent is relatively easy to measure compared to σ and ε'', while ε' is normally known for most materials.

In most dielectrics, the Ohmic losses are relatively small while polarization losses are small in "good" dielectrics. In lossy dielectrics, at microwave frequencies, polarization losses dominate. In high conductivity materials, Ohmic losses are dominant.

1.10.2. Propagation of Waves in Lossy Dielectrics

A lossy dielectric is defined as a material with small but not negligible losses. Loses may include conduction as well as dielectric losses.

Assuming a lossy medium with complex permittivity $\varepsilon=\varepsilon'+\varepsilon''$ and conductivity σ, the general wave equation for the medium can be written directly from Eq. (1.64) by substitution of ε and σ. The equation is

$$\nabla^2\mathbf{E} + \omega^2\mu\varepsilon\left(1-\frac{j\sigma}{\omega\varepsilon}\right)\mathbf{E} = 0$$

where, σ represents the modified conductivity discussed above $(\omega\varepsilon''+\sigma)$ and $\varepsilon=\varepsilon'$. Now, the wave number takes the following form:

$$k = \omega\sqrt{\mu\varepsilon}\sqrt{\left(1-\frac{j\sigma}{\omega\varepsilon}\right)}$$

which, as mentioned before, is a complex quantity. A (complex) propagation constant is defined through the following relation

$$-\gamma^2 = \omega^2\mu\varepsilon\left(1-\frac{j\sigma}{\omega\varepsilon}\right)$$

or

$$\gamma = j\omega\sqrt{\mu\varepsilon}\sqrt{\left(1-\frac{j\sigma}{\omega\varepsilon}\right)} \tag{1.85}$$

This term can now be substituted for jk in Eq. (1.69), using the same assumptions we made for the lossless case. The one dimensional wave equation is now

$$\frac{d^2E_x}{dz^2} - \gamma^2 E_x = 0 \tag{1.86}$$

and, the general solution has the same two wave components: one travelling in the positive z direction, the other in the negative z direction

$$E_x(z) = E_0^+e^{-\gamma z} + E_0^-e^{+\gamma z} \tag{1.87}$$

Similarly, assuming only an outgoing wave, we have

$$E_x(z) = E_0^+e^{-\gamma z} \tag{1.88}$$

Writing

$$\gamma = \alpha + j\beta$$

we have

$$E_x(z) = E_0^+ e^{-(\alpha + j\beta)z} = E_0^+ e^{-\alpha z} e^{-j\beta z} \tag{1.89}$$

In this form, the propagating wave has the same form as Eq. (1.75) where β has replaced k and the exponential term $e^{-\alpha z}$ multiplies the expression. This is therefore a wave, propagating in the positive z direction, with a phase velocity v and with an exponentially decaying amplitude. The phase velocity is now

$$v = \frac{\omega}{\beta}$$

and the wavelength is defined as

$$\lambda = \frac{2\pi}{\beta}$$

The magnetic field intensity can be written directly from Eq. (1.80) as

$$H_y^+(z) = \frac{1}{\eta} E_x^+(z)$$

where the intrinsic impedance is now

$$\eta = \frac{j\omega\mu}{\gamma} \tag{1.90}$$

and is a complex number. The imaginary part of the intrinsic impedance is due to losses

$$\eta = \frac{j\omega\mu}{\gamma} = \sqrt{\frac{\mu}{\varepsilon(1 - \frac{j\sigma}{\omega\varepsilon})}} = \eta_n\left(1 + \frac{j\sigma}{\omega\varepsilon}\right) \tag{1.91}$$

where η_n is the no-loss intrinsic impedance. The imaginary part (reactive) is normally small and is often neglected for low loss dielectrics.

The exponential term $e^{-\alpha z}$ is an attenuation term. The wave is attenuated as it propagates in space. The constant α is therefore an attenuation constant. The attenuation constant is measured in *Nepers/meter*. The *Neper* is a dimensionless constant. A *Neper* is defined as that attenuation constant for which the wave is attenuated to *1/e* of its amplitude in one meter.

1.10.3. Propagation of Waves in Low Loss Dielectrics

A low loss dielectric is a material for which $\sigma/\omega\varepsilon << 1$. The phase constant can now be approximated by

$$\gamma = j\omega\sqrt{\mu\varepsilon}\sqrt{\left(1 - \frac{j\sigma}{\omega\varepsilon}\right)} \approx j\omega\sqrt{\mu\varepsilon}\left(1 - \frac{j\sigma}{2\omega\varepsilon}\right) \tag{1.92}$$

where the following (binomial) expansion was used

$$\left(1 - \frac{j\sigma}{\omega\varepsilon}\right)^{\frac{1}{2}} = 1 - \frac{j\sigma}{2\omega\varepsilon} + \frac{1}{8}\left(\frac{\sigma}{\omega\varepsilon}\right)^2 + H.O.T.$$

and truncated after the second term. The attenuation constant is therefore

$$\alpha \approx \frac{\sigma}{2}\sqrt{\frac{\mu}{\varepsilon}} \tag{1.93}$$

and the phase constant is

$$\beta \approx \omega\sqrt{\mu\varepsilon} \tag{1.94}$$

The phase constant for low loss dielectrics is essentially unchanged from that for the lossless dielectric. We can retain the next term in the binomial expansion and get a more accurate approximation for β as

$$\beta \approx \omega\sqrt{\mu\varepsilon}\left(1 + \frac{1}{8}\left(\frac{\sigma}{\omega\varepsilon}\right)^2\right) \tag{1.95}$$

but, since $\sigma/\omega\varepsilon << 1$, the additional term is negligible and in most practical, low loss cases is not needed.

The intrinsic impedance in low loss dielectrics is complex and can be approximated as

$$\eta = \eta_n\left(1 + \frac{j\sigma}{\omega\varepsilon}\right) \tag{1.96}$$

where η_n is the no-loss intrinsic impedance for the same material (i.e. with the same μ and ε' but $\varepsilon''=0$).The phase velocity remains essentially unchanged from that in the lossless case since β changes very little ($v=\omega/\beta$).

1.10.4. Propagation of Waves in Conductors

In highly conductive materials the losses are high and we can assume that $\sigma >> \omega\varepsilon$ or, that $\varepsilon'' >> \varepsilon'$. This is equivalent to ignoring the displacement currents. Note that high conductivity does not mean perfect conductivity. Perfect conductors do not have losses due to the exclusion of fields from the interior of materials. In the case of high conductivity materials, the complex propagation constant can be written as

$$\gamma \approx j\omega\sqrt{\mu\varepsilon}\sqrt{-\frac{j\sigma}{\omega\varepsilon}} = (1+j)\sqrt{\frac{\omega\mu\sigma}{2}} \tag{1.97}$$

where

$$\sqrt{j} = (1+j)/\sqrt{2}$$

From this we get

$$\alpha = \beta = \sqrt{\frac{\omega\mu\sigma}{2}} = \frac{1}{\delta} \tag{1.98}$$

The attenuation and phase constants are equal and are very large. The wave is attenuated rapidly to the point where propagation in conducting media can only exist within short distances. The propagating wave can now be written as

$$E_x(z) = E_0^+ e^{-z/\delta} e^{-jz/\delta} \tag{1.99}$$

where the term

$$\delta = \sqrt{\frac{2}{\omega\mu\sigma}} \tag{1.100}$$

was used here and in Eq. (1.98). This constant is known as the skin depth or depth of penetration. It is defined as that distance through which the amplitude of a plane wave is attenuated to $1/e$ of its original amplitude. The skin depth in conductors is small. In the microwave range it can be of the order of a few microns (depending on material and frequency).

The phase velocity in good conductors is

$$v = \frac{\omega}{\beta} = \omega\delta = \sqrt{\frac{2\pi}{\mu\sigma}} \tag{1.101}$$

and is obviously small compared to the phase velocity in dielectrics or free space.

The intrinsic impedance is

$$\eta = \sqrt{\frac{\mu}{\varepsilon}} \approx \sqrt{\frac{j\omega\mu}{\sigma}} = (1+j)\frac{1}{\sigma\delta} \tag{1.102}$$

The phase angle of the intrinsic impedance is, therefore, $45°$. This is characteristic of good conductors for which the magnetic field intensity lags behind the electric field intensity by $45°$.

The wavelength also changes dramatically compared to free space or lossless dielectrics. It is given by

$$\lambda = \frac{2\pi}{\beta} = 2\pi\delta \tag{1.103}$$

Since δ is very small, the wavelength is also very short.

We return now to the skin depth. This term is very useful at low frequency applications since it defines the range in which currents or fields can be assumed to exist. In high frequency applications, it defines the depth in which we can have effective propagation of waves. Materials with small skin depths cannot support a propagating wave because of the large attenuation. These can be used for shielding purposes or for absorption of microwaves. It also means that microwaves cannot penetrate effectively in conducting media like sea water and soil. When such penetration occurs it is highly attenuated. At the same time, the very low penetration in conductors allows waveguides to have thin walls and the losses in such walls are minimal. Most conductors in the microwave range behave like perfect conductors and reflect waves rather than absorb them.

Another important point with regard to skin depth, one that is often overlooked is that skin depth is only properly defined for plane waves and only for sinusoidal excitations.

1.11. Electromagnetic Boundary Conditions

Through the constitutive relation we know that the electromagnetic field undergoes changes at the interface between two materials. If we look, for example at the relation $\mathbf{B}=\mu\mathbf{H}$, and assume we have two materials in contact, it is clear that the flux density $\mathbf{B}$ in each material is different. For this reason, a detailed knowledge of the behavior of fields at the interface between materials is important for analysis and for our understanding of electromagnetic phenomena. This is done by applying Maxwell's equations at the interface between two general media. We assume an arbitrarily directed field in one material and calculate the field in the other. For this purpose we use the integral form of Maxwell's equations. For convenience, this is done by separating the field into two components; one tangential and one normal to the interface, and applying Maxwell's equations to each component separately. The conditions for the electric field and magnetic field are calculated separately.

1.11.1. Interface Conditions for the Electric Field

Consider the interface between two general materials shown in Figure 1.5. We apply Maxwell's first equation (in integral form, Eq. (1.19)) to this interface

$$\oint_C \mathbf{E}\cdot d\mathbf{l} = -\int_s \frac{\partial \mathbf{B}}{\partial t}\cdot d\mathbf{s}$$

Since the integral must be performed on a closed contour, we choose the contour *abcda* shown in Figure 1.5. This contour is arbitrary in the sense that we could choose any other contour but is one that allows simple integration. We let the dimension w tend to zero. This makes the area of the loop tend to zero and, therefore, the total magnetic flux through the area is zero. Eq. (1.19) can now be written as

$$\oint_{abcda} \mathbf{E}.d\mathbf{l} = 0$$

Since the distances *bc* and *da* both tend to zero, the integration is along the two segments *ab* and *cd*

$$\int_{ab} E_{1t}dl_1 - \int_{cd} E_{2t}dl_2 = 0$$

where the vector notation has been dropped since the tangential components of the electric field and the direction **dl** are collinear. The minus sign in the second term comes from the fact that E_{2t} and dl_2 are in opposite directions. By assuming that the paths *ab* and *cd* are equal in length and short, the tangential components are constant along their corresponding paths. This gives

$$E_{1t} = E_{2t} \tag{1.104}$$

Thus, the tangential components of the electric field intensity are continuous across material interfaces.

To see how the normal component of the electric field behaves, we use Gauss' law (Maxwell's fourth equation) for the electric flux density:

$$\oint_s \mathbf{D} \cdot \mathbf{ds} = Q$$

where Q is the total charge in the volume enclosed by the surface S. For simplicity, we construct a cylindrical volume, with its axis perpendicular to the interface as shown in Figure 1.6. In this figure, the normal components of the electric flux density and **ds** are collinear in each material. In material *1* they are in opposite directions and in material *2* in the same direction. We allow the height of the cylinder to tend to zero such that all charges enclosed in the volume, if any, are on the interface. The contribution to the integral due to the lateral surface is zero (**D** and **ds** are perpendicular to each other on this surface and therefore their scalar product is zero). Assuming the surface of the base of the cylinder to be S, we have:

$$D_{1n} - D_{2n} = \rho_s \tag{1.105}$$

where ρ_s is the surface charge density at the interface, if any such charge density exists.

In the particular case where the surface charge density is zero, the normal component of the electric flux density is continuous across the interface. For linear, isotropic materials, we can substitute the constitutive relations to obtain interface conditions in terms of **E** or **D**:

$$\frac{D_{1t}}{\varepsilon_1} = \frac{D_{2t}}{\varepsilon_2} \tag{1.106}$$

$$\varepsilon_1 E_{1n} - \varepsilon_2 E_{2n} = \rho_s \tag{1.107}$$

Table 1.3 summarizes the interface conditions for the electric field between two materials.

1.11.2. Interface Conditions for The Magnetic Field

Repeating the process shown above for the magnetic field intensity, we obtain the interface conditions for **H** and **B** using Maxwell's second and fourth equations. To define the interface conditions for the magnetic field intensity **H** consider Figure 1.7. Using the equation

$$\oint_C \mathbf{H} \cdot \mathbf{dl} = \int_s \left(\mathbf{J} + \frac{\partial \mathbf{D}}{\partial t}\right) \cdot \mathbf{ds}$$

and the contour *abcda* we can evaluate **H**. Again, allowing the distances *bc* and *da* to tend to zero, the total electric flux is zero. However, there can still be a current density on the surface at the interface. The equation now is

$$\int_{ab} H_{1t} dl_1 - \int_{ad} H_{2t} dl_2 = \int_{ab} J dl$$

Table 1.3. Interface (boundary) conditions for the electric field.

General conditions at interface	Conditions at interface without surface charge densities
$E_{1t} = E_{2t}$	$E_{1t} = E_{2t}$
$\varepsilon_1 E_{1n} - \varepsilon_2 E_{2n} = \rho_s$	$\varepsilon_1 E_{1n} = \varepsilon_2 E_{2n}$
$\dfrac{D_{1t}}{\varepsilon_1} = \dfrac{D_{2t}}{\varepsilon_2}$	$\dfrac{D_{1t}}{\varepsilon_1} = \dfrac{D_{2t}}{\varepsilon_2}$
$D_{1n} - D_{2n} = \rho_s$	$D_{1n} = D_{2n}$

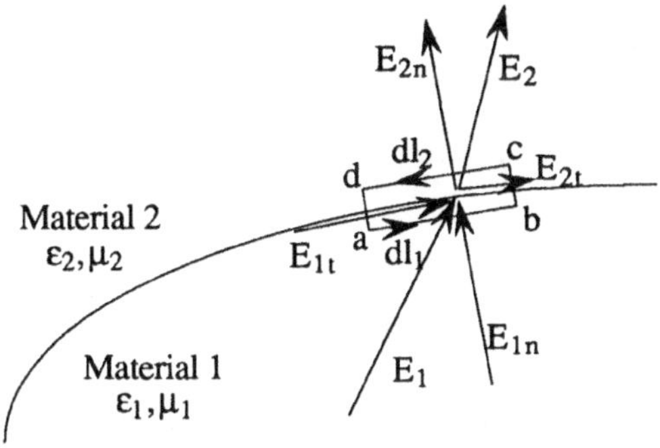

Figure 1.5. Interface conditions for the tangential component of the electric field intensity.

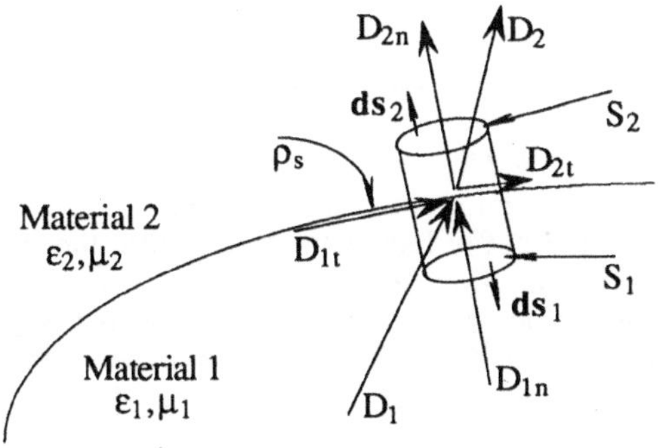

Figure 1.6. Interface conditions for the normal component of the electric field intensity.

In this expression, the vector notation was dropped since the fields are collinear with the path (H_{2t} and dl_2 are in opposite directions, hence the negative sign). The current density is perpendicular to the loop *abcda* and, since it can only depend on the width of the loop, it is integrated over the path *dl*.

Integrating over the two segments *ab* and *cd*, and allowing *ab*=*cd*, we get

$$H_{1t} - H_{2t} = J_s \tag{1.108}$$

From Maxwell's third equation, and using again, the integral form we evaluate the boundary conditions for **B**

$$\oint_s \mathbf{B}.d\mathbf{s} = 0$$

A cylindrical volume is shown in Figure 1.8 where the normal components of **B** are perpendicular to the bases of the cylinder. Thus, B_{n1} is in the same direction as ds_1 and B_{n2} is opposite ds_2. Allowing the volume of the cylinder to tend to zero, we have:

$$\int_{s_1} B_{1n} ds_1 - \int_{s_2} B_{2n} ds_2 = 0$$

With S_1=S_2, the boundary condition becomes:

$$B_{1n} = B_{2n} \tag{1.109}$$

Again, with the use of the constitutive relation in Eq. (1.6), we can write the interface conditions for the normal component of **H** and the tangential component of **B** as

$$\frac{B_{1t}}{\mu_1} - \frac{B_{2t}}{\mu_2} = J_s \tag{1.110}$$

$$\mu_1 H_{1n} = \mu_2 H_{2n} \tag{1.111}$$

These conditions are summarized in Table 1.4.

To obtain these four sets of interface conditions, we used the four Maxwell equations. Since the four equations are not independent, neither are the four interface conditions. Whenever we need to specify interface conditions, we must take care to specify an independent set. As an example, specifying the conditions for tangential components of **H** and **E** or normal components of **B** and **D** is correct but specifying tangential components of **H** and normal components of **D** or tangential components of **E** and normal component of **B** is not since these conditions are equivalent.

Table 1.4. Interface (boundary) conditions for the magnetic field

General conditions at interface	Conditions at interface without surface current densities
$H_{1t} - H_{2t} = J_s$	$H_{1t} = H_{2t}$
$\mu_1 H_{1n} = \mu_2 H_{2n}$	$\mu_1 H_{1n} = \mu_2 H_{2n}$
$\dfrac{B_{1t}}{\mu_1} - \dfrac{B_{2t}}{\mu_2} = J_s$	$\dfrac{B_{1t}}{\mu_1} = \dfrac{B_{2t}}{\mu_2}$
$B_{1n} = B_{2n}$	$B_{1n} = B_{2n}$

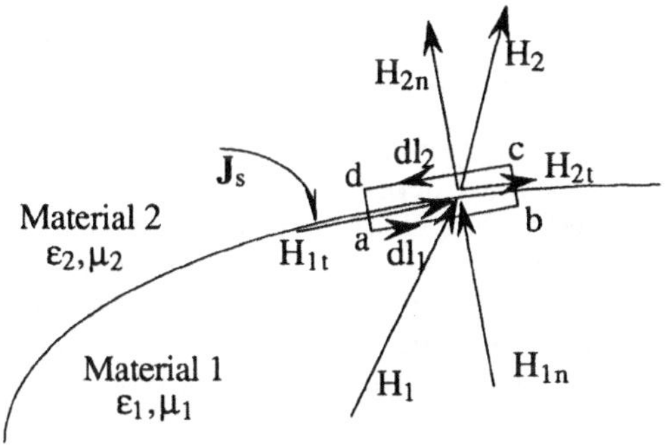

Figure 1.7. Interface conditions for the tangential component of the magnetic field intensity.

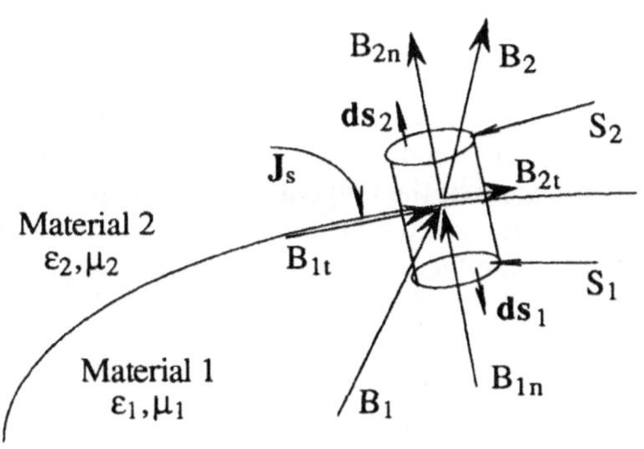

Figure 1.8. Interface conditions for the normal component of the magnetic field intensity.

1.11.3. Interface Conditions Between Two Lossless Dielectrics

At a perfect dielectric interface, current and charge densities cannot exist. This reduces the interface conditions to:

$$E_{1t} = E_{2t} \tag{1.112}$$

$$D_{1n} = D_{2n} \tag{1.113}$$

$$H_{1t} = H_{2t} \tag{1.114}$$

$$B_{1n} = B_{2n} \tag{1.115}$$

In other words, these components of the fields are continuous. The discontinuity in, for example, the normal component of **E** still exists since this is defined by the change in ε across the interface.

1.11.4. Interface Between a Dielectric and a Conductor

A conductor is characterized by its conductivity, which, for practical purposes, is part of the imaginary part of the complex permittivity. In this case, we can expect to have both surface current densities and surface charge densities at the interface. For most conducting metals, the conductivity is high and we may assume they behave as perfect conductors for most practical purposes. The interface conditions between a dielectric and a conductor can be deduced from the general conditions in Eqs. (1.107) through (1.111) by setting the electric and magnetic field intensities inside the conductor to zero. The conditions are:

$$E_{1t} = 0 \tag{1.116}$$

$$D_{1n} = \rho_s \tag{1.117}$$

$$H_{1t} = J_s \tag{1.118}$$

$$B_{1n} = 0 \tag{1.119}$$

This condition is often encountered in microwave applications. It is normally referred to as an "electric wall" to indicate that the tangential components of the electric field are zero. While the argument of a perfect conductor may not apply in other cases, in the microwave range this can be justified by the fact that the skin depth is negligibly small, enforcing this condition implicitly.

1.11.5. Other Interface Conditions

The conditions described above include the general interface conditions and some particularly useful conditions. Depending on the particular condition encountered in practice, other considerations may apply. As an example, at the interface between lossy dielectrics, charge densities may accumulate but surface current densities are usually absent.

One particularly useful approximation often used in microwave engineering is the interface between a dielectric and an infinitely permeable material. Since such materials do not exist, this condition should be viewed only as an approximation.

A magnetic wall, in analogy with the electric wall condition, implies that the tangential component of **H** is zero at the interface. On the other hand, the tangential component of the electric field is now calculated from Eq (1.1) after adding the magnetic current density (magnetization) **M.** In effect we assume that a magnetic wall has a surface magnetic current density or surface magnetization. The conditions at the boundary are:

$$E_{1t} = -M_s \tag{1.120}$$

$$D_{1n} = 0 \tag{1.121}$$

$$H_{1t} = 0 \tag{1.122}$$

$$B_{1n} = 0 \tag{1.123}$$

This condition, together with the electric wall condition are useful in analyzing microwave structures like waveguides, cavities and apertures.

Another aspect of boundary conditions is the behavior of fields at infinity. While we normally do not need to specifically apply these conditions, their existence is implied since we must assume that electromagnetic fields propagate throughout space. These conditions are called radiation or absorption boundary conditions. As an example, an absorption boundary condition, at infinity implies that none of the energy in the electromagnetic field is reflected back. In most cases this can be justified by observing that, in a lossy medium, the only way we might have reflections in an infinite domain is if the energy in the source producing the fields is infinite. For this reason, we will normally assume that we only have an outward propagating wave and an implicit radiation condition at infinity.

In some cases, it is useful to create an artificial radiation boundary condition at a finite boundary. This is often done in computation of electromagnetic fields to ensure that artificially located boundaries do not create reflections which, in the actual infinite domain do not exist. We will expand on this in chapters 9 and 10.

The distinction between absorption and radiation boundary conditions is important. In an absorption boundary condition, the reflections are eliminated by dissipating the energy in the wave. As an example, a lossy material of sufficient thickness will ensure that no significant waves can reflect from the finite boundary. A radiation boundary condition does not dissipate energy but rather radiates it outward.

A related condition that is sometimes useful is an impedance boundary condition. In this type of condition the impedance on the boundary is taken into account. As an example, the resistivity of the conducting boundaries of waveguides and resonators can be used as impedance boundary conditions.

1.12. Bibliography

The material covered in this chapter is a rather common collection of topics in field theory. The topics were selected as needed to support subsequent discussions on microwaves. The list of references below gives a number of good sources on these topics. The list is arranged in three levels, in order of the author preference. In addition to these references, almost any textbook on electromagnetic field covers the material given here

The books by Collin [8] and Pozar [10] were written specifically for microwave engineering.

Maxwell's Treatise [14] is included here as a background reference. Most readers will find the notation difficult to follow but it is always interesting reading.

Smythe's book [15] is particularly interesting in its approach to problem solving.

A very interesting account of Heaviside's contribution to the theory of electromagnetic fields as well as a historical account can be found in Nahin [16].

Textbooks, beginners level: equivalent to a first course in electromagnetics

[1] M.A. Plonus, "Applied Electromagnetics", McGraw-Hill, Inc., 1978.
[2] D.K. Cheng, "Field and Wave Electromagnetics", 2nd ed., Addison-Wesley, Reading, Ma., 1989.
[3] P. Lorain and D. R. Corson, "Electromagnetism Principles and Applications", 2d ed., W. H. Freeman and Company, San Francisco, 1990.
[4] N.N. Rao, "Elements in Engineering Electromagnetics", 3rd ed., Prentice- Hall, Inc., Englewood Cliffs, N.J., 1991.
[5] W.H. Hayt, "Engineering Electromagnetics," 5th ed., McGraw-Hill Book Company, New York, 1989.

Intermediate level: equivalent to a second course in electromagnetics

[6] R. Plosney and R. E. Collin, "Principles and Applications of Electromagnetic Fields," McGraw-Hill Book Company, New York, 1961.
[7] P. Lorain and D. R. Corson, "Electromagnetic Fields and Waves,", 2d ed., W. H. Freeman and Company, San Francisco, 1970.

Advanced level: equivalent to a graduate course in electromagnetics

[8] R.E. Collin, "Electromagnetic Theory of Guided Waves", McGraw-Hill, NY, 1991.
[9] J.A. Kong, "Theory of Electromagnetic Waves," John Wiley & Sons, Inc., New York, 1975.
[10] D.M. Pozar, Microwave Engineering, Addison-Wesley, Reading, MA., 1990.
[11] R.E. Collin, "Foundations for Microwave Engineering", McGraw-Hill, Inc., New York, 1992.
[12] R.F. Harrington, "Time-harmonic Electromagnetic Fields," McGraw- Hill, Inc., New York, 1961.

[13] H.C. Chen, "Theory of Electromagnetic Waves", McGraw-Hill Book Company, New York, 1983.
[14] J. C. Maxwell, "A Treatise on Electricity and Magnetism," Vol. 1 and 2, Dover Publications, Inc., New York, 1954.
[15] W. R. Smythe, "Static and Dynamic Electricity," McGraw-Hill, Inc., New York, 1991.
[16] P.J. Nahin, "Oliver Heaviside: Sage in Solitude", IEEE Press, New York, 1988.

Chapter 2
Transmission Lines, Waveguides, and Resonant Cavities

2.1. Transmission Lines

Transmission of energy from point to point can be accomplished in one of two ways. One is by propagation of fields through free space or through a space filled with some medium. This is the case, for example, of transmission by antennas. The second, one that we will discuss here, is the propagation of energy through a defined structure that physically connects the generator and the load. This is the general definition of a transmission line. We will view the transmission line, the generator, and the load in general terms. For example, a transmission line can be a dielectric slab or two wires, while the generator is a general source - either a true generator or the output from another transmission line.

We make a very important distinction between transmission lines that operate at high frequencies but have their counterpart in low frequency applications. These will be called simply, transmission lines. Examples are the two wire transmission lines and the coaxial line. A second group are the waveguides. The differentiating factor is that waveguides will be viewed as single conductor transmission lines and will be analyzed from a field theory point of view while transmission lines in this definition contain two conductors and will be analyzed from a circuit theory point of view. We note, however, that this distinction is artificial and cannot always be followed. For example, two, infinite parallel plates, even though they can form a transmission line based on this definition, will be analyzed as a waveguide because it is much easier to do so. (However, parallel plates have different properties as waveguides and as transmission lines).

We also make a distinction between low frequency and high frequency transmission lines because of the different methods used for their description and analysis.

At low frequencies we assume that all line properties can be defined in terms of the lumped line parameters and there is no wave propagation on the line. In effect we assume the following:

1. Line length is significantly shorter than the wavelength at the highest frequency the line is required to handle.
2. Line parameters are lumped. The line is characterized by its lumped inductance, resistance, capacitance and conductance, as shown in Figure. 2.1.

As an example, at *50 Hz*, the wavelength in free space is *6,000 km*

$$\lambda = \frac{c}{f} = \frac{3 \times 10^8}{50} = 6 \times 10^6 \ m$$

This means that for any propagation effect to be noticeable, the transmission line must be hundreds of *km* long even if the speed of propagation is lower than the speed of light (that is: the line is made with higher permittivity materials, as is almost always the case).

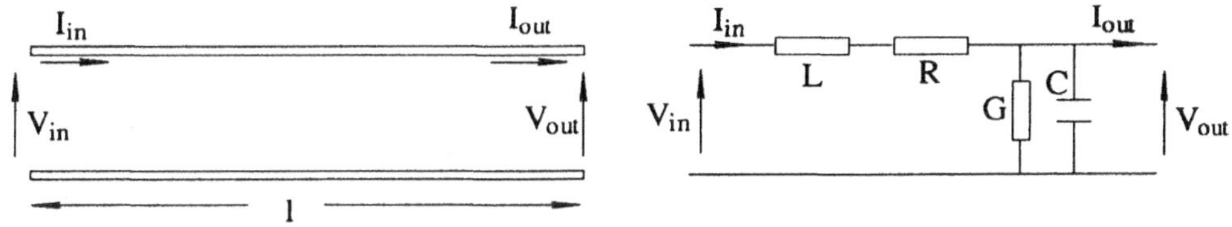

Figure. 2.1. Transmission line and lumped equivalent circuit at low frequencies. Line length is arbitrary.

At high frequencies, say at frequencies for which the wavelength is longer than the transmission line, the applied source energy propagates over the line at a speed of propagation (phase constant) that is dependent on line properties as it does at low frequencies. However, because of the relative size of the wavelength and line, the effect is very important as we shall see shortly.

Before describing the transmission line parameters and transmission line equations, we will use a simple explanation to understand why high frequency transmission lines must be treated differently than low frequency transmission lines.

Consider the transmission line in Figure 2.2. A sinusoidal wave connected to the input, travels along the transmission line at speed v, which depends on the properties of the transmission line. From the definition of the wavelength (see Eq. (1.77)), the phase of the wave changes by 2π in one wavelength. The wave amplitude and phase are shown at a given time, say $t=t_0$ in Figure 2.3. As time increases, a location with constant phase moves to the right and a propagating wave is generated as in Figure 2.3.

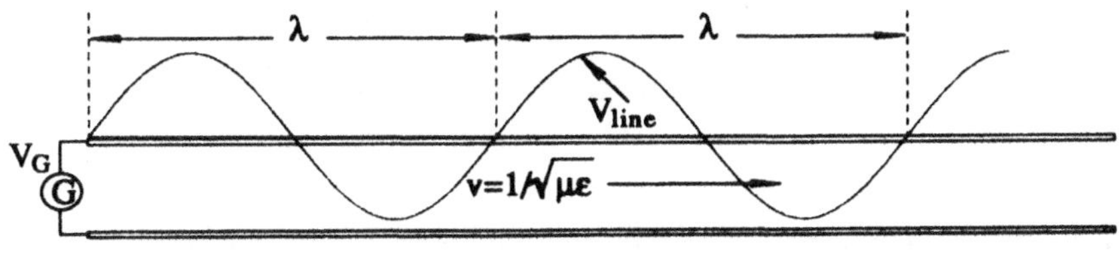

Figure 2.2. A long transmission line under sinusoidal excitation.

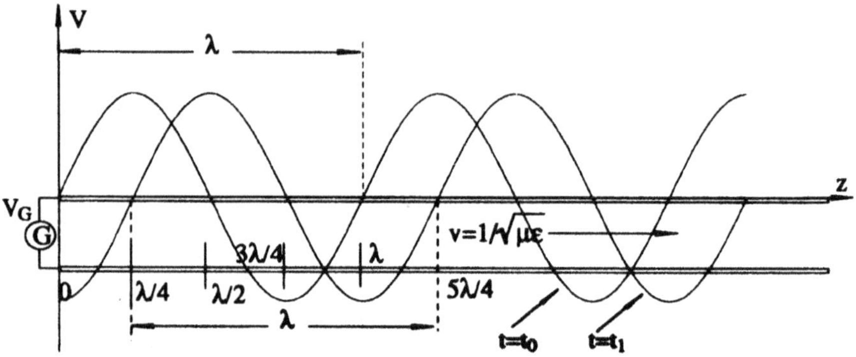

Figure 2.3. Conditions along the line for a traveling wave.

Different locations on the line experience different conditions, indicating that a lumped parameter approach cannot be used. In particular, the electric and magnetic fields along the line vary and a lumped circuit approach does not allow their calculation.

Instead, a distributed parameter approach can be used. A segment of the transmission line of unit length is defined as in Figure 2.4. This unit length line has lumped parameters R, L, C and G as shown. R is the resistance of the line, L the inductance, C is the capacitance and G the conductance of the line. These are given per unit length. The impedance of the line is therefore

$$Z = R + j\omega L$$

and the admittance is

$$Y = G + j\omega C$$

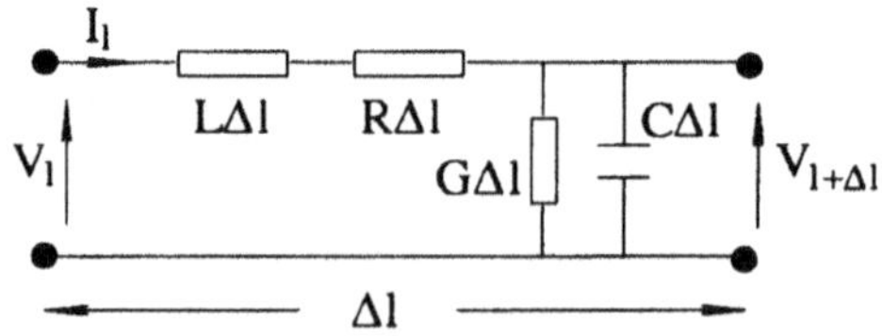

Figure 2.4. Definition of a segment of transmission line. Δ is arbitrary but small. Normally this is taken to be a unit length.

These parameters can now be used to define a transmission line as shown in Figure 2.5. The Δ notation was used to indicate that the segment of line used is arbitrary but must be small compared to wavelength. Now we can write the circuit equations using Kirchoff's laws for one of the segments and find the transmission line equations. Instead, we will use a simpler approach, based on the general solution of the wave equation given in chapter 1. Assuming that the transmission line is uniform and that both a voltage and current wave exist along the line, these waves are solutions to the following scalar wave equations:

$$\frac{\partial^2 V}{\partial z^2} - \gamma^2 V = 0 \tag{2.1}$$

and

$$\frac{\partial^2 I}{\partial z^2} - \gamma^2 I = 0 \tag{2.2}$$

where both V and I are assumed to be phasors. The constant γ is the propagation constant as was defined in chapter 1 (Eq. (1.85)). The solution to these equations has two parts: one propagating in the positive z direction, the other in the negative z direction. Using the notation we used in chapter 1, we get the solution

$$V(z) = V^+ e^{-\gamma z} + V^- e^{\gamma z} \tag{2.3}$$

where V^+ and V^- are the amplitudes (complex in general) of the waves propagating in the positive z and negative z directions respectively. For the current we chose the following solution

$$I(z) = I^+ e^{-\gamma z} - I^- e^{\gamma z} \tag{2.4}$$

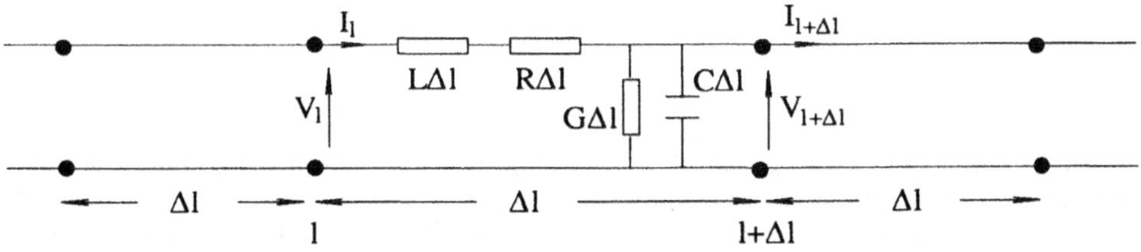

Figure 2.5. A transmission line viewed as a distributed parameter circuit with the segment in Figure 2.4. as the building block.

The reason for the negative sign is not obvious from Eqs. (2.3) and (2.4). However, if we view the first term as a propagating wave in the positive z direction, with voltage V^+ and current I^+, the Poynting vector (the vector product of the two waves) must be in the positive z direction. The Poynting vector for the waves propagating in the negative z directions must be negative. Thus, the need for the negative sign.

Now we define two characteristic quantities of the line:

1. The propagation constant γ, is given by

$$\gamma = \alpha + j\beta = \sqrt{ZY} = \sqrt{(R + j\omega L)(G + j\omega C)} \tag{2.5}$$

where α is the attenuation constant and β the propagation constant.

2. The characteristic impedance Z_0 of the line is given by

$$Z_0 = \sqrt{\frac{Z}{Y}} = \sqrt{\frac{R + j\omega L}{G + j\omega C}} \quad [Ohm] \tag{2.6}$$

Now, the current can be written as

$$I(z) = \frac{V^+}{Z_0} e^{-\gamma z} - \frac{V^-}{Z_0} e^{\gamma z} \tag{2.7}$$

If the transmission line is infinitely long, then the wave propagating in the $-z$ direction cannot exist and the voltage and current waves are

$$V(z) = V^+ e^{-\gamma z} \tag{2.8}$$

and

$$I(z) = \frac{V^+}{Z_0} e^{-\gamma z} \tag{2.9}$$

If the line is not infinite, the line must be terminated in a load Z_l, for the current I to flow (including shorted and open lines). Unless the load is perfectly matched to the impedance of the line $(Z_l = Z_0)$, there will be a current flowing from the load towards the source, since not all energy is transferred to the load. The part flowing towards the source is reflected energy. The voltage at the load is

$$V(l) = V^+ e^{-\gamma l} + V^- e^{\gamma l} \tag{2.10}$$

and the current is

$$I(l) = \frac{V^+}{Z_0} e^{-\gamma l} - \frac{V^-}{Z_0} e^{\gamma l} \tag{2.11}$$

Thus, the impedance of the load is

$$Z_l = Z_0 \frac{V^+ e^{-\gamma l} + V^- e^{\gamma l}}{V^+ e^{-\gamma l} - V^- e^{\gamma l}} \tag{2.12}$$

From this, if the backward propagating wave is zero, $Z_l = Z_0$ as required for matched loads (see Figure 2.6).

Finally, we also mention that the wavelength for any propagating wave is given as

$$\lambda = \frac{2\pi}{\beta} \tag{2.13}$$

and the velocity of propagation on the line is

$$v = \frac{\omega}{\beta} \tag{2.14}$$

The quantity βz is called the electrical length of the line.

2.1.1. The Lossless Transmission Line

A lossless transmission line is a line with zero resistance and zero conductance. If we substitute $R=0$ and $G=0$ in the propagation constant in Eq. (2.5) we get

$$\gamma = j\beta = j\omega\sqrt{LC} \tag{2.15}$$

and the characteristic impedance of the line is real and equal to

$$Z_0 = \sqrt{\frac{L}{C}} \tag{2.16}$$

The other parameters are:

$$\beta = \omega\sqrt{LC} \tag{2.17}$$

$$\alpha = 0 \tag{2.18}$$

$$\lambda = \frac{2\pi}{\omega\sqrt{LC}} \qquad (2.19)$$

$$v = \frac{1}{\sqrt{LC}} \qquad (2.20)$$

In particular, the phase constant and the speed of propagation are only dependent on the inductance and capacitance per unit length. These, in turn, are independent of the shape of the transmission line and only depend on the permeability and permittivity of the materials between the conductors of the line.

2.1.2. Reflection on Transmission Lines and the Reflection Coefficient

2.1.2a. Lossless Transmission Line

To define the reflection coefficient we resort to a lossless transmission line, terminated in a load Z_l as shown in Figure 2.6. The load impedance was calculated in Eq. (2.12) for a general line. For a lossless line, it becomes:

$$Z_l = Z_0 \frac{V^+ e^{-j\beta l} + V^- e^{j\beta l}}{V^+ e^{-j\beta l} - V^- e^{j\beta l}} \qquad (2.21)$$

or, writing this in terms of V^+ and V^- at the load, and using $l=0$ (i.e. we relate the voltages at the load rather than at a distance away), we get

$$\frac{V^-}{V^+} = \frac{Z_l - Z_0}{Z_l + Z_0}$$

This ratio, giving the relation between the backward propagating amplitude V^- and the forward propagating amplitude V^+ is the reflection coefficient at the load. Since it is given as a ratio of voltages it is also called a voltage reflection coefficient. Thus, we can write

$$\Gamma = \frac{V^-}{V^+} = \frac{Z_l - Z_0}{Z_l + Z_0} \qquad (2.22)$$

This can now be used to write the general voltage or current relations at any point on the line, using Eqs. (2.10) and (2.11)

$$V(z) = V^+ e^{-j\beta z} + \Gamma V^+ e^{j\beta z} = V^+\left[e^{-j\beta z} + \Gamma e^{j\beta z}\right] \qquad (2.23)$$

and

$$I(z) = \frac{V^+}{Z_0} e^{-j\beta z} - \Gamma \frac{V^+}{Z_0} e^{j\beta z} = \frac{V^+}{Z_0}\left[e^{-j\beta z} - \Gamma e^{j\beta z}\right] \tag{2.24}$$

The first component of each of these waves is the forward or incident wave and the second is the backward or reflected wave. Again, for perfect matching, Γ is zero and we have only an incident wave.

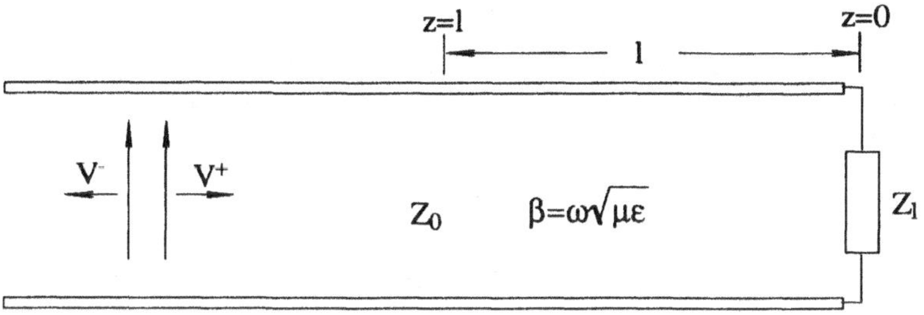

Figure 2.6. A terminated, lossless transmission line with line parameters.

2.1.2b. Lossy Transmission Line

For a lossy line, we start with the same relations for voltage and current as in Eqs. (2.10) and (2.11). The load impedance, which can be complex is

$$Z_l = Z_0 \frac{V^+ e^{-\gamma l} + V^- e^{\gamma l}}{V^+ e^{-\gamma l} - V^- e^{\gamma l}}$$

and, by rewriting this in terms of the forward and backward propagating waves, we can define the reflection coefficient at a point l (e.g. at the load) as

$$\Gamma_l = \frac{V^- e^{\gamma l}}{V^+ e^{-\gamma l}} = \frac{Z_l - Z_0}{Z_l + Z_0} \tag{2.25}$$

This is identical to the previous solution except that now Γ is obviously dependent on γ. If we were to write the reflection coefficient at a distance z_0 in front of the load (see Figure 2.6), we have:

$$\Gamma_{z_0} = \frac{V^- e^{\gamma(l-z_0)}}{V^+ e^{-\gamma(l-z_0)}} = \frac{V^- e^{\gamma l} e^{-\gamma z_0}}{V^+ e^{-\gamma l} e^{\gamma z_0}} = \Gamma_{z_0} e^{-2\gamma z_0} \tag{2.26}$$

or, rewriting in terms of $\gamma = \alpha + \beta$

$$\Gamma_{z_0} = \Gamma e^{-2\alpha z_0} e^{-j2\beta z_0} \tag{2.27}$$

This expression shows that the phase of the reflection coefficient changes as the wave propagates away from the load and the amplitude of the wave attenuates with an attenuation coefficient of 2α. If the line is lossless, Γ remains constant in magnitude but changes in phase.

2.1.3. The Transmission Coefficient

We have defined the reflection coefficient in Eq. (2.27). If a total power is delivered to the transmission line, and some of it is reflected, then the remaining power is transmitted to the load. Thus, we can define a transmission coefficient as:

$$T = 1 + \Gamma \tag{2.28}$$

where we note that the reflection coefficient varies between -1 to $+1$, a result we obtained from Eq. (2.25) by allowing the load impedance to vary from zero to infinity.

In terms of impedances, we can use Eq. (2.25) and write the transmission coefficient at the load

$$T = 1 + \frac{Z_l - Z_0}{Z_l + Z_0} = \frac{2Z_l}{Z_l + Z_0} \tag{2.29}$$

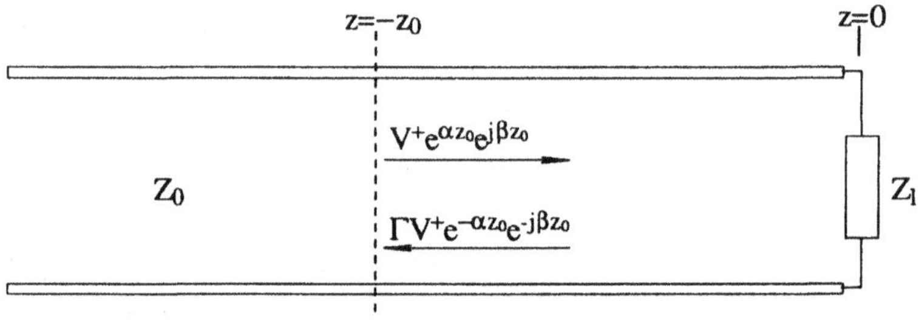

Figure 2.7. Notation used to calculate power on the transmission line.

2.1.4. Power Relations in a Lossy Transmission Line

The power on the line at a distance z_0 from the load, can be calculated by assuming that the load is at $z=0$ and the input at $z=-z_0$ as shown in Figure 2.7. Now, the power entering this section of the transmission line is

$$P_i = \frac{1}{2} Re\left(V_{-z_0} I^*_{-z_0}\right) = \frac{|V^+|^2}{2Z_0}\left[e^{2\alpha z_0} - |\Gamma|^2 e^{-2\alpha z_0}\right] \tag{2.30}$$

The power at the load is obtained by setting $z_0=0$:

$$P_L = \frac{|V^+|^2}{2Z_0}\left[1 - |\Gamma|^2\right] \tag{2.31}$$

2.1.5. Standing Waves on Transmission Lines

If there is a reflected wave from the load propagating along the line, the forward and backward propagating waves will interfere with each other, depending on their phases. If both are in phase, their amplitudes add up and the amplitude is larger. If they are out of phase, their amplitudes subtract and the amplitude of the total wave is lower. Assuming a lossy transmission line, the amplitude of the forward propagating wave is

$$V_f = V^+ e^{-j\beta z} e^{-\alpha z}$$

using the reflection coefficient, the backward voltage wave is

$$V_b = \Gamma V^+ e^{-j\beta z} e^{-\alpha z}$$

Thus, the maximum in the wave occurs when the two waves are in phase and is equal to

$$V_{max} = V^+ e^{-\alpha z}(1 + |\Gamma|) \tag{2.32}$$

and the minimum is equal to

$$V_{min} = V^+ e^{-\alpha z}(1 - |\Gamma|) \tag{2.33}$$

Minima or maxima occur at distances of $\lambda/2$. Since this occurs at fixed locations of z, the wave is said to be fixed in space (as opposed to traveling), or the wave is a standing wave. In other words, if the first minimum occurs at a distance z_1 from the load, the second minimum occurs at $z_1+\lambda/2$ from the load, and so on.

From these we note that if the reflection coefficient is zero, there are no standing waves while, if the reflection coefficient is equal to 1 (total reflection), the maximum amplitude is largest and the minimum amplitude is zero. These relations are shown in Figure 2.8 for a shorted transmission line.

If there is no attenuation of the wave, the maximum is twice the amplitude of the forward wave for total reflection.

We can now define a standing wave ratio. The voltage standing wave ratio *(VSWR)* is defined as

$$VSWR = \frac{V_{max}}{V_{min}} = \frac{1+|\Gamma|}{1-|\Gamma|} \tag{2.34}$$

This is sometimes known as simply the standing wave ratio *(SWR)* and, sometimes, is defined as the ratio of the absolute values of voltages or currents.

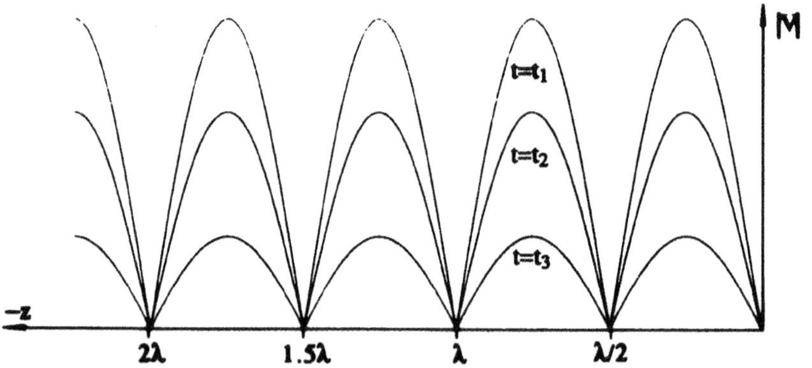

Figure 2.8. Standing wave relations on a transmission line. Complete standing waves without attenuation are shown at three different times.

2.1.6. Line Impedance

The impedance of a line at a point along the line can be viewed as being the input impedance for the line starting at that point and terminating in a load Z_l. Using Figure 2.7, the voltage and current at point z_0, are

$$V(-z_0) = V^{+}\left[e^{\gamma z_0} + \Gamma e^{-\gamma z_0}\right]$$

$$I(-z_0) = \frac{V^{+}}{Z_0}\left[e^{\gamma z_0} - \Gamma e^{-\gamma z_0}\right]$$

The impedance of the line is

$$\frac{V(-z_0)}{I(-z_0)} = Z_{in} = Z_0 \frac{\left[e^{\gamma z_0} + \Gamma e^{-\gamma z_0}\right]}{\left[e^{\gamma z_0} - \Gamma e^{-\gamma z_0}\right]}$$

If we multiply by $e^{-\gamma z_0}$ and divide by $e^{-\gamma z_0}$, we get

$$Z_{in} = Z_0 \frac{\left[1 + \Gamma e^{-2\gamma z_0}\right]}{\left[1 - \Gamma e^{-2\gamma z_0}\right]} \tag{2.35}$$

where

$$\Gamma e^{-2\gamma z_0} = \Gamma_{z_0}$$

is the transmission coefficient at $z=-z_0$ and Γ is the transmission coefficient at the load.

An expression in terms of the load impedance and the characteristic impedance is obtained by substituting the relation for Γ from Eq. (2.25) and using hyperbolic functions instead of the exponent functions (see section 12.2 for a short discussion of hyperbolic and exponential functions). The impedance of the line at distance z_0 from the load is

$$Z_{in} = Z_0 \frac{\left[Z_l + Z_0 tanh \gamma z_0\right]}{\left[Z_0 + Z_l tanh \gamma z_0\right]} \tag{2.36}$$

This line impedance varies along the line and repeats itself every $\lambda/2$. For a lossless transmission line we can replace γ by $j\beta$ and get

$$Z_{in} = Z_0 \frac{\left[Z_l + jZ_0 tan \beta z_0\right]}{\left[Z_0 + jZ_l tan \beta z_0\right]} \tag{2.37}$$

again using the properties of the hyperbolic functions. Either Eq. (2.36) or Eq. (2.37), (or Eq (2.38 for lossless lines) can be used for the calculation of the impedance of the line.

We note here that if the if $Z_l=0$, the input impedance in Eq. (2.37) is that of a shorted transmission line:

$$Z_{in} = jZ_0 tan \beta z_0 \tag{2.38}$$

Thus, the input impedance of a shorted transmission line is purely imaginary and is defined by the tangent function. It can be negative or positive (or zero) and therefore the line behaves either as a capacitive or inductive reactance depending on location on the line. Since the line is shorted, the voltage is zero at the short and maximum at a quarter wavelength from the short. The voltage is again minimum at an additional quarter wavelength from the short.

We also note that if the load in Eq. (2.37) is infinite (open line), the impedance at that point must tend to ∞ and the line is at a voltage maximum (zero current). This condition is identical to a shorted transmission line at voltage maximum or at quarter wavelength from the short. Thus an open transmission line can always be analyzed as a shorted transmission line by adding $\lambda/4$ to, or subtracting $\lambda/4$ from the length of the line.

2.1.7. Impedance Matching

It is often necessary to match a transmission line to a load or to a generator, for the purpose of eliminating standing waves on the line. In general, a transmission line is matched to a load if the load impedance is equal to the characteristic impedance (this does not imply maximum power transfer which requires conjugate matching). Similarly, if the line impedance is equal to the generator impedance, the two are matched. To match a load to a line (or a generator for that matter) a matching network is connected between the line and the load as shown in Figure 2.9b. The location of the matching network depends on the application.

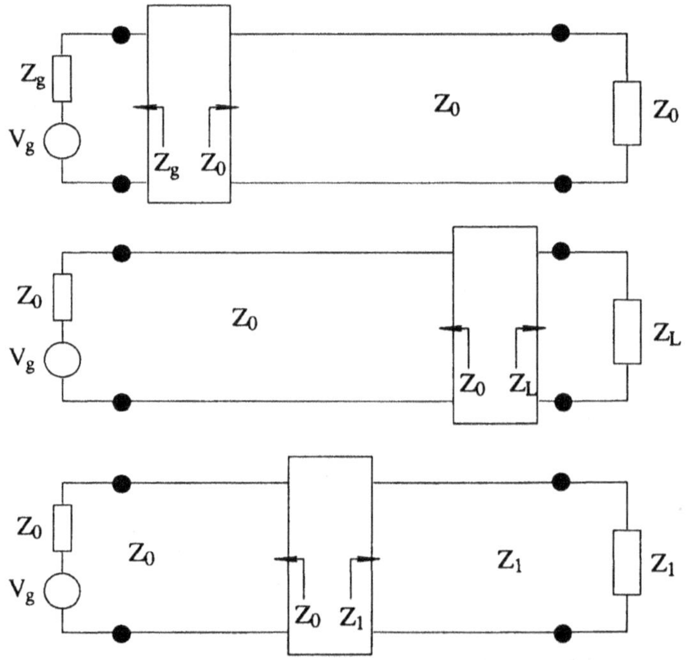

Figure 2.9. Matching networks at the generator side, load side and at an arbitrary location on the line. a) represents a matching stub or stubs at the generator, b) represents matching stubs at the load while c) represents a quarter wavelength transformer.

If we wish to reduce the standing waves on the line, the matching network should be located as closely as possible to the mismatched load. If, however, the line can operate with standing waves, then a more convenient location, at some distance away can be found since the line impedance repeats every $\lambda/2$. Two types of impedance matching networks are discussed here briefly. One is the stub matching which makes use of properties of shorted (or open) transmission lines, the other is the quarter wavelength transformer.

2.1.8. Stub Matching

A section of a transmission line, either open or short circuited is attached either in parallel or in series with the transmission line as shown in Figure 2.10 The impedance of the stub, and therefore its length and location on the line are chosen such that the combined impedance of line and stubs is equal to the characteristic impedance of the line. Although in principle matching to any line impedance is possible, because shorted transmission lines have a purely imaginary impedance, stubs are used to match loads to transmission lines with resistive characteristic impedances. Assuming a characteristic impedance of Z_0 (or admittance Y_0), and a load admittance of Y_0+jX_0, the two can be matched by adding a stub in parallel, at distance l_1 from the load, such that the admittance of the stub is $-jX_0$. The distance l_1 defines the imaginary part of the line admittance from Eq. (2.37). l_2 is then that length of line that cancels the imaginary part of the admittance. The choices of l_1 and l_2 may not be unique, but any practical combination that satisfies the conditions above can be used.

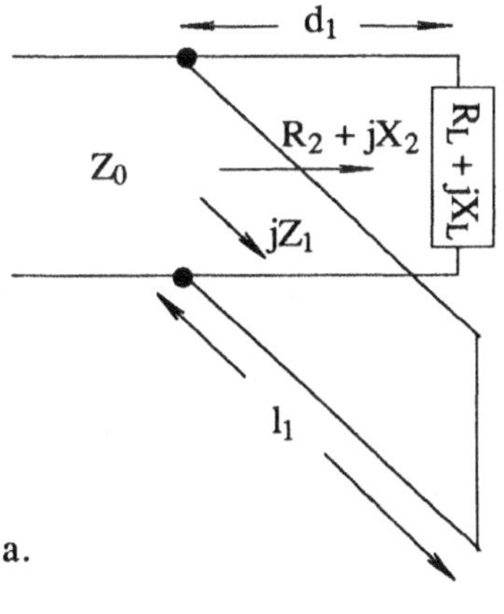

a.

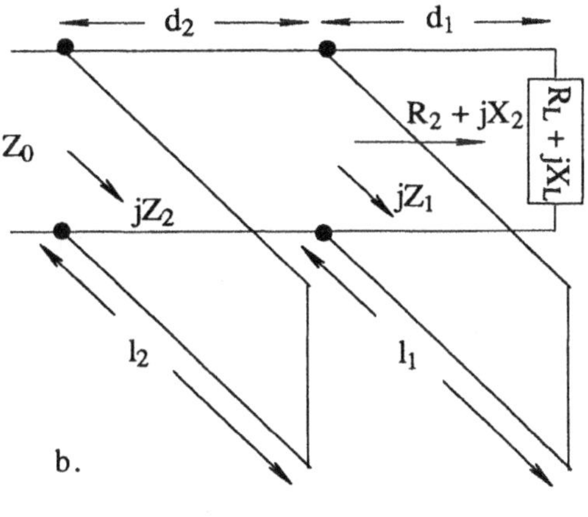

b.

Figure 2.10a. Single stub matching.

Figure 2.10b. Double stub matching

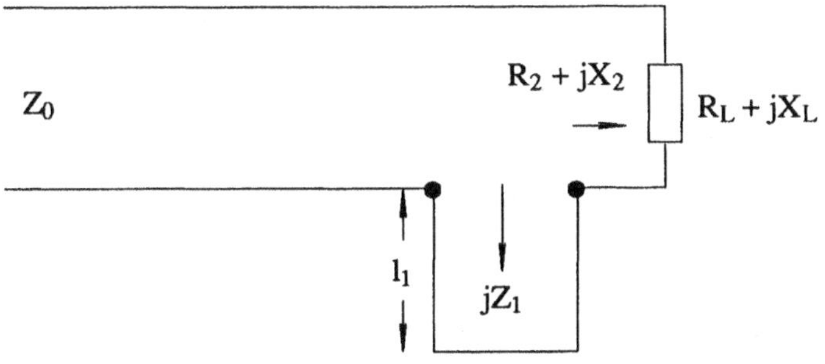

Figure 2.10c. Serial stub matching.

Sometimes the physical conditions of the line do not allow perfect matching with a single stub in which case, two stubs, at different locations may be used. However, two stubs may not be capable of matching all impedances to the line.

The derivation of formulas for impedance matching is rather tedious. Moreover, the design of stubs are most often done using the Smith chart. We will not discuss here the derivations of formulas for stub matching or the Smith chart but rather, give here the formulas for single stub matching for reference. The bibliography section discusses these points further and refers to sources on the others.

For single stub matching: Assuming the stub has the same characteristic impedance as the line, the lengths l_1 and l_2 are calculated as

$$l_1 = \frac{\lambda}{2\pi} \tan^{-1}(A) \qquad for \qquad A \geq 0 \tag{2.39}$$

$$l_1 = \frac{\lambda}{2} + \frac{\lambda}{2\pi} \tan^{-1}(A) \qquad for \qquad A < 0 \tag{2.40}$$

where

$$A = \frac{X_L \pm \sqrt{\dfrac{X_L\left((Z_0 - R_L)^2 + X_L^2\right)}{Z_0}}}{R_L - Z_0} \qquad for \qquad R_L \neq Z_0$$

and

$$A = \frac{-X_L}{2Z_0} \qquad for \qquad R_L = Z_0$$

and all impedances are defined in Figure 2.10. Obviously, any multiples of $\lambda/2$ can be added to l_1 without changing the solution. The length of the stub is:

$$l_2 = \frac{\lambda}{2\pi} \tan^{-1}\left(\frac{1}{Z_0 B}\right) \qquad for \qquad B < 0 \tag{2.41}$$

$$l_2 = \frac{\lambda}{2} + \frac{\lambda}{2\pi} \tan^{-1}\left(\frac{1}{Z_0 B}\right) \qquad for \qquad B > 0 \tag{2.42}$$

where

$$B = \frac{R_L^2 A - (Z_0 - X_L A)(X_L - Z_0 A)}{R_L^2 + (X_L + Z_0 A)^2}$$

2.1.9. Quarter Wavelength Transformer Matching

Stub matching, in effect is capable of removing a mismatch for any load (except a purely reactive load) but it is not an impedance transformer. If different lines must be matched, a transmission line transformer can be used (Figure 2.11).

From Eq. (2.38), it is obvious that the impedance of the line depends on location. Assuming a lossless transmission line, the impedance at a distance z_0 from the load is

$$Z_{in} = Z_0 \frac{\left[Z_l + jZ_0 tan\beta z_0\right]}{\left[Z_0 + jZ_l tan\beta z_0\right]} = Z_0 \frac{\left[Z_l cos\beta z_0 + jZ_0 sin\beta z_0\right]}{\left[Z_0 cos\beta z_0 + jZ_l sin\beta z_0\right]} \tag{2.43}$$

where the following were used:

$$sin\beta z_0 = \frac{e^{-j\beta z_0} - e^{j\beta z_0}}{2j} \qquad cos\beta z_0 = \frac{e^{j\beta z_0} + e^{-j\beta z_0}}{2} \tag{2.44}$$

For a length of line equal to $\lambda/4$, we have from Eq. (2.13)

$$\beta\frac{\lambda}{4} = \frac{\pi}{2}$$

Substituting $\pi/2$ for βz_0, we get

$$Z_{in} = Z_0 \frac{\left[Z_l cos\frac{\pi}{2} + jZ_0 sin\frac{\pi}{2}\right]}{\left[Z_0 cos\frac{\pi}{2} + jZ_l sin\frac{\pi}{2}\right]} = \frac{Z_0^2}{Z_l} \tag{2.45}$$

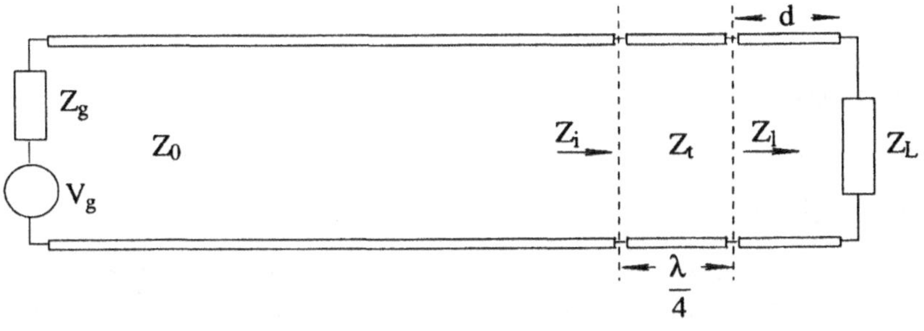

Figure 2.11. A quarter wavelength transformer located at distance *d* from load.

If Z_{in} is the line impedance and Z_l the impedance of a load or a second line impedance, then the transformer must have a characteristic impedance which is equal to

$$Z_t = \sqrt{Z_{in} Z_l} \tag{2.46}$$

where Z_t is the characteristic impedance of the transformer section, Z_{in} the input impedance and Z_l the output impedance. Thus, two different transmission lines or any two impedances may be matched provided a transformer of proper intrinsic impedance can be found. The quarter wavelength transformer is normally connected at a point of maximum or minimum voltage since at that point the impedance of the line is real. The impedance of the line at a point of minimum voltage is

$$Z_l = \frac{Z_0}{VSWR} \tag{2.47}$$

where Z_l is the line impedance and Z_0 is the characteristic impedance of the line. The location of the minimum voltage on the line is at a distance

$$d_{min} = \frac{\lambda}{4\pi}\left(\theta_L + \pi\right) + n\frac{\lambda}{2} \tag{2.48}$$

from the load, where n is any integer, including zero. Thus, the transformer can be located at any of the points in Eq. (2.48). If the characteristic line impedance is Z_0, the characteristic impedance of a transformer, located at a point of minimum voltage must be

$$Z_t = Z_0\sqrt{\frac{1}{\text{VSWR}}} \tag{2.49}$$

Similarly, if the transformer is located at a point of maximum voltage, (by moving it a quarter wavelength in either direction of any of the points in Eq. (2.48)), the characteristic impedance of the transformer for matching is

$$Z_t = Z_0\sqrt{\text{VSWR}} \tag{2.50}$$

These relations are shown in Figure 2.11.

2.2. Waveguides

The propagation of uniform plane waves as discussed in chapter 1 was characterized by the transverse *(TEM)* nature of the wave. The electric and magnetic field intensities were perpendicular to each other and to the direction of propagation. In unbounded space, neither of the fields had a component in the direction of propagation. We will discuss now the propagation of waves in guiding structures. The purpose of these structures is to guide waves, with particular characteristics, along the structure. However, unlike transmission lines these guiding structures are single conductor waveguides, primarily as used with microwaves. As a direct extension to waveguides we will also discuss microwave cavity resonators. These are essentially sections of waveguides shorted at one or both ends to create an open cavity or a closed cavity resonator. A further extension to completely open resonators is discussed from the point of view of complete reflection at boundaries. From this point of view, all that is necessary for resonance is that waves be completely reflected. Thus, a pair of mirrors, concave, flat, or a combination of the two can be viewed as a resonator. This type of open resonator is particularly useful for testing because material handling is simplified considerably.

The uniform plane wave used in section 1.9. was a transverse electromagnetic wave. This implied that neither the electric nor the magnetic field intensity had a component in the direction of propagation. A wave with an electric field perpendicular to the direction of propagation, but with a component of the magnetic field in the direction of propagation is called a transverse electric *(TE)* wave. Similarly, if the magnetic field is perpendicular to the direction of propagation, and the electric field has a component in the direction of propagation, this is a transverse magnetic *(TM)* wave.

Before proceeding with description of waves in waveguides, we need to define the conditions under which the three types of waves can exist. These conditions are then used to define the possible fields or modes in rectangular and cylindrical waveguides.

The starting point are the two curl equations. Each is expanded into three scalar equations by equating the vector components on both sides. By doing so, an equation of one component of one field (electric or magnetic) is written in terms of the transverse components of the other field.
For the electric field:

$$\nabla\times\mathbf{H} = j\omega\epsilon\mathbf{E} \tag{2.51}$$

$$\frac{\partial H_z}{\partial y} - \frac{\partial H_y}{\partial z} = j\omega\epsilon E_x \tag{2.52}$$

$$\frac{\partial H_x}{\partial z} - \frac{\partial H_z}{\partial x} = j\omega\epsilon E_y \tag{2.53}$$

$$\frac{\partial H_y}{\partial x} - \frac{\partial H_x}{\partial y} = j\omega\epsilon E_z \tag{2.54}$$

For the magnetic field:

$$\nabla\times\mathbf{E} = -j\omega\mu\mathbf{H} \tag{2.55}$$

$$\frac{\partial E_z}{\partial y} - \frac{\partial E_y}{\partial z} = -j\omega\mu H_x \tag{2.56}$$

$$\frac{\partial E_x}{\partial z} - \frac{\partial E_z}{\partial x} = -j\omega\mu H_y \tag{2.57}$$

$$\frac{\partial E_y}{\partial x}-\frac{\partial E_x}{\partial y}=-j\omega\mu H_z \tag{2.58}$$

The waves are assumed to propagate in the z direction, with a propagation constant $\gamma=\alpha+j\beta$. Assuming no backward propagating waves (no reflections) in the z direction, the fields have the following general form

$$E_x = E_0 e^{-\gamma z}$$

where E_x was chosen as an example, and the $e^{j\omega t}$ variation is also implied since the fields are all phasors. Because all fields vary with the z parameter, the derivatives with respect to z in Eqs. (2.52, 2.53) and (2.56, 2.57) are

$$\frac{\partial E_y}{\partial z}=-\gamma E_y\,,\qquad \frac{\partial E_x}{\partial z}=-\gamma E_x\,,\qquad \frac{\partial H_y}{\partial z}=-\gamma H_y\,,\qquad \frac{\partial H_x}{\partial z}=-\gamma H_x$$

Now, Eqs. (2.52) through (2.54) and (2.56) through (2.58) are

$$\frac{\partial H_z}{\partial y}+\gamma H_y=j\omega\epsilon E_x \tag{2.59}$$

$$-\gamma H_x-\frac{\partial H_z}{\partial x}=j\omega\epsilon E_y \tag{2.60}$$

$$\frac{\partial H_y}{\partial x}-\frac{\partial H_x}{\partial y}=j\omega\epsilon E_z \tag{2.61}$$

$$\frac{\partial E_z}{\partial y}+\gamma E_y=-j\omega\mu H_x \tag{2.62}$$

$$-\gamma E_x-\frac{\partial E_z}{\partial x}=-j\omega\mu H_y \tag{2.63}$$

$$\frac{\partial E_y}{\partial x}-\frac{\partial E_x}{\partial y}=-j\omega\mu H_z \tag{2.64}$$

Since the nature of the waves is defined by the longitudinal components of the field it is necessary to rewrite these equations such that the transverse components of the fields (those perpendicular to the direction of propagation), in this case E_x, E_y, H_x, and H_y, are written in terms of the longitudinal components (those components in the direction of propagation), E_z, and H_z. As an example, using equations (2.59) and (2.63), we can eliminate H_y, and write the component E_x in terms of H_z and E_z. Substitution of H_y from Eq. (2.63) in Eq. (2.59) gives

$$\frac{\partial H_z}{\partial y} + \frac{\gamma}{j\omega\mu}\left(\gamma E_x + \frac{\partial E_z}{\partial x}\right) = j\omega\varepsilon E_x$$

Multiplying both sides by $j\omega\mu$ and rearranging terms, gives

$$E_x = \frac{1}{\gamma^2 + k^2}\left(-\gamma\frac{\partial E_z}{\partial x} - j\omega\mu\frac{\partial H_z}{\partial y}\right)$$

where $k^2 = \omega^2\mu\varepsilon$.

Repeating the process for the other three transverse components we get a set of four equations

$$E_x = \frac{1}{\gamma^2 + k^2}\left(-\gamma\frac{\partial E_z}{\partial x} - j\omega\mu\frac{\partial H_z}{\partial y}\right) \tag{2.65}$$

$$E_y = \frac{1}{\gamma^2 + k^2}\left(-\gamma\frac{\partial E_z}{\partial y} + j\omega\mu\frac{\partial H_z}{\partial x}\right) \tag{2.66}$$

$$H_x = \frac{1}{\gamma^2 + k^2}\left(-\gamma\frac{\partial H_z}{\partial x} + j\omega\varepsilon\frac{\partial E_z}{\partial y}\right) \tag{2.67}$$

$$H_y = \frac{1}{\gamma^2 + k^2}\left(-\gamma\frac{\partial H_z}{\partial y} - j\omega\varepsilon\frac{\partial E_z}{\partial x}\right) \tag{2.68}$$

These equations are now used to define *TEM*, *TE*, and *TM* propagation of waves by imposing the necessary conditions for each mode of propagation.

2.2.1. TM and TE Modes in Waveguides

As mentioned above, *TEM* waves, require that there be no field in the direction of propagation. Thus, the condition for *TEM* propagation is: $E_z=H_z=0$. Substituting this into Eqs. (2.65) through (2.68) leads to the requirement that all four transverse components are zero, unless $\gamma^2+k^2=0$. Thus, the constant of propagation must be

$$\gamma^2 = -k^2 \qquad or \qquad \gamma = j\omega\sqrt{\mu\varepsilon}$$

which is the condition we obtained for propagation of uniform plane waves in lossless media. Indeed, all the properties we obtained for propagation of plane waves in an unbounded domain, including the definition of intrinsic impedance (or wave impedance) and the speed of propagation (phase velocity) apply here as well. However, *TEM* waves cannot exist in single conductor waveguides (i.e. a rectangular or cylindrical waveguide), a property that can be shown by direct application of Ampere's law (Eq. (7.3)). They can exist in coaxial cables and parallel plate waveguides. For this reason we turn our attention to *TE* and *TM* modes, as they apply to rectangular and cylindrical waveguides.

For *TE* waves to exist, E_z must be zero. Substituting this condition in Eq. (2.65) through (2.68) we get the transverse components for *TE* propagation. These are:

$$E_x = \frac{-j\omega\mu}{\gamma^2 + k^2}\frac{\partial H_z}{\partial y} \tag{2.69}$$

$$E_y = \frac{j\omega\mu}{\gamma^2 + k^2}\frac{\partial H_z}{\partial x} \tag{2.70}$$

$$H_x = \frac{-\gamma}{\gamma^2 + k^2}\frac{\partial H_z}{\partial x} \tag{2.71}$$

$$H_y = \frac{-\gamma}{\gamma^2 + k^2}\frac{\partial H_z}{\partial y} \tag{2.72}$$

It is sometimes convenient to write these as a wave equation in H_z, which is the only longitudinal component in a *TE* wave. Doing so allows representation of fields and properties in terms of a solution to the wave equation. Since these solutions are standard, and we have already obtained them for propagation in free space and in transmission lines, this approach will allow us to build on the existing solutions. We start with the derivative with respect to y of Eq. (2.69) and the derivative with respect to x of Eq. (2.70)

$$\frac{\partial E_x}{\partial y} = \frac{-j\omega\mu}{\gamma^2 + k^2}\frac{\partial^2 H_z}{\partial y^2} \quad \text{and} \quad \frac{\partial E_y}{\partial x} = \frac{j\omega\mu}{\gamma^2 + k^2}\frac{\partial^2 H_z}{\partial x^2}$$

Adding the two equations gives

$$\frac{\partial^2 H_z}{\partial x^2} + \frac{\partial^2 H_z}{\partial y^2} + \frac{\gamma^2 + k^2}{j\omega\mu}\left(-\frac{\partial E_y}{\partial x} + \frac{\partial E_x}{\partial y}\right) = 0$$

Now, using Eq. (2.56)

$$\frac{\partial^2 H_z}{\partial x^2} + \frac{\partial^2 H_z}{\partial y^2} + (\gamma^2 + k^2)H_z = 0 \tag{2.73}$$

This is a wave equation in H_z alone and will be taken from now on as the defining equation for *TE* waves.

Using the condition $H_z=0$ yields the field equations for *TM* waves. The steps are identical to those for the *TE* waves. First, we write the field components by substituting the condition $H_z=0$ in Eqs. (2.65) through (2.68):

$$E_x = \frac{-\gamma}{\gamma^2 + k^2}\frac{\partial E_z}{\partial x} \tag{2.74}$$

$$E_y = \frac{-\gamma}{\gamma^2 + k^2}\frac{\partial E_z}{\partial y} \tag{2.75}$$

$$H_x = \frac{j\omega\varepsilon}{\gamma^2 + k^2}\frac{\partial E_z}{\partial y} \tag{2.76}$$

$$H_y = \frac{-j\omega\varepsilon}{\gamma^2 + k^2}\frac{\partial E_z}{\partial x} \tag{2.77}$$

The wave equation equivalent to these four equations is obtained as above

$$\frac{\partial^2 E_z}{\partial x^2} + \frac{\partial^2 E_z}{\partial y^2} + (\gamma^2 + k^2)E_z = 0 \tag{2.78}$$

The definition of *TE* and *TM* fields in Eqs. (2.69) through (2.72) and (2.74) through (2.78) was done in completely general terms. These apply universally since no other conditions (such as boundary or interface conditions) were required. These equations were written in Cartesian coordinates but this is a minor point since they can be easily modified for cylindrical or any other system of coordinates. However, our purpose was to define these for analysis of fields and waves in waveguides and resonators. The additional conditions we impose are boundary conditions on the waveguide boundaries. Definition of *TE* and *TM* modes in rectangular and cylindrical waveguides is our next goal.

2.2.2. Rectangular Waveguides

A rectangular waveguide is shown in Figure 2.12. The dimensions of the waveguide are assumed to be the internal dimensions and the walls to be perfectly conducting.

To see how the electric and magnetic fields in a waveguide of the type shown in Figure 2.12 behave, we must solve Eq. (2.73) or Eq. (2.78) subject to boundary conditions on the conducting boundaries. We start with *TM* waves because the boundary conditions are straight forward.

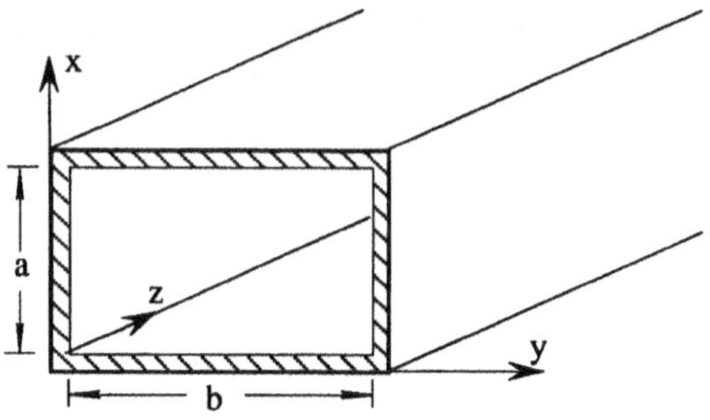

Figure 2.12. Structure and dimensions of a rectangular waveguide.

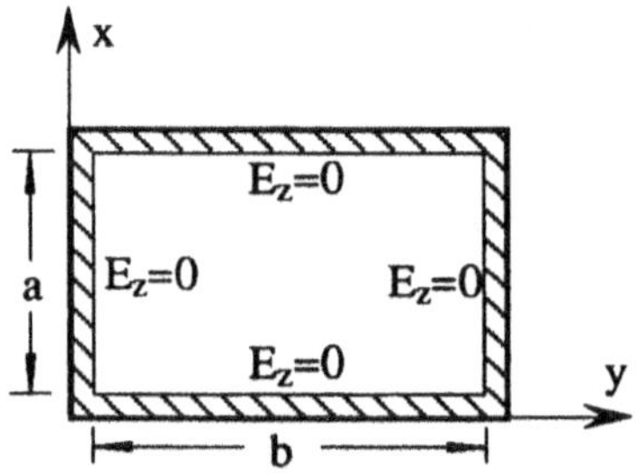

Figure 2.13. Boundary conditions on the electric field for *TM* modes in a rectangular waveguide.

2.2.3. TM Modes in Rectangular Waveguides

To find the fields for *TM* modes we must solve the following equation (Eq. (2.78)

$$\frac{\partial^2 E_z}{\partial x^2} + \frac{\partial^2 E_z}{\partial y^2} + (\gamma^2 + k^2)E_z = 0$$

subject to the condition that the electric field must vanish on the conducting boundaries. These conditions (shown in Figure 2.13) are:

$E_z(0,y)=0.$ *and* $E_z(a,y)=0.$ *in the x direction*

$E_z(x,0)=0.$ *and* $E_z(x,b)=0.$ *in the y direction*

Eq. (2.78) can be solved directly using separation of variables. This leads to two second order ordinary differential equations, the solutions to which are sinusoidal functions. Instead of proceeding in this fashion, we will simply assume a solution in terms of sinusoids, and evaluate the amplitude and spatial frequencies of the sinusoids so that the solution satisfies the boundary conditions. The solution we seek is of the form

$$E_z(x,y) = E_0 \sin(k_x x)\sin(k_y y)$$

This is in fact, the solution one obtains through separation of variables. To satisfy the boundary conditions, k_x must be such that the *sin* function is zero at $x=0$ and $x=a$. Similarly, k_y forces the second *sin* function to zero at $y=0$ and $y=b$. These conditions are satisfied if

$$k_x = \frac{m\pi}{a} \quad \text{and} \quad k_y = \frac{n\pi}{b} \quad \text{where } m,n \text{ are integers}$$

The solution now is

$$E_z(x,y) = E_0 \sin\left(\frac{m\pi}{a}x\right) \sin\left(\frac{n\pi}{b}y\right) \tag{2.79}$$

If we now substitute this solution into Eqs. (2.74) and (2.75) and evaluate the derivatives of E_z, we obtain the transverse components of **E** and **H**:

$$E_x(x,y) = \frac{-\gamma}{\gamma^2 + k^2} E_0 \frac{m\pi}{a} \cos\left(\frac{m\pi}{a}x\right) \sin\left(\frac{n\pi}{b}y\right) \tag{2.80}$$

$$E_y(x,y) = \frac{-\gamma}{\gamma^2 + k^2} E_0 \frac{n\pi}{b} \sin\left(\frac{m\pi}{a}x\right) \cos\left(\frac{n\pi}{b}y\right) \tag{2.81}$$

$$H_x(x,y) = \frac{j\omega\varepsilon}{\gamma^2 + k^2} E_0 \frac{n\pi}{b} \sin\left(\frac{m\pi}{a}x\right) \cos\left(\frac{n\pi}{b}y\right) \tag{2.82}$$

$$H_z(x,y) = \frac{-j\omega\varepsilon}{\gamma^2 + k^2} E_0 \frac{m\pi}{a} \cos\left(\frac{m\pi}{a}x\right) \sin\left(\frac{n\pi}{b}y\right) \tag{2.83}$$

To obtain a relation between the constants k_x, k_y, and γ, we substitute the general solution given in Eq. (2.79) into Eq. (2.78). This gives a relation called a dispersion relation

$$\gamma^2 + k^2 = \left(\frac{m\pi}{a}\right)^2 + \left(\frac{n\pi}{b}\right)^2$$

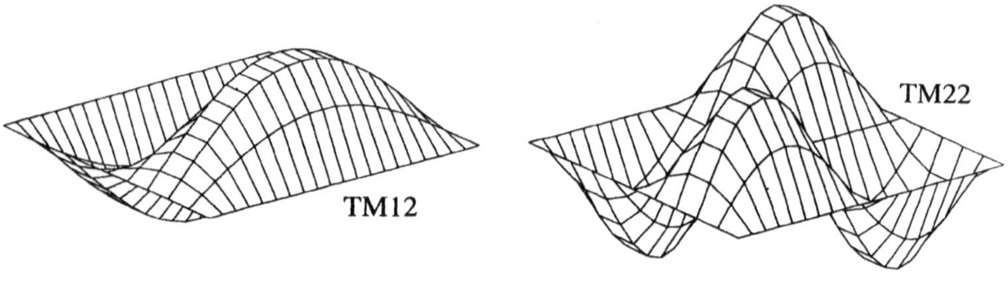

Figure 2.14. The electric field distribution in a waveguide for the TM_{12} and TM_{22} modes.

Or, in a more convenient form

$$\gamma^2 = -k^2 + \left(\frac{m\pi}{a}\right)^2 + \left(\frac{n\pi}{b}\right)^2$$

Since there are no losses (perfect conductors, lossless dielectrics), we can write

$$\gamma = j\beta = j\sqrt{\omega^2\mu\varepsilon - \left(\frac{m\pi}{a}\right)^2 - \left(\frac{n\pi}{b}\right)^2} \qquad (2.84)$$

Because m and n are discrete values, the only fields that can exist are those defined by combinations of integer values of m and n. There is an infinite number of possible combinations of the integers, and, correspondingly, an infinite number of possible fields in the waveguide. Each possible combination is called a mode. The modes defined here are for *TM* waves, therefore they are called *TM* modes. The integers m and n are the indices of the mode. Thus, if $m=1$, $n=3$, this defines a TM_{13} mode. The indices indicate the number of half cycles in the respective direction of the waveguide.

From Eq. (2.84) we note that the propagation constant γ can be zero. This clearly indicates that the waves cannot propagate if this condition is encountered. The frequency at which this happens is called a cutoff frequency and it is obviously different for each mode. The cutoff frequency is given by the condition

$$\omega^2\mu\varepsilon - \left[\left(\frac{m\pi}{a}\right)^2 + \left(\frac{n\pi}{b}\right)^2\right] = 0$$

or

$$f_c = \frac{1}{2\pi\sqrt{\mu\varepsilon}}\sqrt{\left(\frac{m\pi}{a}\right)^2 + \left(\frac{n\pi}{b}\right)^2} \tag{2.85}$$

Below this frequency there is no propagation in the waveguide.

Also, we note that neither m nor n can be zero since that would force all field components to be zero. The lowest possible mode (i.e. the one with lowest cutoff frequency) is the TM_{11} mode.

The electric field distribution (E_z) in a waveguide for the TM_{12} and TM_{22} modes is shown in Figure 2.14. for a waveguide for which $a=2b$. The electric field is zero on the boundaries and has a half cycle variation in the transverse directions for which the index is 1 and a full cycle variation in directions in which the index is 2.

2.2.4. TE Modes in Rectangular Waveguides

For TE waves to exist, E_z must be zero. The wave equation to solve is now

$$\frac{\partial^2 H_z}{\partial x^2} + \frac{\partial^2 H_z}{\partial y^2} + (\gamma^2 + k^2)H_z = 0$$

subject to the condition that the x and y components of the electric field must vanish on the conducting boundaries. These conditions are shown in Figure 2.15 and can be written in terms of the magnetic field intensity H_z as

$$\frac{\partial H_z}{\partial x}(0,y) = 0. \quad \text{and} \quad \frac{\partial H_z}{\partial x}(a,y) = 0. \quad \text{in the x direction}$$

$$\frac{\partial H_z}{\partial y}(x,0) = 0. \quad \text{and} \quad \frac{\partial H_z}{\partial y}(x,b) = 0. \quad \text{in the y direction}$$

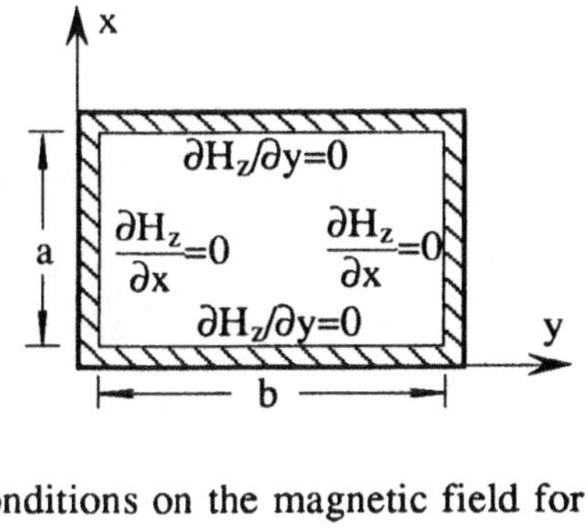

Figure 2.15. Boundary conditions on the magnetic field for TE modes in a rectangular waveguide.

The solution is obtained in an analogous manner to that for *TM* waves and is

$$H_z(x,y) = H_0 \cos\left(\frac{m\pi}{a}\right)x \cos\left(\frac{n\pi}{b}\right)y \tag{2.86}$$

Substitution of this solution in Eqs. (2.69) through (2.72) provides the transverse components of the fields as:

$$E_x(x,y) = \frac{j\omega\mu}{\gamma^2 + k^2} H_0 \frac{n\pi}{b} \cos\left(\frac{m\pi}{a}\right)x \sin\left(\frac{n\pi}{b}\right)y \tag{2.87}$$

$$E_y(x,y) = \frac{-j\omega\mu}{\gamma^2 + k^2} H_0 \frac{m\pi}{a} \sin\left(\frac{m\pi}{a}\right)x \cos\left(\frac{n\pi}{b}\right)y \tag{2.88}$$

$$H_x(x,y) = \frac{\gamma}{\gamma^2 + k^2} H_0 \frac{m\pi}{a} \sin\left(\frac{m\pi}{a}\right)x \cos\left(\frac{n\pi}{b}\right)y \tag{2.89}$$

$$H_y(x,y) = \frac{\gamma}{\gamma^2 + k^2} H_0 \frac{n\pi}{b} \cos\left(\frac{m\pi}{a}\right)x \sin\left(\frac{n\pi}{b}\right)y \tag{2.90}$$

The dispersion relation for *TE* waves is obtained by substituting the solution in Eq.(2.86) into the wave equation (Eq. (2.73)), and is identical to that for the *TM* modes. The cutoff frequency is also the same as that for *TM* modes.

However, unlike *TM* modes, in *TE* modes either *m* or *n* can be zero. This indicates that the lower resonant mode is a TE_{0n} or TE_{m0}, depending on the dimensions *a* and *b*. If $a>b$, the lowest cutoff frequency is for a TE_{10} mode. Also, *TM* and *TE* modes with the same indices have the same frequency.

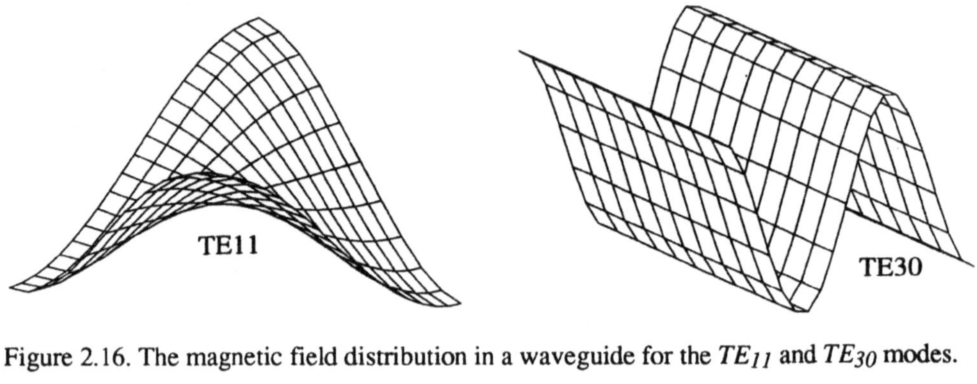

Figure 2.16. The magnetic field distribution in a waveguide for the TE_{11} and TE_{30} modes.

The magnetic field *(H_z)* distribution in a waveguide with $a=2b$, for the TE_{11} and TE_{30} modes are shown in Figure 2.16. The field is maximum on the surfaces and has two half cycle variations in the x direction and no variation in the y direction.

The discussion on waveguides was limited here to rectangular cross section waveguides. Other shapes can be analyzed in a similar fashion. However, since the analysis entails the solution of the wave equation, only simple shapes can be solved exactly. In particular, cylindrical waveguides can be analyzed using steps identical to those presented above.

2.2.5. Cylindrical Waveguides

Electromagnetic waves will also propagate in waveguides that are not rectangular in cross section. For conducting boundary waveguides, the basic requirement is that reflections from the boundary must occur. However, waves can also propagate along structures that are not conducting, provided that total reflection from internal surfaces occurs.

In this section we discuss cylindrical or circular cross section waveguides, the conditions for propagation and the modes the waveguide can support.

The solution is similar to that conducted for rectangular waveguides, with the obvious need for solution in cylindrical coordinates. A cylindrical waveguide is shown in Figure 2.17.

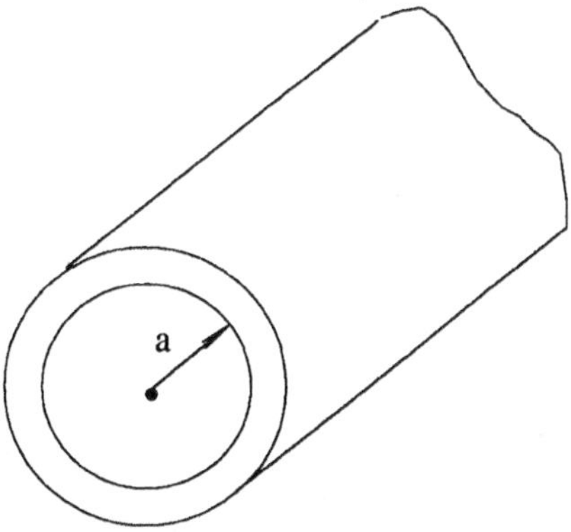

Figure 2.17. Structure and dimensions of a cylindrical waveguide.

We start with the two curl equations given in Eq. (2.51) and (2.55) and expand them in components in cylindrical coordinates. Without repeating the whole process, which only differs in details we get the transverse components of the electric and magnetic fields in cylindrical coordinates as a function of the longitudinal components E_z and H_z as

$$E_r = \frac{1}{\gamma^2 + k^2}\left(-\gamma\frac{\partial E_z}{\partial r} - j\omega\mu\frac{1}{r}\frac{\partial H_z}{\partial \phi}\right) \qquad (2.91)$$

$$E_\phi = \frac{1}{\gamma^2 + k^2}\left(-\gamma\frac{1}{r}\frac{\partial E_z}{\partial f} + j\omega\mu\frac{\partial H_z}{\partial r}\right) \tag{2.92}$$

$$H_r = \frac{1}{\gamma^2 + k^2}\left(-\gamma\frac{\partial H_z}{\partial r} + j\omega\varepsilon\frac{1}{r}\frac{\partial E_z}{\partial f}\right) \tag{2.93}$$

$$H_\phi = \frac{1}{\gamma^2 + k^2}\left(-\gamma\frac{1}{r}\frac{\partial H_z}{\partial f} - j\omega\varepsilon\frac{\partial E_z}{\partial r}\right) \tag{2.94}$$

These equations are completely equivalent to Eqs. (2.65) through (2.68) except for the obvious use of cylindrical coordinates.

2.2.6. *TM Modes in Cylindrical Waveguides*

To find the *TM* modes in a cylindrical waveguide we set $H_z=0$ in Eqs. (2.91) through (2.94). The following wave equation must be solved

$$\frac{1}{r}\frac{\partial}{\partial r}\left[\frac{r\partial E_z}{\partial r}\right] + \frac{1}{r^2}\frac{\partial^2 E_z}{\partial \phi^2} + (\gamma^2 + k^2)E_z = 0 \tag{2.95}$$

subject to the boundary condition:

$E_z(a,\phi)=0$ *on the circular boundary at* $r=a$

The solution to Eq. (2.95) is obtained by direct separation of variables. As we have done for the rectangular waveguide, we will not follow the solution, which is quite tedious but rather, present the solution here and discuss the conditions under which the solution is valid. The solution for the electric field is

$$E_z = c_n J_n(r\sqrt{\gamma^2 + k^2})\cos(n\phi) \tag{2.96}$$

where J_n is the Bessel function of the first kind and nth order, and n is an integer and indicates the number of integer variations of ϕ. The term $(\gamma^2+k^2)^{1/2}r$ is the argument of the Bessel function and c_n is a constant that depends on the excitation level.

The transverse components of the fields can now be calculated from Eqs. (2.91) through (2.94) by direct substitution

For the required boundary conditions to be satisfied, the Bessel function must satisfy

$$J_n a\sqrt{\gamma^2 + k^2} = 0$$

If we define the roots of the Bessel function as those that satisfy the relation

$$J_n(x_{nm}) = 0$$

where m indicates the number of integer variations in r. From this we get the following dispersion relation

$$x_{nm} = \sqrt{(\gamma^2 + k^2)}\,a$$

and we view this as the mth root of the Bessel function. The possible roots define the existent mode in the waveguide. As an example, the first root is $x_{01}=2.405$, the second is $p_{02}=5.52$ and $x_{11}=3.832$. The lowest propagating mode is the TM_{01} mode while, as an example, the TM_{10} does not exist since $m=1,2,....$.

To calculate the propagation constant, we use the lossless case:

$$\gamma = j\beta$$

or, using the relation above

$$\left(\frac{x_{nm}}{a}\right)^2 = (j\beta)^2 + k^2 = -\beta^2 + \omega^2\mu\varepsilon$$

This gives for β

$$\beta = \sqrt{\omega^2\mu\varepsilon - \left(\frac{x_{nm}}{a}\right)^2}$$

The cutoff frequency for mode mn is obtained by setting $\beta=0$. Doing so gives the cutoff frequency as

$$f_c = \frac{x_{nm}}{2\pi a\sqrt{\mu\varepsilon}} \qquad (2.97)$$

The electric and magnetic field distribution for a TM_{11} mode is shown in Figure 2.18.

The transverse components of the electric and magnetic fields are found by substituting E_z as found in Eq. (2.96) above and $H_z=0$ in Equations (2.91) through (2.94). This gives

$$E_r = -\frac{\gamma}{\gamma^2 + k^2}\frac{\partial E_z}{\partial r} = -\frac{\gamma}{\sqrt{\gamma^2 + k^2}}\,c_n J_n'(r\sqrt{\gamma^2 + k^2})\cos(n\phi) \qquad (2.98)$$

$$E_\phi = -\frac{\gamma}{\gamma^2 + k^2}\frac{1}{r}\frac{\partial E_z}{\partial \phi} = -\frac{\gamma n}{\left(\gamma^2 + k^2\right)r}\,c_n J_n(r\sqrt{\gamma^2 + k^2})\sin(n\phi) \qquad (2.99)$$

$$H_r = -\frac{j\omega\varepsilon}{(\gamma^2 + k^2)r}\frac{\partial E_z}{\partial \phi} = -\frac{j\omega\varepsilon n}{(\gamma^2 + k^2)r} c_n J_n(r\sqrt{\gamma^2 + k^2})\sin(n\phi) \qquad (2.100)$$

$$H_\phi = -\frac{j\omega\varepsilon}{\gamma^2 + k^2}\frac{\partial E_z}{\partial r} = -\frac{j\omega\varepsilon}{\sqrt{(\gamma^2 + k^2)}} c_n J_n'(r\sqrt{\gamma^2 + k^2})\cos(n\phi) \qquad (2.101)$$

In these relations, J' indicates the derivative of the corresponding Bessel function with respect to its argument $(\gamma^2+k^2)^{1/2}r$.

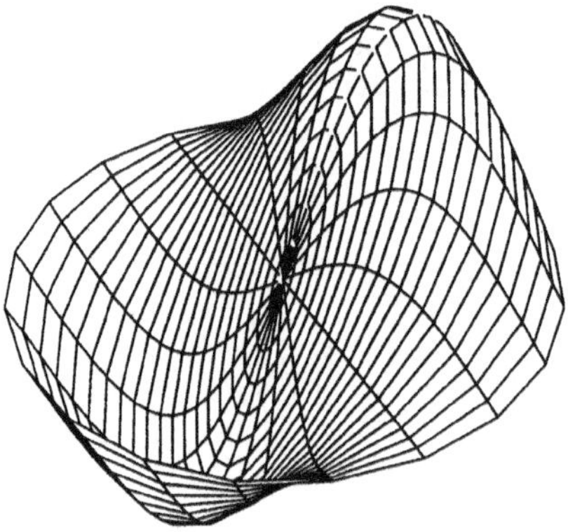

Figure 2.18. The electric field distribution in a cylindrical waveguide for the TM_{11} mode.

2.2.7. TE Modes in Cylindrical Waveguides

TE modes are found in an analogous manner, by setting $E_z=0$ in Eqs. (2.91) through (2.94) and defining the equivalent wave equation

$$\frac{1}{r}\frac{\partial}{\partial x}\left[\frac{r\partial H_z}{\partial x}\right] + \frac{1}{r^2}\frac{\partial^2 H_z}{\partial \phi^2} + (\gamma^2 + k^2)H_z = 0 \qquad (2.102)$$

The solution to this equation is identical in form to the solution for *TM* modes

$$H_z = c_n J_n(r\sqrt{\gamma^2+k^2})\cos(n\phi) \qquad (2.103)$$

The necessary boundary condition now is for the normal derivative of H_z to vanish:

$$\frac{\partial H_z}{\partial n} = \frac{\partial H_z}{\partial r} = 0$$

Or, since $E_z=0$, from Eq. (2.99) it is sufficient to require that the tangential component of E be zero, that is: $E_\phi(r,\phi,z)=0$ at $r=a$. Calculating this we get

$$E_\phi(r,\phi,z) = \frac{j\omega\mu}{\sqrt{\gamma^2+k^2}} c_n J_n'(r\sqrt{\gamma^2+k^2})\cos(nf) \tag{2.104}$$

where J_n' indicates the derivative of the Bessel function with respect to the argument $(\gamma^2+k^2)^{1/2}r$ and the required condition is that

$$J_n'\left(\sqrt{\gamma^2+k^2}\,a\right) = 0$$

We define the roots of this function as those that satisfy the relation

$$J_n'(x_{nm}') = 0$$

where m indicates the number of integer variations in r. From this the dispersion relation is

$$x_{nm}' = \sqrt{(\gamma^2+k^2)}\,a$$

and we view this as the mth root of the Bessel function. The possible roots define the existent mode in the waveguide. As an example, the first root is $x_{01}=3.832$, the second is $x_{02}=7.016$ and $x_{11}=1.841$. The lowest propagating mode is the TE_{11} mode while, as an example, the TE_{10} does not exist since $m=1,2,\ldots..$

The propagation constant is given by

$$\beta = \sqrt{\omega^2\mu\varepsilon - \left(\frac{x_{nm}'}{a}\right)^2}$$

The cutoff frequency for mode mn is obtained by setting $b=0$. This gives the cutoff frequency as

$$f_c = \frac{x_{nm}'}{2\pi a\sqrt{\mu\varepsilon}}$$

The transverse components of the fields are found similar to the TM case except that the condition $E_z=0$ and H_z from Eq. (2.102) are used. The transverse fields are given by

$$H_r = -\frac{\gamma}{\sqrt{\gamma^2+k^2}} c_n J_n'(r\sqrt{\gamma^2+k^2})\cos(n\phi) \tag{2.105}$$

$$H_\phi = \frac{\gamma n}{\left(\gamma^2 + k^2\right) r} C_n J_n(r\sqrt{\gamma^2 + k^2})\sin(n\phi) \tag{2.106}$$

$$E_r = \frac{j\omega\mu n}{\left(\gamma^2 + k^2\right) r} C_n J_n(r\sqrt{\gamma^2 + k^2})\sin(n\phi) \tag{2.107}$$

$$E_f = \frac{j\omega\mu}{\sqrt{\left(\gamma^2 + k^2\right)}} C_n J'_n(r\sqrt{\gamma^2 + k^2})\cos(n\phi) \tag{2.108}$$

2.3. Cavity Resonators

A cavity resonator is built by adding two walls inside a waveguide. A rectangular cavity is built by adding two conducting walls at $z=0$ and $z=d$ as shown in Figure 2.19. This can be viewed as a modified waveguide, in which there are standing waves in the z direction as well as in the x and y directions. The main difference between cavities and waveguides is that in cavities, the z direction imposes additional boundary conditions and there is no propagation of waves as in waveguides. The cavity acts as a resonant structure in which there is exchange of energy between the electric and magnetic field at given (resonant) frequencies. This is equivalent to resonant *LC* circuits in the case of lossless cavities and to *RLC* circuits in the case of lossy cavities.

The analysis of fields in a cavity requires the solution of the full three dimensional wave equation with the required boundary conditions. The procedure here will be to take the *TM* and *TE* waves we have already defined and to modify them to satisfy the additional boundary conditions imposed by the additional conducting walls.

TM fields satisfy the following equation

$$\frac{\partial^2 E_z}{\partial x^2} + \frac{\partial^2 E_z}{\partial y^2} + \frac{\partial^2 E_z}{\partial z^2} + (\gamma^2 + k^2)E_z = 0 \tag{2.109}$$

while*TE* fields satisfy

$$\frac{\partial^2 H_z}{\partial x^2} + \frac{\partial^2 H_z}{\partial y^2} + \frac{\partial^2 H_z}{\partial z^2} + (\gamma^2 + k^2)H_z = 0 \tag{2.110}$$

where the term in brackets (γ^2+k^2) is also known as the cut-off wave number.

Comparing these with Eqs. (2.78) and (2.73), it is obvious that the fields E_z and H_z are those in Eqs. (2.79) and (2.86) multiplied by an additional product due to the z dependence of the field.

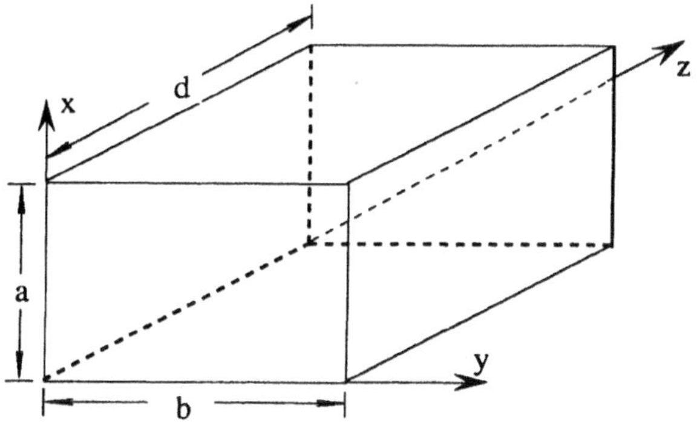

Figure 2.19. Structure and dimensions of a rectangular cavity resonator.

2.3.1. TM and TE Modes in Cavity Resonators

For *TM* modes, the additional boundary conditions are:

$$E_x(x,y,0) = 0. \qquad E_x(x,y,d) = 0. \qquad E_y(x,y,0) = 0. \qquad E_y(x,y,d) = 0.$$

The electric field that satisfies Eq. (2.109) and these boundary conditions is

$$E_z(x,y,z) = E_0 \sin\left(\frac{m\pi}{a}\right)x \sin\left(\frac{n\pi}{b}\right)y \cos\left(\frac{p\pi}{d}\right)z \tag{2.111}$$

Now we can substitute this into Eqs. (2.74) through (2.77) to obtain

$$E_x(x,y,z) = \frac{-\gamma}{\gamma^2+k^2} E_0 \frac{m\pi}{a} \cos\left(\frac{m\pi}{a}\right)x \sin\left(\frac{n\pi}{b}\right)y \cos\left(\frac{p\pi}{d}\right)z \tag{2.112}$$

$$E_y(x,y,z) = \frac{-\gamma}{\gamma^2+k^2} E_0 \frac{n\pi}{b} \sin\left(\frac{m\pi}{a}\right)x \cos\left(\frac{n\pi}{b}\right)y \cos\left(\frac{p\pi}{d}\right)z \tag{2.113}$$

$$H_x(x,y,z) = \frac{j\omega\varepsilon}{\gamma^2+k^2} E_0 \frac{n\pi}{b} \sin\left(\frac{m\pi}{a}\right)x \cos\left(\frac{n\pi}{b}\right)y \cos\left(\frac{p\pi}{d}\right)z \tag{2.114}$$

$$H_y(x,y,z) = \frac{-j\omega\varepsilon}{\gamma^2+k^2} E_0 \frac{m\pi}{a} \cos\left(\frac{m\pi}{a}\right)x \sin\left(\frac{n\pi}{b}\right)y \cos\left(\frac{p\pi}{d}\right)z \tag{2.115}$$

The first of these equation can be written, in general terms, as

$$E_x = E_0 e^{-\gamma z}$$

Therefore,

$$\frac{\partial E_x}{\partial z} = -\gamma E_x \tag{2.116}$$

Using this derivative in Eqs. (2.112) and (2.113), the fields are now

$$E_x(x,y,z) = \frac{-1}{\gamma^2 + k^2} E_0 \frac{p\pi}{d}\frac{m\pi}{a} \cos\left(\frac{m\pi}{a}\right)x \sin\left(\frac{n\pi}{b}\right)y \sin\left(\frac{p\pi}{d}\right)z \tag{2.117}$$

$$E_y(x,y,z) = \frac{-1}{\gamma^2 + k^2} E_0 \frac{p\pi}{d}\frac{n\pi}{b} \sin\left(\frac{m\pi}{a}\right)x \cos\left(\frac{n\pi}{b}\right)y \sin\left(\frac{p\pi}{d}\right)z \tag{2.118}$$

$$H_x(x,y,z) = \frac{j\omega\varepsilon}{\gamma^2 + k^2} E_0 \frac{n\pi}{b} \sin\left(\frac{m\pi}{a}\right)x \cos\left(\frac{n\pi}{b}\right)y \cos\left(\frac{p\pi}{d}\right)z \tag{2.119}$$

$$H_y(x,y,z) = \frac{-j\omega\varepsilon}{\gamma^2 + k^2} E_0 \frac{m\pi}{a} \cos\left(\frac{m\pi}{a}\right)x \sin\left(\frac{n\pi}{b}\right)y \cos\left(\frac{p\pi}{d}\right)z \tag{2.120}$$

Similarly, we can write the dispersion relation as

$$k^2 = \left(\frac{m\pi}{a}\right)^2 + \left(\frac{n\pi}{b}\right)^2 + \left(\frac{p\pi}{d}\right)^2 \tag{2.121}$$

or

$$f_{mnp} = \frac{1}{2\pi\sqrt{\mu\varepsilon}} \sqrt{\left(\frac{m\pi}{a}\right)^2 + \left(\frac{n\pi}{b}\right)^2 + \left(\frac{p\pi}{d}\right)^2} \tag{2.122}$$

where the indices m,n,p, indicate the mode at which the cavity resonates.

2.3.2. *TE Modes in a Rectangular Cavity Resonator*

A discussion similar to that for the *TM* waves can be followed here. The z component of the magnetic field is now

$$H_z(x,y,z) = H_0 \cos\left(\frac{m\pi}{a}\right)x \cos\left(\frac{n\pi}{b}\right)y \sin\left(\frac{p\pi}{d}\right)z \tag{2.123}$$

and substitution in the *TE* field equation (2.69) through (2.72) gives

$$E_x(x,y,z) = \frac{j\omega\mu}{\gamma^2 + k^2} H_0 \frac{n\pi}{b} \cos\left(\frac{m\pi}{a}\right)x \sin\left(\frac{n\pi}{b}\right)y \sin\left(\frac{p\pi}{d}\right)z \tag{2.124}$$

$$E_y(x,y,z) = \frac{-j\omega\mu}{\gamma^2 + k^2} H_0 \frac{m\pi}{a} \sin\left(\frac{m\pi}{a}\right)x \cos\left(\frac{n\pi}{b}\right)y \sin\left(\frac{p\pi}{d}\right)z \tag{2.125}$$

$$H_x(x,y,z) = \frac{\gamma}{\gamma^2 + k^2} H_0 \frac{m\pi}{a} \sin\left(\frac{m\pi}{a}\right)x \cos\left(\frac{n\pi}{b}\right)y \cos\left(\frac{p\pi}{d}\right)z \tag{2.126}$$

$$H_y(x,y,z) = \frac{\gamma}{\gamma^2 + k^2} H_0 \frac{n\pi}{b} \cos\left(\frac{m\pi}{a}\right)x \sin\left(\frac{n\pi}{b}\right)y \cos\left(\frac{p\pi}{d}\right)z \tag{2.127}$$

The dispersion relation is the same as in Eq. (2.121) and the resonant frequency for *TE* modes is also given by Eq. (2.122).

From the equations for *TM* and *TE* modes we note that for *TM* modes, neither *m* and *n* can be zero while *p* can be zero. For *TE* modes, either *m* or *n* can be zero (but not both) while *p* must be nonzero.

The various modes possible are defined by *m,n,* and *p*. Some of the modes may have the same resonant frequency even though they are different modes. As an example, for a cubic cavity *(a=b=d)*, TM_{110}, TE_{011}, TE_{101} all have the same resonant frequency. These are called degenerate modes. The mode with lowest frequency (again, it depends on *a, b,* and *d*) is called the dominant mode.

2.3.3. Cylindrical Cavity Resonators

The procedure for computation of fields and modes in cylindrical cavity resonators is identical to that used for the rectangular cavity. We will not do so here since most of the steps are obvious. We will only mention here that the cutoff frequencies of resonant cylindrical cavities are given by

$$f_{mnp} = \frac{1}{2\pi\sqrt{\mu\varepsilon}} \sqrt{\left(\frac{x_{nm}}{a}\right)^2 + \left(\frac{p\pi}{d}\right)^2} \tag{2.128}$$

for *TM* modes and by

$$f_{mnp} = \frac{1}{2\pi\sqrt{\mu\varepsilon}} \sqrt{\left(\frac{x'_{nm}}{a}\right)^2 + \left(\frac{p\pi}{d}\right)^2} \tag{2.129}$$

for *TE* modes.

In these relations, p is the third mode index as for rectangular cavities and d is the length of the cavity. A cylindrical cavity resonator is shown in Figure 2.20. The values for x_{nm} and x'_{nm} were given in sections 2.2.6. and 2.2.7. as

$$x_{nm} = \sqrt{(\gamma^2 + k^2)}\, a$$

$$x'_{nm} = \sqrt{(\gamma^2 + k^2)}\, a$$

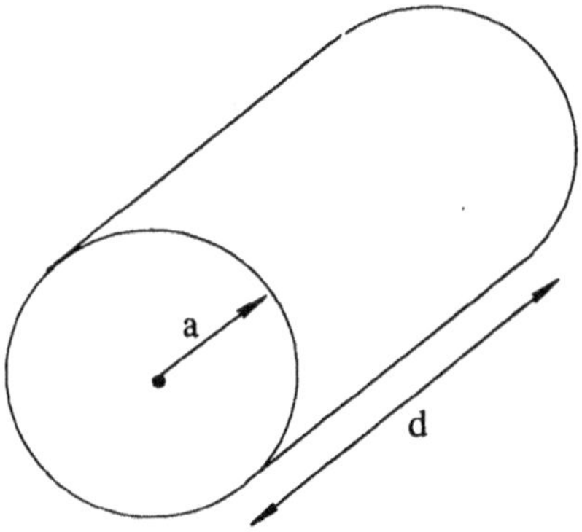

Figure 2.20. A cylindrical resonant cavity.

2.3.4. *Energy in a Cavity Resonator*

Power and energy relations in cavity resonators are defined by the Poynting theorem. Since there is a certain amount of energy stored in the fields of a cavity, the calculation of this power is an important aspect of analysis. This is particularly obvious if we recall that in a resonant device, these relations change dramatically at, or near resonance. This was true with resonant circuits and is certainly true with resonant cavities. The stored and dissipated energy in a cavity define the basic qualities of the cavity. A lossless cavity is not practically realizable, therefore we also define a quantity called "quality factor" of the cavity which is a measure of losses in the cavity. A shift in the resonant frequency of the cavity can also be described in terms of energy. These relations can then be used to characterize a cavity and for measurements in the cavity.

The Poynting vector was defined in chapter 1 as

$$\mathbf{P} = \mathbf{E} \times \mathbf{H}$$

where, **E** and **H** are the electric and magnetic field intensities in the cavity. **P** is the Poynting vector. Its direction indicates the direction of propagation of energy and its magnitude is the power density. For our purposes here it is more convenient to write the total power through a surface

$$\int_s (\mathbf{E}\times\mathbf{H})\cdot d\mathbf{s} = -\frac{\partial}{\partial t}\int_v \left(\frac{1}{2}\varepsilon E^2 + \frac{1}{2}\mu H^2\right)dv - \int_v \sigma E^2 dv$$

Or, in a simplified form as

$$\int_s (\mathbf{E}\times\mathbf{H})\cdot d\mathbf{s} = -\frac{\partial}{\partial t}\int_v (w_e + w_m)dv - \int_v p_d dv \qquad (2.130)$$

where w_e is the stored electric energy density, w_m the stored magnetic energy density and p_d the dissipated power density. The average power density is

$$P_{av} = \frac{1}{2} Re(\mathbf{E}\times\mathbf{H})$$

The stored electric and magnetic energy in the cavity can then be written as:

$$W_0 = \int_v \left(\frac{\varepsilon|E|^2}{4} + \frac{\mu|H|^2}{4}\right)dv \qquad (2.131)$$

where E and H are the fields in the cavity and v the volume of the cavity. This relation is correct at any frequency regardless of resonance. The dissipated power is

$$P_d = \frac{1}{2}\int_v \sigma|E|^2 dv \qquad (2.132)$$

where σ may represent conducting materials in the cavity, including lossy dielectrics or, more often, conductivity of the walls.

2.3.5. Quality Factor of a Cavity Resonator

The quality factor of the cavity resonator is defined as the ratio between the stored energy in the cavity and the dissipated power per unit time

$$Q = \frac{\omega_0 W_0}{P_d} \qquad (2.133)$$

where ω_0 is the resonant frequency. Since the higher the Q factor, the more selective the cavity is, Q is a measure of the bandwidth of the cavity. It also defines, indirectly, the amount of energy needed to couple into the cavity to maintain an energy balance. Ideal cavities have an infinite quality factor.

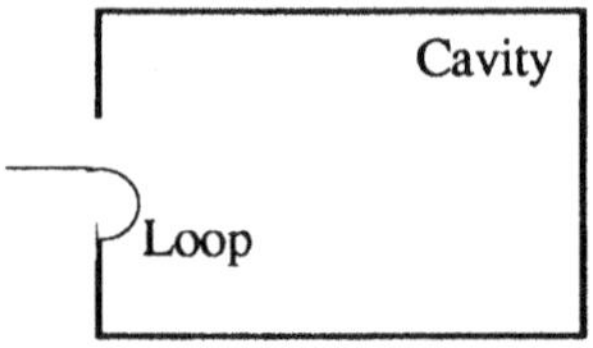

Figure 2.21. Coupling to a cavity by a small loop in the cavity.

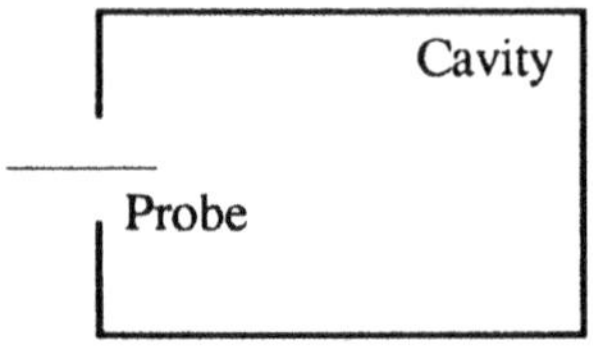

Figure 2.22. Coupling to a cavity by a small probe in the cavity.

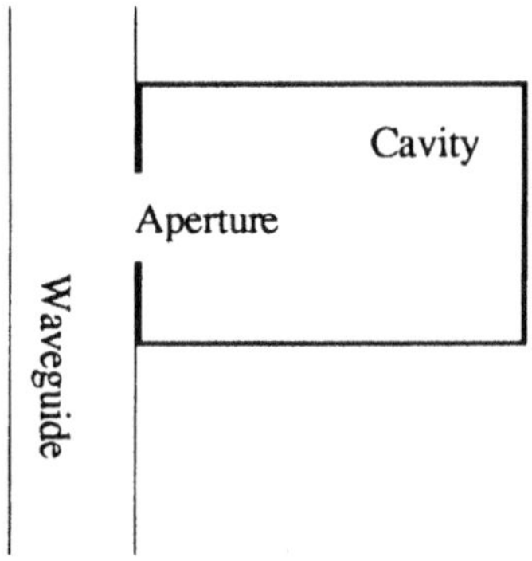

Figure 2.23. Coupling to a cavity by a small aperture in the wall of the cavity. The figure shows coupling to a waveguide.

2.3.6. Coupling to Cavities

The coupling of electromagnetic energy to cavities has not been addressed specifically in the discussion above. If the cavity was to be ideal, the fields within the cavity would be of infinite amplitude, provided that the necessary modes can be excited. In realistic cavities, there are always some losses but these are usually small. The fields are large and the amount of stored energy is also large. However, the small amount of power dissipated has to be compensated for by external sources otherwise the cavity would cease to oscillate. This is done by coupling energy into the cavity. A cavity for which the lost energy is balanced is called a critically coupled cavity.

The introduction of energy into the cavity can be done in a number of ways. The most obvious of these is to have within the cavity a source that generates the necessary fields. A small loop (Figure 2.21) or a simple probe excitation (Figure 2.22) can be used. Similarly, the cavity can be coupled through a small aperture through which a small amount of energy "leaks" into the cavity. (Figure 2.23). The only requirement of these methods of coupling is that they excite the required modes. The three coupling methods in Figure 2.21 through 2.23 can excite different modes.

2.4. Perturbation Techniques

Introduction of any material in a microwave cavity disturbs the field pattern inside and changes both the resonant frequency and the Q-factor of the cavity. In the extreme, when the cavity is loaded with conducting materials or lossy dielectrics, the modes of the cavity may also change. Often, for the purpose of measurement of material properties, a small sample is introduced inside a microwave cavity. This affects the cavity properties and the changes in resonant frequency and Q-factor can be used to measure various properties of the material.

Under certain conditions, these changes in cavity properties can be calculated directly from the properties of the empty cavity by a perturbation method. The guiding principle is that as long as the perturbation or "disturbance" of the original cavity is small, the changes in properties are also small and can be superimposed on those of the original cavity. Therefore, separation between the two is possible and any change in cavity parameters can be attributed to the perturbing material. Although, for testing purposes, perturbation normally means insertion of a piece of material, it can also mean removal of material, changes in material properties or, for that matter, distortion in the cavity boundaries.

The assumptions needed for application of perturbation techniques are:

1. The perturbation is small compared to the dimensions of the cavity. While most often the dimensions are meant to be geometrical dimensions, the electric and magnetic properties must also be considered. As an example, a small sample with very large permittivity may have a larger effect on the fields than a much larger sample with low permittivity. Since the perturbation is in the field not in geometry, the size of the sample and its properties must be such that the field perturbation is small
2. The fields throughout the cavity remain essentially the same, with the exception of the location of the perturbation and its vicinity.

As a consequence, it is implicitly understood that the changes in resonant frequency and Q-factor are small.

The use of perturbation methods is particularly useful for validation of measurements, or for validation of numerical calculations. Used with care, the method can also be used for estimation of properties when the perturbation is not necessarily small. One should, however, use caution since with large perturbations, the modes in the cavity may change. In addition, the solution obtained is not accurate and accuracy deteriorates with the size of the perturbation.

We discuss here three specific situations. The first is that of volume perturbation shown in Figure 2.24. Here the volume is reduced (or increased) by a small amount due to, for example, metallic deposits on the cavity wall, carburation on the walls or a "bump" in the cavity due to externally or internally induced damage. The second type of perturbation is that of changes in the material properties in the cavity. Either ε or μ or both may change. This form of perturbation is useful in monitoring or measurement of material properties since, in general, these changes are relatively small. This situation is depicted in Figure 2.25. The third situation is that of an insertion (or removal) of a small sample from the cavity. The testing equivalent is the insertion of a sample for testing for material properties or testing for small inclusions in a sample that fills the cavity. This is shown in Figure 2.26.

2.4.1. Volume Perturbation

We assume the cavity has perfectly conducting walls and that a perturbation has been introduced such that the cavity volume is increased by a small volume ΔV as shown in Figure 2.24. Also, for simplicity, we assume there are no losses in the cavity and the sources of the fields are outside the cavity (as is almost always the case)

The energy relationships in the cavity can be defined in a formal manner starting with Maxwell's field equations. Starting with Faraday's and Ampere's laws, the fields in the unperturbed cavity are:

$$\nabla\times\mathbf{E}_0 = -j\omega_0\mu\mathbf{H}_0 \tag{2.134}$$

$$\nabla\times\mathbf{H}_0 = j\omega_0\varepsilon\mathbf{E}_0 \tag{2.135}$$

and in the perturbed cavity:

$$\nabla\times\mathbf{E} = -j\omega\mu\mathbf{H} \tag{2.136}$$

$$\nabla\times\mathbf{H} = j\omega\varepsilon\mathbf{E} \tag{2.137}$$

Here $\mathbf{E}$ and $\mathbf{H}$ are the fields in the perturbed cavity and $\mathbf{E}_0$ and $\mathbf{H}_0$ are the fields in the unperturbed cavity. ω_0 is the resonant frequency of the unperturbed cavity and ω the resonant frequency in the perturbed cavity. These equations were written in the time harmonic forms as in Eqs. (1.37) through (1.40) in chapter 1.

The complex conjugates of Eqs. (2.134) and (2.135) are

$$\nabla\times\mathbf{E}_0^* = j\omega_0\mu\mathbf{H}_0^* \tag{2.138}$$

$$\nabla\times\mathbf{H}_0^* = -j\omega_0\varepsilon\mathbf{E}_0^* \tag{2.139}$$

The dot product between **H** and Eq. (2.138) is

$$\mathbf{H}\cdot(\nabla\times\mathbf{E}_0^*) = j\omega_0\mu\mathbf{H}\cdot\mathbf{H}_0^* \tag{2.140}$$

The dot product between $\mathbf{E}_0^*$ and Eq. (2.137) is

$$\mathbf{E}_0^*\cdot(\nabla\times\mathbf{H}) = j\omega\varepsilon\mathbf{E}_0^*\cdot\mathbf{E} \tag{2.141}$$

Subtracting equation (2.141) from Eq. (2.140) gives

$$\mathbf{H}\cdot(\nabla\times\mathbf{E}_0^*) - \mathbf{E}_0^*\cdot(\nabla\times\mathbf{H}) = j\omega_0\mu\mathbf{H}\cdot\mathbf{H}_0^* - j\omega\varepsilon\mathbf{E}_0^*\cdot\mathbf{E} \tag{2.142}$$

the left hand side can be rewritten by using the following vector identity

$$\nabla\cdot(\mathbf{A}\times\mathbf{B}) = \mathbf{B}\cdot(\nabla\times\mathbf{A}) - \mathbf{A}\cdot(\nabla\times\mathbf{B})$$

as

$$\nabla\cdot(\mathbf{E}_0^*\times\mathbf{H}) = j\omega_0\mu\mathbf{H}\cdot\mathbf{H}_0^* - j\omega\varepsilon\mathbf{E}_0^*\cdot\mathbf{E} \tag{2.143}$$

Similarly, the dot product between **E** and Eq. (2.139) is

$$\mathbf{E}\cdot(\nabla\times\mathbf{H}_0^*) = -j\omega_0\varepsilon\mathbf{E}_0^*\cdot\mathbf{E} \tag{2.144}$$

and the dot product between $\mathbf{H}_0^*$ and Eq. (2.136) is

$$\mathbf{H}_0^*\cdot(\nabla\times\mathbf{E}) = -j\omega\mu\mathbf{H}_0^*\cdot\mathbf{H} \tag{2.145}$$

Subtracting Eq. (2.144) from Eq. (2.145) and using the same vector identity we get

$$\nabla\cdot(\mathbf{E}\times\mathbf{H}_0^*) = -j\omega\mu\mathbf{H}_0^*\cdot\mathbf{H} + j\omega_0\varepsilon\mathbf{E}_0^*\cdot\mathbf{E} \tag{2.146}$$

We now integrate Eqs. (2.143) and (2.146) over the volume V of the perturbed cavity. For Eq. (2.143) we have

$$\int_V \nabla\cdot(\mathbf{E}_0^*\times\mathbf{H})dv = \int_V \left(j\omega_0\mu\mathbf{H}\cdot\mathbf{H}_0^* - j\omega\varepsilon\mathbf{E}_0^*\cdot\mathbf{E}\right)dv$$

Using the divergence theorem we get

$$\oint_S (\mathbf{E}_0^* \times \mathbf{H}) \cdot \mathbf{ds} = \int_V \left(j\omega_0 \mu \mathbf{H} \cdot \mathbf{H}_0^* - j\omega\varepsilon \mathbf{E}_0^* \cdot \mathbf{E} \right) \mathrm{dv} \tag{2.147}$$

Similarly, for Eq. (2.146)

$$\oint_S (\mathbf{E} \times \mathbf{H}_0^*) \cdot \mathbf{ds} = \int_V \left(-j\omega\mu \mathbf{H}_0^* \cdot \mathbf{H} + j\omega_0 \varepsilon \mathbf{E}_0^* \cdot \mathbf{E} \right) \mathrm{dv} \tag{2.148}$$

Summing Eqs. (2.147) and (2.148):

$$\oint_S (\mathbf{E}_0^* \times \mathbf{H} + \mathbf{E} \times \mathbf{H}_0^*) \cdot \mathbf{ds} = -j(\omega - \omega_0) \int_V \left(\mu \mathbf{H} \cdot \mathbf{H}_0^* + \varepsilon \mathbf{E}_0^* \cdot \mathbf{E} \right) \mathrm{dv}$$

Because we assume perfectly conducting cavity walls, the second term on the left hand side vanishes (i.e. the tangential component of **E** vanishes on the unperturbed cavity boundary and therefore the closed surface integral of this term is zero). We get

$$\oint_S (\mathbf{E}_0^* \times \mathbf{H}) \cdot \mathbf{ds} = -j(\omega - \omega_0) \int_V \left(\mu \mathbf{H} \cdot \mathbf{H}_0^* + \varepsilon \mathbf{E}_0^* \cdot \mathbf{E} \right) \mathrm{dv}$$

Similarly, if we assume that the perturbed cavity surface is $S=S_0+\Delta S$, the following can be written:

$$\oint_{\Delta S} (\mathbf{E}_0^* \times \mathbf{H}) \cdot \mathbf{ds} = -j\omega_0 \int_{\Delta V} \left(-\mu \mathbf{H} \cdot \mathbf{H}_0^* + \varepsilon \mathbf{E}_0^* \cdot \mathbf{E} \right) \mathrm{dv}$$

Thus, we can write for the shift in resonant frequency due to a change in the volume of the cavity by ΔV

$$\frac{\omega - \omega_0}{\omega_0} = -\frac{\int_{\Delta V} \left(\varepsilon \mathbf{E}_0^* \cdot \mathbf{E} - \mu \mathbf{H} \cdot \mathbf{H}_0^* \right) dv}{\int_V \left(\mu \mathbf{H} \cdot \mathbf{H}_0^* + \varepsilon \mathbf{E}_0^* \cdot \mathbf{E} \right) dv} \tag{2.149}$$

Finally, based on the basic assumptions above (that the fields vary little due to perturbation) we replace **H** by $\mathbf{H}_0$, and **E** by $\mathbf{E}_0$ and get

$$\frac{\omega - \omega_0}{\omega_0} = -\frac{\int_{\Delta V} \left(\varepsilon \mathbf{E}_0^* \cdot \mathbf{E}_0 - \mu \mathbf{H}_0 \cdot \mathbf{H}_0^* \right) dv}{\int_V \left(\mu \mathbf{H}_0 \cdot \mathbf{H}_0^* + \varepsilon \mathbf{E}_0^* \cdot \mathbf{E}_0 \right) dv} \tag{2.149}$$

Also, performing the dot products we can write in a simpler form:

$$\frac{\omega-\omega_0}{\omega_0}=-\frac{\int_{\Delta V}\left(\varepsilon|\mathbf{E}_0|^2-\mu|\mathbf{H}_0|^2\right)dv}{\int_V\left(\mu|\mathbf{H}_0|^2+\varepsilon|\mathbf{E}_0|^2\right)dv} \tag{2.150}$$

Thus, the shift in resonant frequency depends on the change in the volume of the cavity. To see this relationship we rewrite this as

$$\frac{\omega-\omega_0}{\omega_0}=-\frac{\Delta W_\varepsilon-\Delta W_\mu}{W_\varepsilon+W_\mu} \tag{2.151}$$

where W_m is the stored magnetic energy, W_e the stored electric energy, and Δ indicates a shift in these energies.

The shift in resonant frequency in Eq. (2.150) depends on where the volume changes. If this is at a point of minimum electric field, only the change in magnetic energy is significant and an increase in magnetic energy increases the resonant frequency. If this happens at a point of maximum electric field, an increase in cavity volume decreases the resonant frequency. From the testing point of view it is also important to realize that locating a sample that decreases or increases the effective volume of the cavity (like a highly conductive sample) in a location of low field may mean that no shift in resonant frequency may be detected. With small samples this is particularly important.

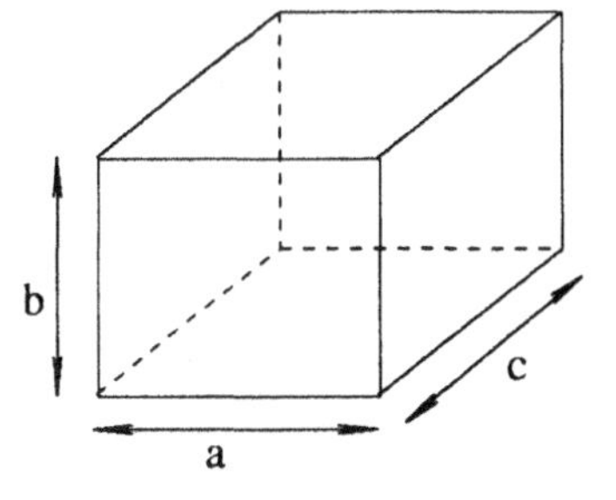

Figure 2.24a. Unperturbed rectangular cavity resonator.

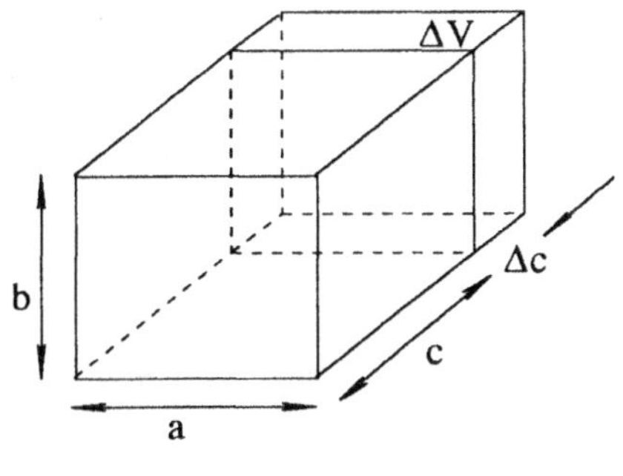

Figure 2.24b. Volume perturbation of the cavity.

2.4.2. Material Perturbation

The resonant frequency of a cavity is also shifted by changes in material properties. This situation is shown in Figure 2.25. To define these changes we start with the same relations as in Eqs (2.134) through (2.137). In Eqs. (2.136) and (2.137) the permittivity and permeability are varied by Δ_ε and Δ_μ respectively. The fields in the unperturbed cavity are the same as in Eqs. (2.134) and Eq. (2.135). In the perturbed cavity we have

$$\nabla\times\mathbf{E} = -j\omega(\mu + \Delta\mu)\mathbf{H} \tag{2.152}$$

$$\nabla\times\mathbf{H} = j\omega(\varepsilon + \Delta\varepsilon)\mathbf{E} \tag{2.153}$$

The dot product between **H** and Eq. (2.138) is

$$\mathbf{H}\cdot(\nabla\times\mathbf{E}_0^*) = j\omega_0\mu\mathbf{H}\cdot\mathbf{H}_0^* \tag{2.154}$$

The dot product between $\mathbf{E}_0^*$ and Eq. (2.137) is

$$\mathbf{E}_0^*\cdot(\nabla\times\mathbf{H}) = j\omega(\varepsilon + \Delta\varepsilon)\mathbf{E}_0^*\cdot\mathbf{E} \tag{2.155}$$

Subtracting Eq. (2.155) from Eq. (2.154) gives

$$\mathbf{H}\cdot(\nabla\times\mathbf{E}_0^*) - \mathbf{E}_0^*\cdot(\nabla\times\mathbf{H}) = j\omega_0\mu\mathbf{H}\cdot\mathbf{H}_0^* - j\omega(\varepsilon + \Delta\varepsilon)\mathbf{E}_0^*\cdot\mathbf{E} \tag{2.156}$$

and following the same process as above we get

$$\nabla\cdot(\mathbf{E}_0^*\times\mathbf{H}) = j\omega_0\mu\mathbf{H}\cdot\mathbf{H}_0^* - j\omega(\varepsilon + \Delta\varepsilon)\mathbf{E}_0^*\cdot\mathbf{E} \tag{2.157}$$

Similarly, the dot product between **E** and Eq. (2.139) is

$$\mathbf{E}\cdot(\nabla\times\mathbf{H}_0^*) = -j\omega_0\varepsilon\mathbf{E}_0^*\cdot\mathbf{E} \tag{2.158}$$

and the dot product between $\mathbf{H}_0^*$ and Eq. (2.136) is

$$\mathbf{H}_0^*\cdot(\nabla\times\mathbf{E}) = -j\omega(\mu + \Delta\mu)\mathbf{H}_0^*.\mathbf{H} \tag{2.159}$$

Subtracting Eq. (2.158) from Eq. (2.159) and using the same vector identity gives

$$\nabla\cdot(\mathbf{E}\times\mathbf{H}_0^*) = -j\omega(\mu + \Delta\mu)\mathbf{H}_0^*\cdot\mathbf{H} + j\omega_0\varepsilon\mathbf{E}_0^*\cdot\mathbf{E} \tag{2.160}$$

We now integrate Eqs. (2.157) and (2.160) over the volume V_0 of the cavity. For Eq. (2.157) we have

$$\int_{V_0} \nabla\cdot(\mathbf{E}_0^*\times\mathbf{H})dv = j\int_{V_0} \omega_0\mu\mathbf{H}\cdot\mathbf{H}_0^* dv - j\int_{V_0} \omega(\varepsilon+\Delta\varepsilon)\mathbf{E}_0^*\cdot\mathbf{E}dv$$

Using the divergence theorem we get

$$\int_{S_0} (\mathbf{E}_0^*\times\mathbf{H})\cdot d\mathbf{s} = j\int_{V_0} \omega_0\mu\mathbf{H}\cdot\mathbf{H}_0^* dv - j\int_{V_0} \omega(\varepsilon+\Delta\varepsilon)\mathbf{E}_0^*\cdot\mathbf{E}dv \qquad (2.161)$$

Similarly, for Eq. (2.160)

$$\int_{S_0} (\mathbf{E}\times\mathbf{H}_0^*)\cdot d\mathbf{s} = -j\int_{V_0} \omega(\mu+\Delta\mu)\mathbf{H}_0^*\cdot\mathbf{H}dv + j\int_{V_0} \omega_0\varepsilon\mathbf{E}_0^*\cdot\mathbf{E}dv \qquad (2.162)$$

Summing Eqs. (2.161) and (2.162)

$$\int_{S_0} \left[\mathbf{E}\times\mathbf{H}_0^* + \mathbf{E}_0^*\times\mathbf{H}\right]\cdot d\mathbf{s} = j\int_{V_0} \left[\omega_0\mu - \omega(\mu+\Delta\mu)\right]\mathbf{H}_0^*\cdot\mathbf{H}dv + j\int_{V_0} \left[\omega_0\varepsilon - \omega(\varepsilon+\Delta\varepsilon)\right]\mathbf{E}_0^*\cdot\mathbf{E}dv$$

The left hand side is equal to zero since the tangential component of **E** is zero on S_0. Thus

$$\int_{V_0} \left[\omega_0\mu - \omega(\mu+\Delta\mu)\right]\mathbf{H}_0^*\cdot\mathbf{H}dv + \int_{V_0} \left[\omega_0\varepsilon - \omega(\varepsilon+\Delta\varepsilon)\right]\mathbf{E}_0^*\cdot\mathbf{E}dv = 0$$

Rearranging terms we can write

$$\frac{\omega-\omega_0}{\omega} = -\frac{\int_{V_0}\left[\Delta\mu\mathbf{H}\cdot\mathbf{H}_0^* + \Delta\varepsilon\mathbf{E}\cdot\mathbf{E}_0^*\right]dv}{\int_{V_0}\left[\mu\mathbf{H}\cdot\mathbf{H}_0^* + \varepsilon\mathbf{E}\cdot\mathbf{E}_0^*\right]dv}$$

and, as above, by performing the dot product we get

$$\frac{\omega-\omega_0}{\omega} = -\frac{\int_{V_0}\left[\Delta\mu|\mathbf{H}|^2 + \Delta\varepsilon|\mathbf{E}|^2\right]dv}{\int_{V_0}\left[\mu|\mathbf{H}|^2 + \varepsilon|\mathbf{E}|^2\right]dv} \qquad (2.163)$$

In terms of energy in the cavity this is written as

$$\frac{\omega - \omega_0}{\omega_0} = -\frac{\Delta W_\varepsilon + \Delta W_\mu}{W_\varepsilon + W_\mu} \tag{2.164}$$

but now the change in energy is due to changes in material properties.

This relation clearly indicates that either a change in permittivity or a change in permeability or both change the resonant frequency, provided that these changes occur at a location of nonzero fields. Any increase in these properties will cause a decrease in the resonant frequency, providing a powerful tool for detection and testing. The result in Eq. (2.164) assumes that the permittivity and/or the permeability of the whole cavity has changed. The same result can be obtained if only part of the cavity contains material but the calculation is more complicated. However, the same conclusions apply and any material inserted in the cavity will change the resonant frequency.

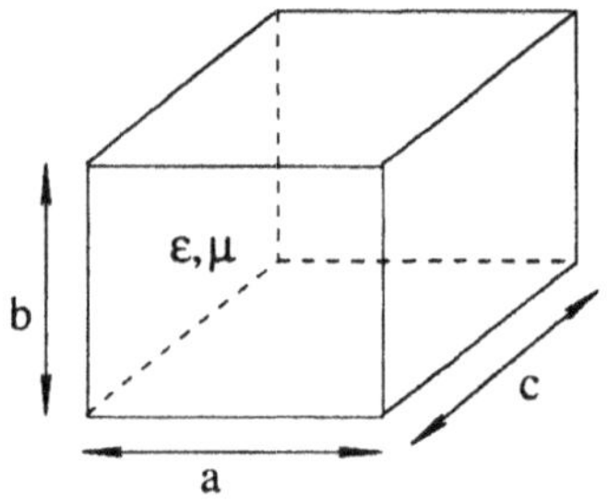

Figure 2.25a. Unperturbed rectangular cavity resonator.

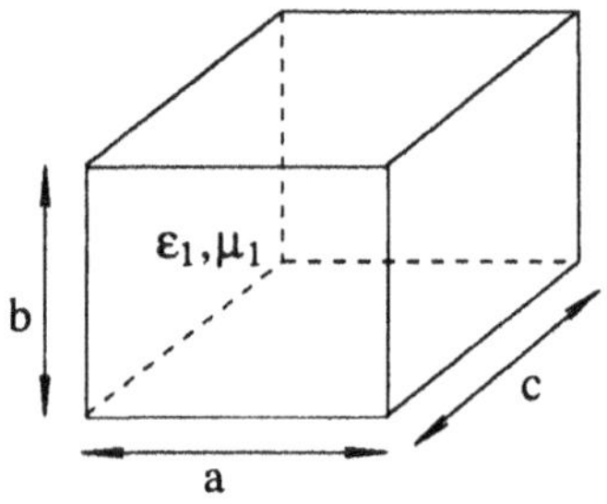

Figure 2.25b. Material perturbation of the cavity.

The perturbation calculations above were based on small changes either in material properties or in the cavity geometry. If the changes are large, the simple results obtained above do not apply. The resonant frequency changes but in different, more complicated relations. Therefore, large changes can also be measured but not necessarily calculated because of the large changes in fields.

2.4.3. Perturbation by Material Insertion

Figure 2.26 shows a variation of the material perturbation method. The analysis is similar to the previous analysis except that the perturbation is only taken over the volume of the insertion which is considered to be small for perturbation to apply. Thus, Eq. (2.164) applies directly but in Eq. (163) the integration is only on V_i which stands for insertion volume

$$\frac{\omega-\omega_0}{\omega}=-\frac{\int_{V_i}\left[\Delta\mu|\mathbf{H}|^2+\Delta\varepsilon|\mathbf{E}|^2\right]dv}{\int_{V_0}\left[\mu|\mathbf{H}|^2+\varepsilon|\mathbf{E}|^2\right]dv} \tag{2.165}$$

or

$$\frac{\omega-\omega_0}{\omega_0}=-\frac{\Delta W_\varepsilon+\Delta W_\mu}{W_\varepsilon+W_\mu} \tag{2.166}$$

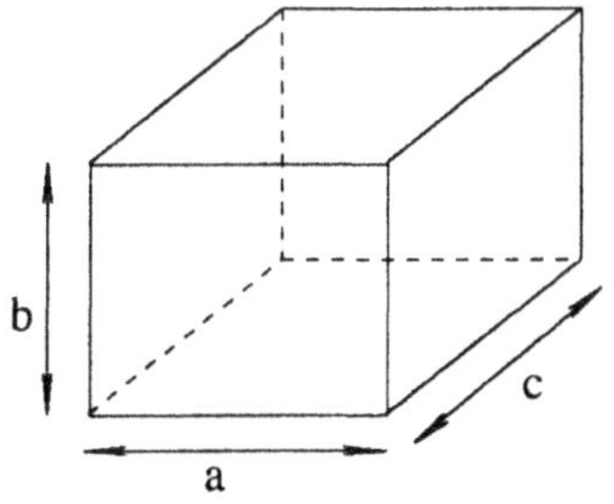

Figure 2.26a. Unperturbed rectangular cavity resonator.

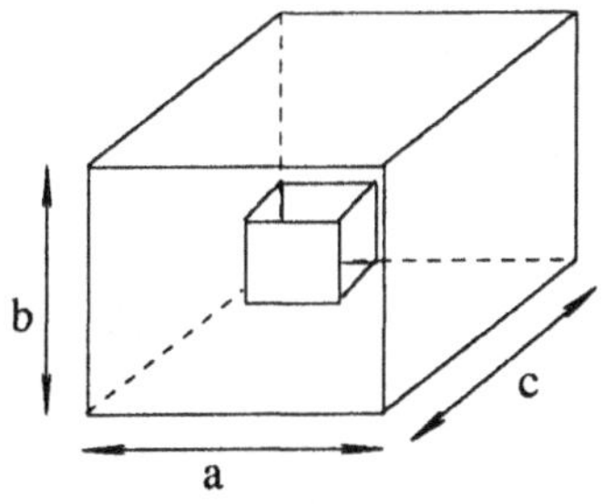

Figure 2.26b. Insertion of material in a cavity.

Although this represents a minor variation of the perturbation technique, it is often used in testing. The insertion can, in principle be a conductor in which case, the perturbation is a volume perturbation. However, this is not often done as the perturbation might not be small, violating the basic assumptions involved.

2.5. Bibliography

Transmission lines are part of most textbooks on electromagnetics. In particular, Collin [1,2], Balanis [3], and Pozar [4] discuss these with a view on microwave applications. There are also specialized books on transmission lines which discuss low frequency transmission lines as well. Examples are [5] and [6]. A particularly clear and simple presentation can be found in Plonus [7]. Finally, a short but clear and complete section on transmission lines is given by Adam [8].

The very important subject of impedance matching is also treated by most authors. It is worth noting however, that most treat the subject in conjunction with the Smith chart. As examples, Pozar [4] uses normalized impedances while Cheng [9] uses both normalized impedances and normalized admittances. Unlike most other sources, Plonus [7] calculates locations of stub matching from the location of a voltage minimum rather than from the load. This produces particularly simple analytical solutions, without the need for Smith charts.

The theory of waveguides, as well as perturbation techniques can be found in any of the textbooks mentioned. A particularly extensive treatment can be found in [1], [2] and [3]. Specialized sources on waveguides and cavity resonators are Lewin [10], Argence and Kahan [11], and Waldron [12]. A considerable amount of information can be found in Harvey [13]. and in Thomas [14]. Although not discussed here, there are a variety of open cavity resonators. We will discuss these later in chapter 8 but a very thorough text on the subject is the book by Weinstein [15]. He also discusses apertures and open waveguides. Also, to be noted is that there are many types of transmission lines that we have not discussed explicitly but which are often used for testing. These include the open transmission lines and microstrip lines. Some good references on the subject are Gupta [16], Itoh [17] and Wadell [18].

[1] R.E. Collin, "Electromagnetic Theory of Guided Waves", McGraw-Hill, NY, 1991.
[2] R.E. Collin, "Foundations for Microwave Engineering", McGraw-Hill, Inc., New York, 1992.
[3] C.A. Balanis, "Advanced Engineering Electromagnetics", John Wiley & Sons, New York, 1989.
[4] D.M. Pozar, Microwave Engineering, Addison-Wesley, Reading, MA., 1990.
[5] R.L. Liboff and G.C. Dalman, "Transmission Lines, Waveguides, and Smith Charts", Macmillan, London, 1985.
[6] L.N. Dworsky, "Modern Transmission Line Theory and Applications", John Wiley & Sons, New York, 1979.
[7] M.A. Plonus, "Applied Electromagnetics", McGraw-Hill, Inc., 1978.
[8] S.F. Adam, "Microwave Theory and Applications", Prentice Hall, Englewood Cliffs, NJ, 1969.
[9] D.K. Cheng, "Field and Wave Electromagnetics", 2nd ed., Addison-Wesley, Reading, Ma., 1989.
[10] L. Lewin, "Theory of Waveguides", John Wiley & Sons, New York, 1975.
[11] E. Argence and T. Kahan, "Theory of Waveguides and Cavity Resonators", Blackie and Sons, Ltd., London, 1967.
[12] R.A. Waldron, "The Theory of Waveguides and Cavities", Maclaren, London, 1967.
[13] A.F. Harvey, "Microwave Engineering", Academic Press, London, 1963.
[14] H.E. Thomas, "Handbook of Microwave Techniques and Equipment", Prentice-Hall, Inc., Englewood Cliffs, NJ., 1972.
[15] L.A. Weinstein, "Open Resonators and Open Waveguides", The Golem Press, Boulder, Colorado, 1969.
[16] K.C. Gupta, R.L. Garg and I.J. Bahl, "Microstrip Lines and Slotlines", Artech House, Dedham, MA., 1979.
[17] I. Itoh, Ed., "Planar Transmission Line Structures" IEEE, New York, 1987.
[18] B.C. Wadell, "Transmission Line Design Handbook", Artech House, Boston, 1991.

Chapter 3.
Reflection, Transmission, and Scattering of Waves

3.1. Introduction

In this chapter we discuss behavior of waves at dielectric and conductor interfaces as well as scattering from bodies. The problem of conditions at the interfaces between two materials is a fundamental problem in testing. Many of the methods we use later depend on measurement of reflected or transmitted waves from a single interface or from multiple interfaces. To define the reflected fields, we define the reflection and transmission coefficients for a variety of interfaces.

The first topic is polarization of plane waves. This is defined here because the reflection and transmission conditions are dependent on the polarization of the incident wave. After this, a number of important conditions are defined. These include reflection and transmission at lossy and lossless interfaces as well as at conductor interfaces. Both oblique and normal incidence are used. The reflection and transmission due to multiple surfaces is also discussed as well as from conductor backed dielectrics. These will be used later (in chapter 6 through 8) to define testing arrangements.

Scattering here is defined as Rayleigh scattering (scatterer is much smaller than the wavelength at which testing takes place) and the scatterer is in the far field of the source. This is a somewhat limited view of scattering but, within the limitations of space, it should give a good background on most scattering applications in *NDT*. Remaining topics that cannot be covered are referred to in the bibliography section.

3.2. Polarization of Plane Waves

The electric (or magnetic) field intensity of a uniform plane wave has a direction in space. This direction may either be constant or may change as the wave propagates. Polarization of a plane wave refers to the direction of the electric field intensity vector in space. Polarization is defined for the electric field vector only, since the magnetic field is always obtainable from the electric field by use of the intrinsic impedance and there is no need to define its polarization separately. The reason why polarization of plane waves is important is because of the properties associated with polarization. As for polarized light, the transmission and refraction of the wave are affected by the polarization. A simple example is the direction of an antenna. If the electric field is, say, horizontally directed in space (polarized in this direction), then a receiving antenna must also be directed horizontally if the electric field (and therefore the current) in the antenna is to be maximized. Thus, wave properties are closely linked with polarization. Polarization of plane waves can be intentional or as a result of propagation through materials.

We will treat linear polarization first as the simplest polarization and then extend linear polarization to include the important cases of circular and elliptical polarization.

3.2.1. Linear Polarization

The electric field

$$\mathbf{E} = \hat{\mathbf{y}} E_y(z) \tag{3.1}$$

of a wave propagating in the z direction and directed in the y direction is "linearly polarized" in the y direction. The electric field intensity varies in the y direction (its amplitude is constant but, as it propagates, the phase changes). This linearly polarized wave is shown in Figure 3.1a. where the electric field is shown as it changes with time.

Similarly, the electric field

$$\mathbf{E} = \hat{\mathbf{x}} E_x(z) \tag{3.2}$$

of a wave propagating in the z direction and directed in the x direction is "linearly polarized" in the x direction as shown in Figure 3.1b.

Taking the two linearly polarized waves above, and superimposing the two, we obtain polarization in an arbitrary direction in space, depending on the two amplitudes. As an example, if E_x and E_y are equal, the electric field is polarized at $45°$ to the x axis as shown in Figure 3.2. In general the superposition of two fields each with a linearly polarized wave produce a linearly polarized wave in the direction $tan^{-1}(E_y/E_x)$.

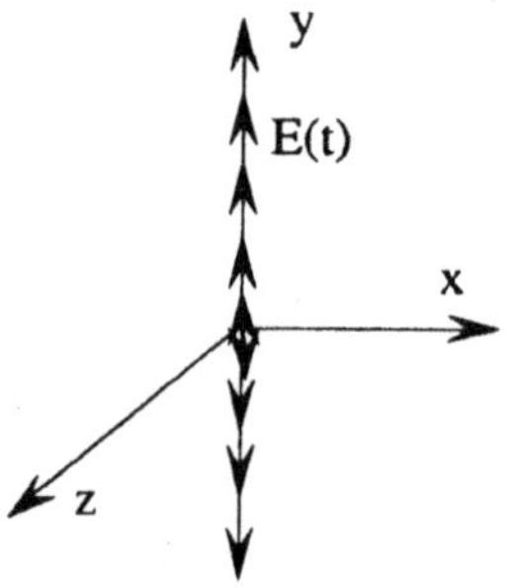

Figure 3.1a. Time variation of a linearly polarized wave in the y direction.

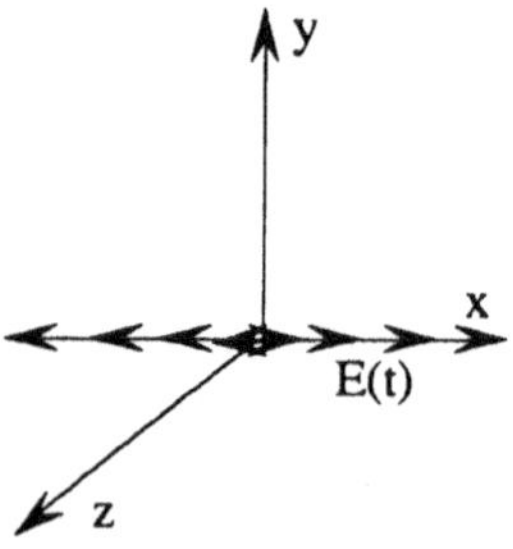

Figure 3.1b. Time variation of a linearly polarized wave in the x direction.

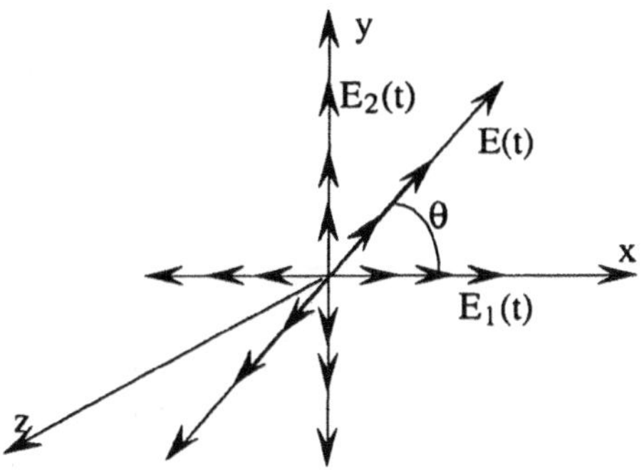

Figure 3.2. General linear polarization obtained by superposition of two linearly polarized waves.

3.2.2. Elliptical and Circular Polarization

If one wave is polarized in a given direction, say the x direction and the other at $90°$ degrees to the first, say in the y direction but also lagging or leading in phase by $90°$ degrees we obtain elliptical polarization. To see this, consider the following wave, consisting of the two electric fields in Eqs. (3.1) and (3.2) but with E_y lagging by $90°$ with respect to E_x. The electric field of the superposed wave is

$$\mathbf{E} = \hat{\mathbf{x}}E_1(z) + \hat{\mathbf{y}}jE_2(z) = \hat{\mathbf{x}}E_1e^{-jkz} - \hat{\mathbf{y}}jE_2e^{-jkz} \tag{3.3}$$

where the amplitudes E_1 and E_2 are real and E_2 lags behind E_1 by $90°$. Taking the real part of Eq. (3.3) we obtain the time domain representation as

$$\mathbf{E}(z,t) = \hat{\mathbf{x}}E_1cos(\omega t - kz) + \hat{\mathbf{y}}E_2cos(\omega t - kz - \frac{\pi}{2}) \tag{3.4}$$

To examine the change of $\mathbf{E}$, at a point in space, say $z=0$, we write

$$\mathbf{E}(0,t) = \hat{\mathbf{x}}E_1cos(\omega t) + \hat{\mathbf{y}}E_2cos(\omega t - \frac{\pi}{2}) = \hat{\mathbf{x}}E_1cos(\omega t) + \hat{\mathbf{y}}E_2sin(\omega t) \tag{3.5}$$

The vector $\mathbf{E}(0,t)$ now describes an elliptical path as t changes. The elliptical path is due to the different amplitudes of the component E_1 and E_2. The rotation of the vector is in the counterclockwise direction. The wave in Eq. (3.3) is said to be elliptically polarized (see Figure 3.3a). Because the rotation is in the counterclockwise direction (positive direction for phase) the wave is said to be right elliptically polarized. If E_y leads E_x by $90°$ the rotation is in the clockwise direction and the wave is said to be left elliptically polarized. This is shown in Figure 3.3b. This can be written from Eqs. (3.3) through (3.5) as

$$\mathbf{E} = \hat{\mathbf{x}} E_1 e^{-jkz} + \hat{\mathbf{y}} j E_2 e^{-jkz}$$

or, in the time domain at $z=0$:

$$\mathbf{E}(0,t) = \hat{\mathbf{x}} E_1 cos(\omega t) + \hat{\mathbf{y}} E_2 cos(\omega t + \frac{\pi}{2}) = \hat{\mathbf{x}} E_1 cos(\omega t) - \hat{\mathbf{y}} E_2 sin(\omega t) \qquad (3.6)$$

Circular polarization is a special case of elliptical polarization and is obtained if $E_1 = E_2$ as in Figure 3.4. If E_1 lags behind E_2, the rotation of the field is clockwise.

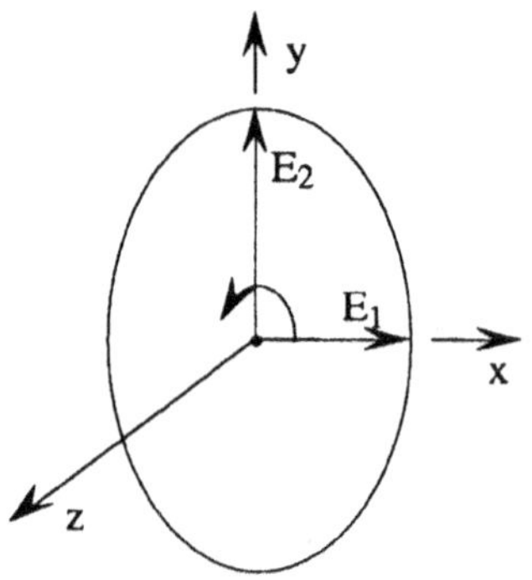

Figure 3.3a. Right elliptically polarized wave: E_2 lags behind E_1.

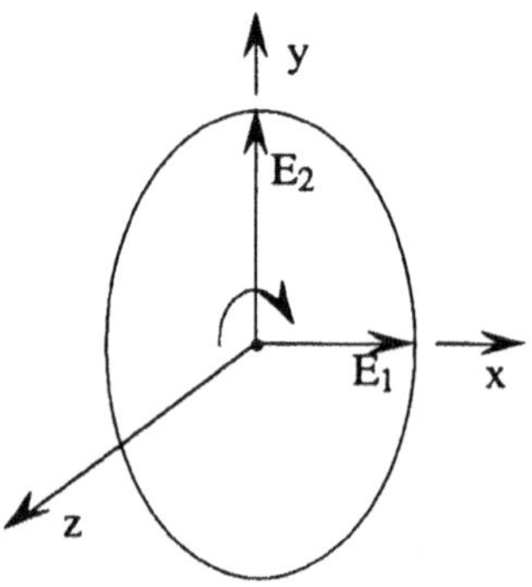

Figure 3.3b. Left elliptically polarized wave: E_1 lags behind E_2.

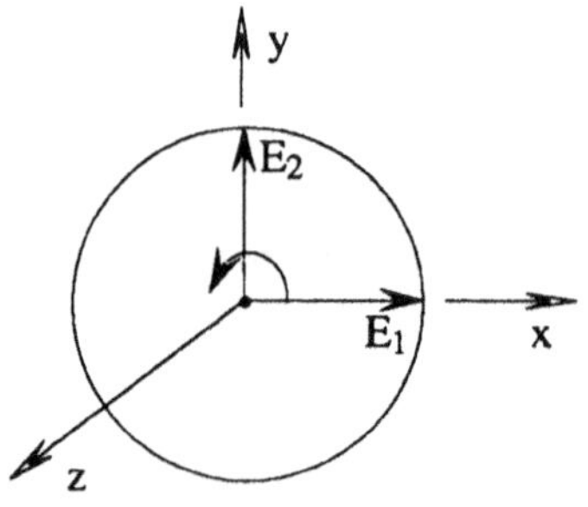

Figure 3.4a. Right circularly polarized wave: E_2 lags behind E_1.

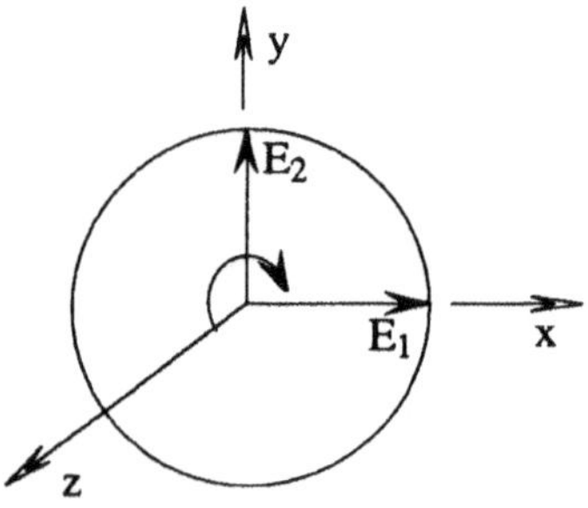

Figure 3.4b. Left circularly polarized wave: E_1 lags behind E_2.

The lead or lag in Eqs. (3.5) and (3.6) were set to *90°* because the results lead to simple expressions. However, the phase difference between two waves can be arbitrary. If this is the case, the waves are elliptically polarized (or circularly polarized if the amplitudes are the same). The only difference is the ellipse the vector **E** describes is rotated in space (its axes do not coincide with the *x* and *y* axes). A general, elliptically polarized wave is shown in Figure 3.5.

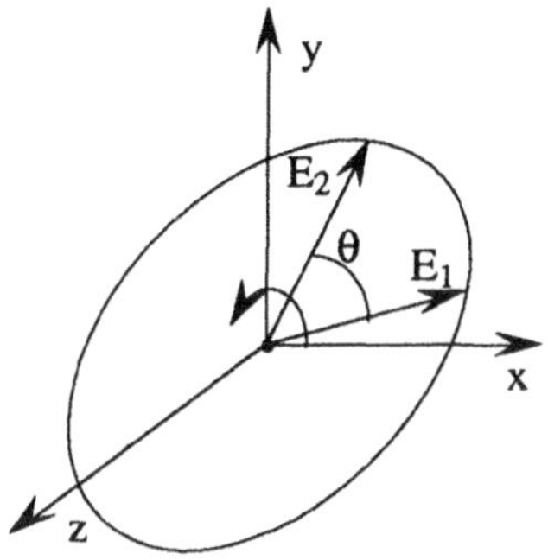

Figure 3.5. Elliptically polarized wave produced by two linearly polarized fields that are *45°* out of phase.

3.3. Reflection and Transmission of Plane Waves

The propagation of waves in free space and in materials was discussed at some length in chapter 1. In chapter 2 we discussed propagation along transmission lines where, among other aspects, we defined reflection from a mismatch on the line as well as transmission to a load. The same basic effects occur in propagation through free space and in propagation from one material into another. We discuss here propagation properties of waves as they propagate through different materials and changes in their amplitudes and directions as they propagate through the interfaces between materials. This aspect of propagation of waves is quite complex and many of the properties of waves are defined by materials and their interfaces. Reflection from interfaces, transmission through materials and refraction of waves are fundamental properties. As an examples waves are reflected from conducting surfaces giving rise to standing waves. Propagation through lossy (dispersive) materials may change the nature of waves from uniform to nonuniform plane waves. These properties are discussed next for plane waves. The various properties depend on the materials involved, direction of propagation, and polarization of the waves. A general discussion, taking into account all aspects of propagation at interfaces between materials is not possible under these constraints since this would require use of nonuniform waves. To keep the discussion simple, and within the context of plane waves, we will discuss a number of simple interface conditions. These include perpendicular and oblique incidence on conducting and dielectric interfaces, conditions often encountered in scattering.

The basic principle is to write the waves on both sides of the interface and to match the components of the fields (electric and magnetic) on the boundary. In general, this means applying the continuity conditions of the fields at the surface. From these, the fields on both sides of the interface are found. The general conditions are shown in Figure 3.6. The wave in material *(1)*, propagates at an angle to the interface. The tangential components of the electric field intensity are continuous at the interface. The normal components are discontinuous although, in general we will assume that there are no charge densities or current densities at the interface. Thus, the wave is partly transmitted into material *(2)* and partly reflected at an angle equal to the angle of incidence. The transmission angle is different than the incidence angle as one would expect from two different materials. Because there is reflection, there are also standing waves to the left of the interface.

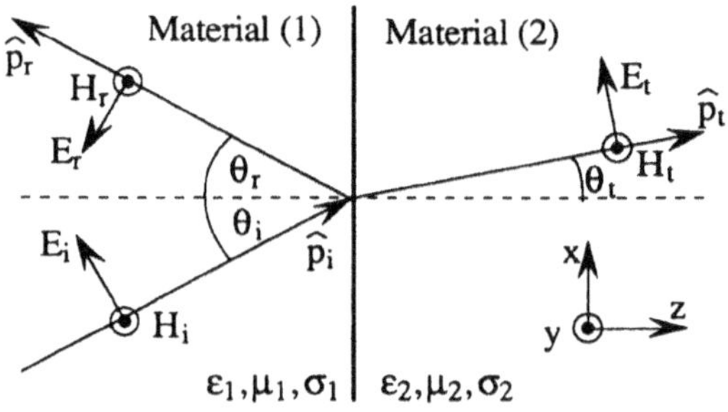

Figure 3.6. The general relationships between fields at an interface between two materials.

The use of boundary conditions for the electric and magnetic field was discussed at some length in section 1.11. These conditions define the continuity (or jump) of the components of the fields. The behavior of the electromagnetic wave at the interfaces between materials is discussed next as it has important applications in the microwave domain. By analogy to light waves, we expect similar behavior, including reflection, transmission and refraction of waves at the interface. For generality, we start with a general, lossy dielectric interface and then proceed to discuss lossless, low loss dielectrics, and conductors. Normal incidence of the wave on the interface is considered first followed by oblique incidence with polarized waves. Some sources on material on he more general case of oblique incidence in lossy dielectrics are given in the bibliography section.

3.3.1. Reflection and Transmission at a General Dielectric Interface: Normal Incidence

Figure 3.7 shows an incident wave $\mathbf{E}_i$ propagating in a medium with permittivity ε_1 and permeability μ_1. The wave encounters the interface between material *(1)* and material *(2)*, which is a general lossy dielectric with permittivity ε_2 and permeability μ_2. From our previous discussion of losses and skin depth, we know that there is a wave propagating into material *(2)* where, at least part of it is absorbed. There is also a reflected wave, as shown in Figure 3.7.

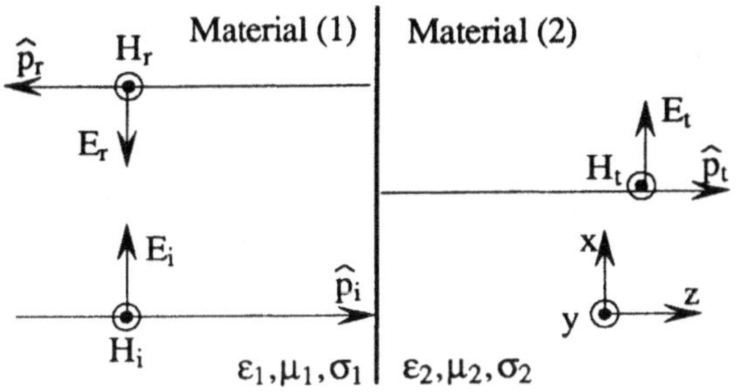

Figure 3.7. Reflection and transmission at a general interface: normal incidence.

To simplify the discussion we assume a wave propagating in the positive *z* direction, as in Eq. (1.73). The use of a single component vector for **E** does not detract from the generality of the discussion. The incident electric field is therefore,

$$\mathbf{E}_i = \hat{\mathbf{x}} E_{i1} e^{-\gamma_1 z} \tag{3.7}$$

where E_{1i} is the amplitude of the incident wave and the propagation constant is

$$\gamma = \alpha + j\beta \tag{3.8}$$

as discussed in chapter 1.

Note that we have chosen to use general media for both materials. If, for example, material *(1)* is free space, γ is replaces by $j\beta$. The wave reflected from the interface propagates backward in material *(1)*. Its form is

$$\mathbf{E}_r = \hat{\mathbf{x}} E_{r1} e^{+\gamma_1 z}$$

The backward propagation is indicated by the positive sign in the exponent (the wave is propagating in the negative *z* direction). The amplitude E_{r1} is, in general different than E_{i1} and may be zero (no reflection at the interface). Thus, the wave in material *(1)* has two components; one propagating in the positive *z* direction and one in the negative *z* direction

$$\mathbf{E}_1 = \hat{\mathbf{x}} E_{i1} e^{-\gamma_1 z} + \hat{\mathbf{x}} E_{r1} e^{+\gamma_1 z} \tag{3.9}$$

The magnetic field also has two components and can be calculated from Eq. (1.1) as

$$\mathbf{H}_1 = \hat{\mathbf{y}}\frac{E_{i1}}{\eta_1}e^{-\gamma_1 z} - \hat{\mathbf{y}}\frac{E_{r1}}{\eta_2}e^{+\gamma_1 z} \tag{3.10}$$

Note that the backward propagating magnetic field is negative. This sign is found directly from Maxwell's first equation (Faraday's law) but, here we simply point out that the sign must be negative since the propagation of this term is in the negative z direction. Thus, from the Poynting vector we have

$$\mathbf{P}_1^+ = \left(\hat{\mathbf{x}}E_{i1}e^{-\gamma_1 z}\right)\times\left(\hat{\mathbf{y}}\frac{E_{i1}}{\eta_1}e^{-\gamma_1 z}\right) = \hat{\mathbf{x}}\times\hat{\mathbf{y}}\frac{E_{i1}^2}{\eta_1}e^{-2\gamma_1 z} = \hat{\mathbf{z}}\frac{E_{i1}^2}{\eta_1}e^{-2\gamma_1 z} \tag{3.11}$$

which propagates in the positive z direction. Similarly, the Poynting vector for the backward propagating wave is

$$\mathbf{P}_1^- = \left(\hat{\mathbf{x}}E_{r1}e^{-\gamma_1 z}\right)\times\left(-\hat{\mathbf{y}}\frac{E_{r1}}{\eta_1}e^{-\gamma_1 z}\right) = -\hat{\mathbf{x}}\times\hat{\mathbf{y}}\frac{E_{r1}^2}{\eta_1}e^{2\gamma_1 z} = -\hat{\mathbf{z}}\frac{E_{r1}^2}{\eta_1}e^{2\gamma_1 z} \tag{3.12}$$

In these relations, the intrinsic impedance η_1 and the propagation constant are completely general and given as (see chapter 1)

$$\eta_1 = \sqrt{\frac{j\omega\mu_1}{\sigma_1 + j\omega\varepsilon_1}} \tag{3.13}$$

$$\gamma_1 = \sqrt{j\omega\mu_1(\sigma_1 + j\omega\varepsilon_1)} \tag{3.14}$$

We will assume here that part of the wave propagates into material *(2)* and call this a transmitted wave while part of it is reflected. Then we find the conditions for this to happen and the magnitudes of the transmitted and reflected waves. To do so, we first define the electric and magnetic fields in material *(2)*. In general, we may assume that **E** and **H** in material *(2)* are of the same form as in material *(1)* except, that in this case, the backward propagating wave in material *(2)* cannot exist. This is because material *(2)* extends to infinity and there is no mechanism for a reflected wave in material *(2)* to exist. Thus we get

$$\mathbf{E}_2 = \hat{\mathbf{x}}E_2e^{-\gamma_2 z} \tag{3.15}$$

$$\mathbf{H}_2 = \hat{\mathbf{y}}\frac{E_2}{\eta_2}e^{-\gamma_2 z} \tag{3.16}$$

and, as expected, this wave propagates in the positive z direction. The propagation constant and the intrinsic impedance in material *(2)* are

$$\eta_2 = \sqrt{\frac{j\omega\mu_2}{\sigma_2 + j\omega\varepsilon_2}} \tag{3.17}$$

$$\gamma_2 = \sqrt{j\omega\mu_2(\sigma_2 + j\omega\varepsilon_2)} \tag{3.18}$$

By defining a reflection coefficient, denoted by Γ, the reflected wave can be written in terms of the incident wave as

$$\mathbf{E}_r = \hat{\mathbf{x}}\Gamma E_{i1} e^{+\gamma_1 z} \tag{3.19}$$

where $E_{r1} = \Gamma E_{i1}$.

Similarly, the transmitted wave is $\mathbf{E}_2$ given in Eq. (3.15). By defining a transmission coefficient T, we can write

$$\mathbf{E}_2 = \hat{\mathbf{x}} T E_{i1} e^{-\gamma_2 z} \tag{3.20}$$

since $E_{t2} = E_{i1} T$

Since the magnetic field and the electric field are related through the intrinsic impedance we can write the following for the magnetic field

$$\mathbf{H}_i = \hat{\mathbf{y}} \frac{E_{i1}}{\eta_1} e^{-\gamma_1 z} \tag{3.21}$$

$$\mathbf{H}_r = -\hat{\mathbf{y}} \Gamma \frac{E_{i1}}{\eta_1} e^{+\gamma_1 z} \tag{3.22}$$

$$\mathbf{H}_2 = \hat{\mathbf{y}} T \frac{E_{i1}}{\eta_1} e^{-\gamma_2 z} \tag{3.23}$$

The transmission and reflection coefficients can now be evaluated from the relations at the boundary. To do so we set $z=0$ and write the relations between the incident, reflected, and transmitted waves. These are

$$\mathbf{E}_i + \mathbf{E}_r = \mathbf{E}_2$$

and

$$\mathbf{H}_i + \mathbf{H}_r = \mathbf{H}_2$$

Substitution of the expressions for the fields $\mathbf{E}_i$, $\mathbf{E}_r$, $\mathbf{E}_2$, $\mathbf{H}_i$, $\mathbf{H}_r$, and $\mathbf{H}_2$ at the interface ($z=0$) gives the following two equations

$$1 + \Gamma = T \tag{3.24}$$

$$\frac{1}{\eta_1} - \frac{\Gamma}{\eta_1} = \frac{T}{\eta_2}$$

or

$$\Gamma = \frac{\eta_2 - \eta_1}{\eta_1 + \eta_2} \tag{3.25}$$

and

$$T = \frac{2\eta_2}{\eta_1 + \eta_2} \tag{3.26}$$

We note here that since η_1 and η_2 are in general complex, Γ and T are also complex. Γ can be negative (depending on the relative values of the intrinsic impedances) but T is always positive. Complex transmission and reflection coefficients cause a phase shift in the reflected and transmitted waves at the interface. Both coefficients are ratios of impedances and, therefore, are dimensionless.

For the transmited wave to be nonzero, the intrinsic impedance η_2 must be nonzero. This is in general the case, except for perfect conductors. In a perfect conductor the transmitted wave is zero ($T=0$) and the reflected wave is equal and opposite in sign to the incident wave ($\Gamma=-1$). Finite conductivity materials behave similar to this, especially at very high frequencies. In most cases in the microwave domain we will assume that this condition is satisfied.

The electric and magnetic fields in material *(1)* can be calculated as the sum of the incident and reflected waves

$$\mathbf{E}_1(z) = \mathbf{E}_i(z) + \mathbf{E}_r(z) = \hat{\mathbf{x}}E_{i1}\left(e^{-\gamma_1 z} + \Gamma e^{+\gamma_1 z}\right) \tag{3.27}$$

By adding and subtracting the term $\Gamma e^{-j\gamma_2}$ to the right hand side, we can write

$$\mathbf{E}_1(z) = \hat{\mathbf{x}}E_{i1}\left((1 + \Gamma)e^{-\gamma_1 z} + \Gamma\left(e^{+\gamma_1 z} - e^{-\gamma_1 z}\right)\right)$$

or:

$$\mathbf{E}_1(z) = \hat{\mathbf{x}}E_{i1}\left(Te^{-\gamma_1 z} + \Gamma(j2\sin\gamma_1 z)\right) \tag{3.28}$$

Similarly, for the magnetic field

$$\mathbf{H}_1(z) = \hat{\mathbf{y}}\frac{E_{i1}}{\eta_1}\left(e^{-\gamma_1 z} - \Gamma e^{+\gamma_1 z}\right)$$

or

$$\mathbf{H}_1(z) = \hat{\mathbf{y}}\frac{E_{i1}}{\eta_1}\left(Te^{-\gamma_1 z} - \Gamma(j2\sin\gamma_1 z)\right) \qquad (3.29)$$

The fields in material (2) are

$$\mathbf{E}_2 = \hat{\mathbf{x}}TE_{i1}e^{-\gamma_2 z} \qquad (3.30)$$

and

$$\mathbf{H}_2 = \hat{\mathbf{x}}T\frac{E_{i1}}{\eta_2}e^{-\gamma_2 z} \qquad (3.31)$$

Eqs. (3.28) and (3.29) are very useful since they indicate that the sum of the incident and reflected wave consist of a wave component propagating in the positive z direction and a non-propagating wave component. This later (second term in Eq. (3.28) or Eq. (3.29)) is responsible for the standing waves we often encounter. We also observe that if the transmission coefficient is zero, only a standing wave exists while if the reflection coefficient is zero, there are no standing waves. Standing waves will be discussed separately in conjunction with reflection from conducting surfaces but we point out here that standing waves may exist any time there is reflection of a wave.

3.3.2. Reflection and Transmission at a Lossy Dielectric Interface: Normal Incidence

The discussion above is completely general and assumes two general materials. In many cases, material *(1)* is taken as free space. The only difference is that η_1 becomes η_0 while in the expressions for $\mathbf{E}_i$ and $\mathbf{E}_r$, γ_1 is replaced by $j\beta_1$. This situation is shown in Figure 3.7 if $\sigma_1=0$, $\varepsilon_1=\varepsilon_0$, and $\mu_1=\mu_0$.

Thus, for example, if material *(1)* is free space, and material *(2)* is a lossy dielectric, the reflection and transmission coefficients are

$$\Gamma = \frac{\eta_2 - \eta_0}{\eta_0 + \eta_2} \qquad (3.32)$$

$$T = \frac{2\eta_2}{\eta_0 + \eta_2} \qquad (3.33)$$

The electric and magnetic fields are

$$\mathbf{E}_1(z) = \hat{\mathbf{x}} E_{i1}\left(Te^{-\beta_1 z} + \Gamma(j2\sin\beta_1 z)\right) \tag{3.34}$$

$$\mathbf{H}_1(z) = \hat{\mathbf{y}} \frac{E_{i1}}{\eta_0}\left(Te^{-\beta_1 z} - \Gamma(j2\sin\beta_1 z)\right) \tag{3.35}$$

$$\mathbf{E}_2 = \hat{\mathbf{x}} T E_{i1} e^{-\gamma_2 z} \tag{3.36}$$

$$\mathbf{H}_2 = \hat{\mathbf{x}} T \frac{E_{i1}}{\eta_2} e^{-\gamma_2 z} \tag{3.37}$$

where

$$\beta_1 = \omega\sqrt{\mu_0 \varepsilon_0}$$

$$\eta_1 = \sqrt{\frac{\mu_0}{\varepsilon_0}}$$

and γ_2, η_2 remain as in Eqs. (3.30) and (3.31). Note that even though $\mathbf{E}_2$ and $\mathbf{H}_2$ are the same as in Eqs. (3.17) and (3.18), the transmission coefficient is different and therefore the fields in material *(2)* are also different.

Obviously, for any other lossless dielectric, the permittivity and permeability of free space are replaced by those of the dielectric.

3.3.3. Reflection and Transmission at a Lossless Dielectric Interface: Normal Incidence

A common combination of material at an interface is free space and a lossless dielectric. Using again Figure 3.7, material *(1)* is characterized by ε_1, μ_1, $\sigma_1=0$ and material *(2)* is characterized by ε_2, μ_2, $\sigma_2=0$. If material *(1)* is free space and material *(2)* is a lossless dielectric, the transmission and reflection coefficients can be written in terms of the relative permeability and relative permittivity of material *(2)* alone

$$\Gamma = \frac{\eta_2 - \eta_0}{\eta_0 + \eta_2} = \frac{\sqrt{\dfrac{\mu_{r2}}{\varepsilon_{r2}}} - 1}{\sqrt{\dfrac{\mu_{r2}}{\varepsilon_{r2}}} + 1} \tag{3.38}$$

$$T = \frac{2\sqrt{\dfrac{\mu_{r2}}{\varepsilon_{r2}}}}{\sqrt{\dfrac{\mu_{r2}}{\varepsilon_{r2}}} + 1} \tag{3.39}$$

The electric and magnetic fields are

$$\mathbf{E}_1(z) = \hat{\mathbf{x}}E_{i1}\left(Te^{-\beta_1 z} + \Gamma(j2sin\beta_1 z)\right) \tag{3.40}$$

$$\mathbf{H}_1(z) = \hat{\mathbf{y}}\frac{E_{i1}}{h_0}\left(Te^{-\beta_1 z} - \Gamma(j2sin\beta_1 z)\right) \tag{3.41}$$

$$\mathbf{E}_2 = \hat{\mathbf{x}}TE_{i1}e^{-j\beta_2 z} \tag{3.42}$$

$$\mathbf{H}_2 = \hat{\mathbf{x}}T\frac{E_{i1}}{\eta_2}e^{-j\beta_2 z} \tag{3.43}$$

where

$$\beta_1 = \omega\sqrt{\mu_0\varepsilon_0} \qquad \beta_2 = \omega\sqrt{\mu_2\varepsilon_2}$$

$$\eta_1 = \sqrt{\frac{\mu_0}{\varepsilon_0}} \qquad \eta_2 = \sqrt{\frac{\mu_2}{\varepsilon_2}}$$

Note that while η_1 and β_1 are real, η_2 and β_2 may be complex, depending on permittivity. If material *(1)* is not free space but a general lossless dielectric, the permittivity and permeability of this material are substituted for those of free space.

We note from these expressions that if we are interested in reducing the reflection at the interface, the ratio of μ_r and ε_r must tend to *1*. This is the same as to say that the closer the materials are in their ratio, the lower the reflection. If the ratio $\mu_{r2}/\varepsilon_{r2}$ is equal to that of free space (i.e. $\mu_{r0}/\varepsilon_{r0}=1$), there is no reflection at the interface and the whole wave is transmitted across the boundary. We can considered this to be the case of perfect impedance matching since both materials have the same intrinsic impedance. While this situation is certainly not common in dielectrics, some ferrites can be made to closely resemble this condition by adjusting their permeability. Note that this case, that of total transmission, is characterized by $\Gamma=0$ and $T=1$.

Similarly, if we are interested in large reflections, the permittivity must be made as large as possible. In the limit, with permittivity tending to infinity, the reflection coefficient tends to *1* and the transmission coefficient tends to zero. Metals behave close to this condition and in most cases we consider a conductor's interface to be totally reflecting.

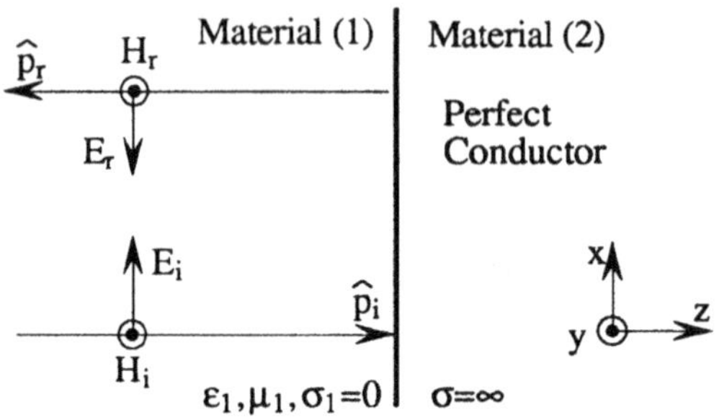

Figure 3.8. Reflection and transmission at a dielectric/conductor interface: normal incidence.

3.3.4. Reflection and Transmission at an Air Conductor Interface: Normal Incidence

We consider next the propagation of a wave from free space (or any lossless dielectric), impinging on a perfect or high conductivity conductor as in Figure 3.8.

For perfect conductors, we can write directly

$$\Gamma = -1 \qquad and \qquad T = 0 \tag{3.44}$$

From Eqs. (3.40) and (3.41), the electric and magnetic fields are:

$$\mathbf{E}_l(z) = -\hat{\mathbf{x}} j 2 E_{il} sin\beta_l z \tag{3.45}$$

$$\mathbf{H}_l(z) = \hat{\mathbf{y}} 2 \frac{E_{il}}{\eta_0} cos\beta_l z \tag{3.46}$$

The wave formed by these fields is unique in that it does not propagate. This a standing wave, similar to the standing wave on transmission lines we discussed in chapter 2. The principle of a standing wave is shown in Figure 3.9 where the amplitude of the wave changes with time but the location of the peaks and zeroes (nodes) is constant in space. To see that this is the case, it is most convenient to calculate the Poynting vector

$$\mathbf{P} = \mathbf{E}_l(z) \times \mathbf{H}_l^*(z) = \hat{\mathbf{z}} j \frac{4}{\eta_0} |E_{il}|^2 sin\beta_l z \; cos\beta_l z$$

where * denotes the complex conjugate vector. Since **P** is purely imaginary no power is transferred into the conductor.

This property of total reflection at conducting boundary is both a nuisance and an indispensable property. Sometimes, it manifests itself as a problem, primarily as an indication of mismatch either in transmission lines, waveguides, and in free space. We will use it extensively to measure thickness of conductors and dielectrics. In waveguides, total reflection at conducting interfaces defines the boundary conditions and therefore the modes of propagation, as well as losses in the conductors.

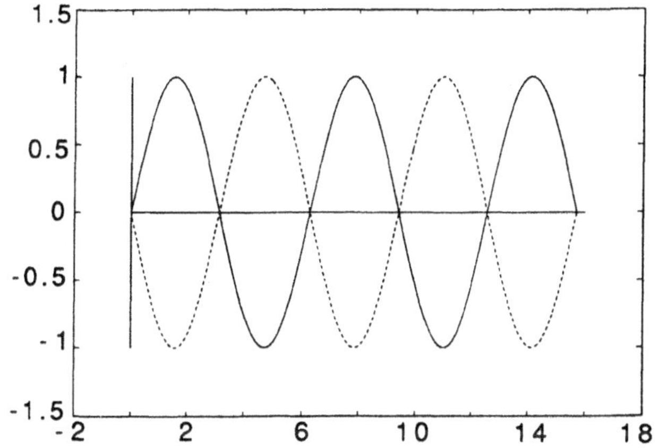

Figure 3.9. Standing waves due to reflection at a conducting interface. Total reflection (complete standing waves) without attenuation is shown.

For a good (but not perfect) conductor, we can either use the general, lossy dielectric results, in Eqs. (3.27) through (3.31) or we may use the results in section 1.10.4. and write the expressions in terms of the skin depth. The latter approach is used here. The reflection and transmission coefficients in terms of the skin depth are

$$\Gamma = \frac{\eta_2 - \eta_0}{\eta_0 + \eta_2} = \frac{(1+j)\dfrac{1}{\sigma\delta} - \eta_0}{(1+j)\dfrac{1}{\sigma\delta} + \eta_0} \tag{3.47}$$

$$T = \frac{2(1+j)\dfrac{1}{\sigma\delta}}{\eta_0 + (1+j)\dfrac{1}{\sigma\delta}} \tag{3.48}$$

These expression are both complex, and according to Eq. (1.95) are only approximations. The smaller the skin depth, the closer Γ is to -1 and T to zero, as one would expect for a perfect conductor.

The electric field and magnetic fields in free space are now

$$\mathbf{E}_1(z) = \hat{\mathbf{x}}E_{i1}\left(\left(\frac{2(1+j)\frac{1}{\sigma\delta}}{\eta_0+(1+j)\frac{1}{\sigma\delta}}\right)e^{-j\beta_1 z}+\left(\frac{(1+j)\frac{1}{\sigma\delta}-\eta_0}{(1+j)\frac{1}{\sigma\delta}+\eta_0}\right)(j2\sin\beta_1 z)\right) \quad (3.49)$$

$$\mathbf{H}_1(z) = \hat{\mathbf{y}}\frac{E_{i1}}{\eta_0}\left(\left(\frac{2(1+j)\frac{1}{\sigma\delta}}{\eta_0+(1+j)\frac{1}{\sigma\delta}}\right)e^{-j\beta_1 z}+\left(\frac{(1+j)\frac{1}{\sigma\delta}-\eta_0}{(1+j)\frac{1}{\sigma\delta}+\eta_0}\right)(j2\sin\beta_1 z)\right) \quad (3.50)$$

and the fields in material *(2)* are

$$\mathbf{E}_2 = \hat{\mathbf{x}}TE_{i1}e^{-j\gamma_2 z} = \hat{\mathbf{x}}\left(\frac{2(1+j)\frac{1}{\sigma\delta}}{\eta_0+(1+j)\frac{1}{\sigma\delta}}\right)e^{-j\gamma_2 z} \quad (3.51)$$

$$\mathbf{H}_2 = \hat{\mathbf{y}}T\frac{E_{i1}}{(1+j)\frac{1}{\sigma\delta}}e^{-j\gamma_2 z} = \hat{\mathbf{y}}\frac{E_{i1}}{(1+j)\frac{1}{\sigma\delta}}\left(\frac{2(1+j)\frac{1}{\sigma\delta}}{\eta_0+(1+j)\frac{1}{\sigma\delta}}\right)e^{-j\gamma_2 z} \quad (3.52)$$

These are rather small, as expected for a good conductor, but they are not zero.

The expressions in Eqs. (3.49) through (3.52) are not different from than those in Eqs. (3.27) through (3.31) except for the fact that the standing wave component is large, as it should be for a good conductor. Calculation of the Poynting vector at the interface, shows a small real part (propagating wave into the conductor, or transmitted wave) and a large imaginary part (standing waves). A useful calculation at this point is the power flow into the conductor. This can be calculated from the transmitted electric and magnetic fields. The Poynting vector for the transmitted wave is

$$\mathbf{E}\times\mathbf{H}^* = \hat{\mathbf{z}}T^2\frac{|E_{i1}|^2}{\eta_2}e^{-j2\gamma_2 z} = \hat{\mathbf{z}}\frac{4\eta_2^2}{(\eta_2+\eta_0)^2}|E_{i1}|^2e^{-j2\gamma_2 z}$$

Assuming that $|\eta_2| << |\eta_0|$ (this is a good approximation in a good conductor), the total power flow into the conductor is minimal. The real part of the Poynting vector represents the dissipated energy in the conductors and this decays as $e^{-2\alpha z}$, indicating very rapid attenuation.

3.3.5. Reflection and Transmission at an Interface: Oblique Incidence

A plane wave, obliquely incident on an interface between two materials undergoes changes similar to those for normal incidence. Part of the wave is transmitted and part of it reflected. In some cases, there is either only transmission or only reflection. In oblique incidence, the continuity of the field components is taken into account as a boundary condition at the interface. The behavior of the wave at the interface depends on the polarization of the wave. A plane of incidence is defined by two vectors: the direction of propagation of the incident wave (i.e. the Poynting vector), and the normal to the surface of the interface, as shown in Figure 3.10. The direction of the electric or magnetic field intensities for uniform plane waves is always normal to the direction of propagation. The electric field can be either in the plane of incidence, perpendicular to it or at any angle to this plane. If the electric field is parallel to the plane of incidence, we refer to this as parallel polarization (sometimes called H polarization). Perpendicular polarization refers to the case of the electric field perpendicular to the plane of incidence, (sometimes called E polarization). The general case of arbitrary polarization can always be treated by separating the electric field into its normal and parallel components as a combination of parallel and perpendicular polarization.

We treat the problem of oblique incidence by looking at propagation of waves obliquely incident at an interface for parallel and perpendicular incidence separately for dielectrics and conductors. The case of oblique incidence for perpendicular polarization is shown explicitly while the results for parallel polarization are given as simple modifications of the perpendicular polarization results, without the details. The case of incidence on conducting interfaces is given first, followed by incidence on lossless dielectrics. The more general case of incidence on lossy dielectrics is not treated here because it normally results in nonuniform plane waves. However the principles used here are directly applicable.

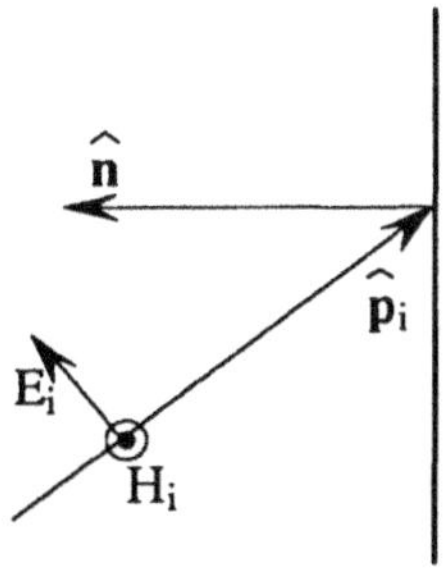

Figure 3.10. Definition of the plane of incidence. The electric field in this figure is parallel to the plane of incidence, the magnetic field is perpendicular.

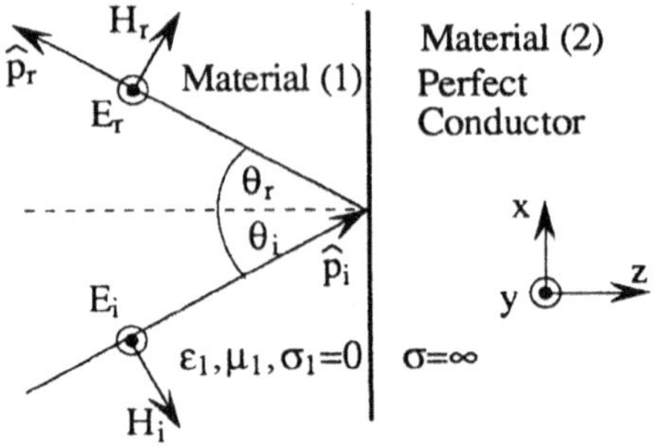

Figure 3.11. Incident and reflected waves for oblique incidence at a dielectric-conductor interface. The electric field is polarized perpendicular to the plane of incidence.

3.3.6. Oblique Incidence on a Conducting Interface: Perpendicular Polarization

In this configuration the electric field is perpendicular to the plane of incidence as shown in Figure 3.11. We assume incidence from a dielectric with permittivity ε_1 on a perfect conductor plane surface.

The direction of propagation is now defined as $\hat{\mathbf{p}}_i = \hat{\mathbf{x}}sin\theta_i + \hat{\mathbf{z}}cos\theta_i$ for the incident wave and $\hat{\mathbf{p}}_r = \hat{\mathbf{x}}sin\theta_r - \hat{\mathbf{z}}cos\theta_r$ for the reflected wave. The variable z in the exponent of the wave is now replaced by $xsin\theta_i + zcos\theta_i$ for the incident wave and by $xsin\theta_r - zcos\theta_r$ for the reflected wave. The electric field propagates in the (x,z) plane and is directed perpendicular to the plane of incidence, in the y direction. Using the directions of propagation, the incident electric and magnetic fields are

$$\mathbf{E}_i(x,z) = \hat{\mathbf{y}}E_{i1}e^{-j\beta_1(xsin\theta_i + zcos\theta_i)} \tag{3.53}$$

$$\mathbf{H}_i(x,z) = \frac{E_{i1}}{\eta_1}(-\hat{\mathbf{x}}cos\theta_i + \hat{\mathbf{z}}sin\theta_i)e^{-j\beta_1(xsin\theta_i + zcos\theta_i)} \tag{3.54}$$

where the material index number was dropped as the waves only exist in material *(1)*. The reflected fields are

$$\mathbf{E}_r(x,z) = \hat{\mathbf{y}}E_{r1}e^{-j\beta_1(xsin\theta_r - zcos\theta_r)}$$

$$\mathbf{H}_r(x,z) = \frac{E_{r1}}{\eta_1}(\hat{\mathbf{x}}cos\theta_r + \hat{\mathbf{z}}sin\theta_r)e^{-j\beta_1(xsin\theta_r - zcos\theta_r)}$$

In a perfect conductor there is no transmitted wave, therefore, at the interface $(z=0)$, the total electric field intensity must be zero:

$$\mathbf{E}_i(x,0) + \mathbf{E}_r(x,0) = \hat{\mathbf{y}}\left[E_{i1}e^{-j\beta_1(x\sin\theta_i)} + E_{r1}e^{-j\beta_1(x\sin\theta_r)}\right] = 0$$

For this to be satisfied, the following must hold

$$E_{r1} = -E_{i1} \qquad \text{and} \qquad \theta_r = \theta_i$$

The amplitudes of the reflected and incident waves are the same and the angle of incidence and reflection are the same (Snell's law). The reflected electric and magnetic fields can be written in terms of the incident field as

$$\mathbf{E}_r(x,z) = -\hat{\mathbf{y}}E_{i1}e^{-j\beta_1(x\sin\theta_i - z\cos\theta_i)}$$

$$\mathbf{H}_r(x,z) = -\frac{E_{i1}}{\eta_1}(\hat{\mathbf{x}}\cos\theta_i + \hat{\mathbf{z}}\sin\theta_i)e^{-j\beta_1(x\sin\theta_i - z\cos\theta_i)}$$

It is now obvious that the reflection coefficient in this case is equal to -1, since the reflection coefficient is the ratio of the amplitudes of $E_r(x,z)$ and $E_i(x,z)$ and these are equal in absolute values(see Eq. (3.54)).

The total field is the sum of the incident and reflected waves. The total electric and magnetic field intensities are (after rearranging terms):

$$\mathbf{E}_1(x,z)=\hat{\mathbf{y}}E_{i1}\left[e^{-j\beta_1(z\cos\theta_i)}-e^{j\beta_1(z\cos\theta_i)}\right]e^{-j\beta_1(x\sin\theta_i)}= -\hat{\mathbf{y}}j2E_{i1}\sin(\beta_1 z\cos\theta_i)e^{-j\beta_1(x\sin\theta_i)}$$

$$\mathbf{H}_1(x,z) = -2\frac{E_{i1}}{\eta_1}\left[\hat{\mathbf{x}}\cos\theta_i\cos(\beta_1 z\cos\theta_i) + \hat{\mathbf{z}}j\sin\theta_i\sin(\beta_1 z\cos\theta_i)\right]e^{-j\beta_1 x\sin\theta_i}$$

where the properties of exponential functions were used to obtain the following identities (see chapter 12)

$$j2\sin(\beta_1 z\cos\theta_i) = e^{j\beta_1 z\cos\theta_i} - e^{-j\beta_1 z\cos\theta_i}$$

$$2\cos(\beta_1 z\cos\theta_i) = e^{j\beta_1 z\cos\theta_i} + e^{-j\beta_1 z\cos\theta_i}$$

Note that the wave in the z direction is a standing wave, but the wave in the x direction is propagating.

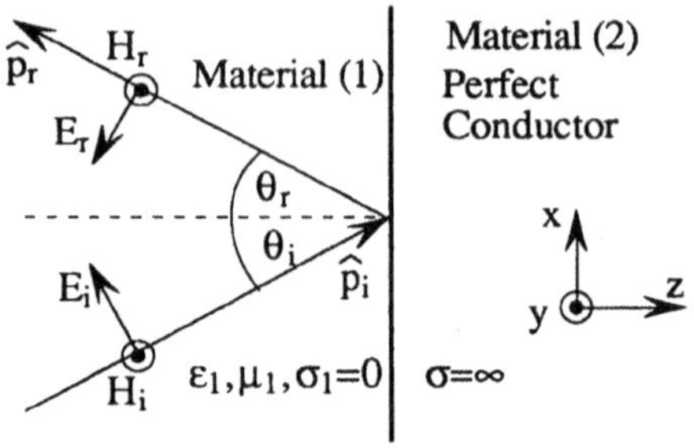

Figure 3.12. Incident and reflected waves for oblique incidence at a dielectric-conductor interface. The electric field is polarized parallel to the plane of incidence.

3.3.7. Oblique Incidence on a Conducting Interface: Parallel Polarization

The discussion of the previous section applies here as well, but the components of the fields are different. The electric field now lies in the incidence plane and therefore will have x and z components while the magnetic field, is perpendicular to the plane of incidence and therefore is in the y direction (see Figure 3.12). The incident electric and magnetic fields are

$$\mathbf{E}_i(x,z) = E_{il}(\hat{\mathbf{x}}cos\theta_i - \hat{\mathbf{z}}sin\theta_i)e^{-j\beta_l(xsin\theta_i + zcos\theta_i)}$$

$$\mathbf{H}_i(x,z) = \hat{\mathbf{y}}\frac{E_{il}}{\eta_i}e^{-j\beta_l(xsin\theta_i + zcos\theta_i)}$$

Following similar considerations we obtain for the electric and magnetic fields

$$\mathbf{E}_l(x,z) = -2E_{il}\left[\hat{\mathbf{x}}jcos\theta_i\ sin(\beta_{lz}cos\theta_i) + \hat{\mathbf{z}}sin\theta_i\ cos(\beta_{lz}cos\theta_i)\right]e^{-j\beta_l xsin\theta_i}$$

$$\mathbf{H}_l(x,z) = \hat{\mathbf{y}}2\frac{E_{il}}{\eta_l}\ cos(\beta_{lz}cos\theta_i)e^{-j\beta_l xsin\theta_i}$$

As with perpendicular polarization, the wave propagating in the z direction, consists of E_{1x} and H_{1y} and these are out of phase. Therefore, we have a standing wave, oscillating exactly as for the perpendicular polarization. The wave in the x direction is a propagating wave as for perpendicular polarization.

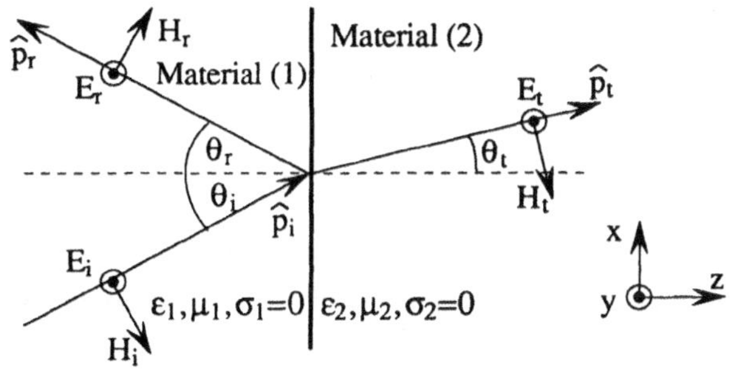

Figure 3.13. Incident, reflected, and refracted waves for oblique incidence at a dielectric-dielectric interface. The electric field is polarized perpendicular to the plane of incidence.

3.3.8. Oblique Incidence on a Dielectric Interface: Perpendicular Polarization

The case considered here is shown in Figure 3.13. The two dielectrics are lossless and either one can be free space. As with normal incidence on dielectric boundaries, part of the wave will be reflected and part will be transmitted as shown. The transmitted wave is refracted as it passes into the second dielectric, depending on the relative values of the dielectric constants,

Before continuing, we recall two well known relations from optics relating the reflection and refraction angles to the incident angles. These are

$$\theta_i = \theta_r \tag{3.55}$$

and

$$\frac{\sin\theta_i}{\sin\theta_t} = \frac{n_2}{n_1} \tag{3.56}$$

The first indicates that the reflection angle and the incident angle are the same (Snell's law). The second relates the refraction angle to the indices of refraction of the two materials, where the index of refraction of a material is

$$n=\sqrt{\mu_r\varepsilon_r} \tag{3.57}$$

In the particular case in which material *(1)* is free space, $n_1=1$ and we have

$$\frac{\sin\theta_i}{\sin\theta_t} = n_2 = \sqrt{\mu_{r2}\varepsilon_{r2}}$$

The electric and magnetic fields are the same as Eqs. (3.53) and (3.54)

$$\mathbf{E}_i(x,z) = \hat{\mathbf{y}} E_{i1} e^{-j\beta_1(x\sin\theta_i + z\cos\theta_i)} \tag{3.58}$$

$$\mathbf{H}_i(x,z) = \frac{E_{i1}}{\eta_1}(-\hat{\mathbf{x}}\cos\theta_i + \hat{\mathbf{z}}\sin\theta_i) e^{-j\beta_1(x\sin\theta_i + z\cos\theta_i)} \tag{3.59}$$

where again, the direction of propagation of the incident wave is given by $\hat{\mathbf{p}}_i = \hat{\mathbf{x}}\sin\theta_i + \hat{\mathbf{z}}\cos\theta_i$.

Analogous to the results we obtained for oblique incidence on a conducting boundary, we write the reflected fields in terms of the reflection coefficient and the incident waves as

$$\mathbf{E}_r(x,z) = \hat{\mathbf{y}} \Gamma E_{i1} e^{-j\beta_1(x\sin\theta_i - z\cos\theta_i)} \tag{3.60}$$

$$\mathbf{H}_r(x,z) = \frac{E_{i1}\Gamma}{\eta_1}(\hat{\mathbf{x}}\cos\theta_i + \hat{\mathbf{z}}\sin\theta_i) e^{-j\beta_1(x\sin\theta_i - z\cos\theta_i)} \tag{3.61}$$

where Γ remains to be determined.

Similarly, the transmitted wave is written directly from the incident wave by means of the transmission coefficient T as

$$\mathbf{E}_t(x,z) = \hat{\mathbf{y}} T E_{i1} e^{-j\beta_2(x\sin\theta_t + z\cos\theta_t)} \tag{3.62}$$

$$\mathbf{H}_t(x,z) = \frac{E_{i1}T}{\eta_2}(-\hat{\mathbf{x}}\cos\theta_t + \hat{\mathbf{z}}\sin\theta_t) e^{-j\beta_2(x\sin\theta_t + z\cos\theta_t)} \tag{3.63}$$

To determine the transmission and reflection coefficients the tangential components of the electric field intensity and those of the magnetic field intensity on both sides of the interface (i.e. at $z=0$) are equated. From Figure 3.13, the following two relations express the continuity of the magnetic and electric fields

$$E_i + E_r = E_t \qquad and \qquad H_i\cos\theta_i - H_r\cos\theta_i = H_t\cos\theta_t$$

where the relation $\theta_i = \theta_r$ was used. From the relations

$$H_i = \frac{E_i}{\eta_1}, \qquad H_r = \frac{E_r}{\eta_1}, \qquad and \qquad H_t = \frac{E_t}{\eta_2}$$

we have

$$E_i + E_r = E_t \qquad and \qquad \frac{E_i}{\eta_1}\cos\theta_i - \frac{E_r}{\eta_1}\cos\theta_i = \frac{E_t}{\eta_2}\cos\theta_t$$

Solution of these two equations for E_r and E_t gives

$$E_r = E_i\frac{\eta_2\cos\theta_i - \eta_1\cos\theta_t}{\eta_2\cos\theta_i + \eta_1\cos\theta_t}$$

and

$$E_t = E_i\frac{2\eta_2\cos\theta_i}{\eta_2\cos\theta_i + \eta_1\cos\theta_t}$$

From which the reflection and transmission coefficients are

$$\Gamma = \frac{\eta_2\cos\theta_i - \eta_1\cos\theta_t}{\eta_2\cos\theta_i + \eta_1\cos\theta_t} \tag{3.64}$$

$$T = \frac{2\eta_2\cos\theta_i}{\eta_2\cos\theta_i + \eta_1\cos\theta_t} \tag{3.65}$$

Now, the total fields in each material can be written directly. In medium *(2)*, where we only have a transmitted wave, Eqs. (3.62) and (3.63) describe the wave. In material *(1)*, the fields are the sum of the transmitted and reflected waves (from Eqs. (3.60) and (3.58) for E_1 and from Eqs. (3.61) and (3.59) for H_1):

$$\mathbf{E}_1(x,z) = \hat{\mathbf{y}}E_{i1}\left[e^{-j\beta_1(x\sin\theta_i + z\cos\theta_i)} + \Gamma e^{-j\beta_1(x\sin\theta_i - z\cos\theta_i)}\right]$$

$$\mathbf{H}_1(x,z) = -\hat{\mathbf{x}}\frac{E_{i1}}{\eta_1}(\cos\theta_i + \Gamma\cos\theta_i)e^{-j\beta_1(x\sin\theta_i + z\cos\theta_i)} +$$

$$\hat{\mathbf{z}}(\sin\theta_i + \Gamma\sin\theta_i)e^{-j\beta_1(x\sin\theta_i - z\cos\theta_i)}$$

E is in the *y* direction but *H* has a component in the *x* and *z* directions. As was the case for conducting interfaces, we have a propagating wave in the *x* direction and a standing wave in the *z* direction.

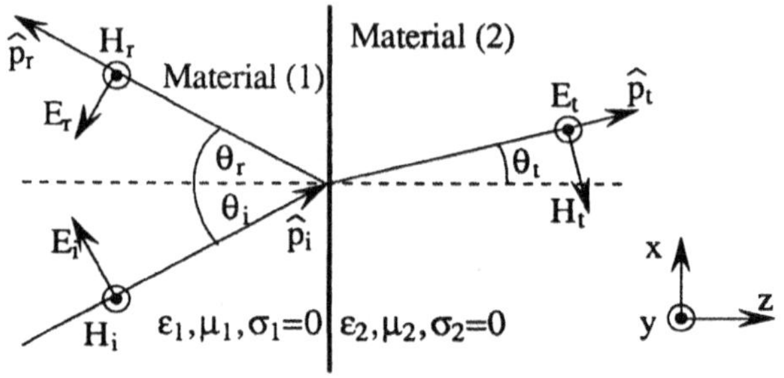

Figure 3.14. Incident, reflected, and refracted waves for oblique incidence at a dielectric-dielectric interface. The electric field is polarized parallel to the plane of incidence.

3.3.9. Oblique Incidence on a Dielectric Interface: Parallel Polarization

The situation considered here is shown in Figure 3.14. The electric field is parallel to the plane of incidence. The electric and magnetic field intensities are as given in section 3.3.7.

Following essentially the same procedures we have for the reflection and transmission coefficients

$$\Gamma = \frac{\eta_2 \cos\theta_t - \eta_1 \cos\theta_i}{\eta_2 \cos\theta_t + \eta_1 \cos\theta_i} \tag{3.66}$$

$$T = \frac{2\eta_2 \cos\theta_i}{\eta_2 \cos\theta_t + \eta_1 \cos\theta_i} \tag{3.67}$$

The total fields in medium 1 are calculated as in the previous section:

$$E_1(x,z) = -\hat{\mathbf{x}}E_{i1}(\cos\theta_i + \Gamma\cos\theta_i)e^{-j\beta_1(x\sin\theta_i + z\cos\theta_i)} +$$

$$\hat{\mathbf{z}}(\sin\theta_i + \Gamma\sin\theta_i)e^{-j\beta_1(x\sin\theta_i + z\cos\theta_i)}$$

$$\mathbf{H}_1(x,z) = \hat{\mathbf{y}}\frac{E_{i1}}{\eta_1}\left[e^{-j\beta_1(x\sin\theta_i + z\cos\theta_i)} + \Gamma e^{-j\beta_1(x\sin\theta_i + z\cos\theta_i)}\right]$$

3.4. Brewster Angle, Total Reflection, and Surface Waves

The reflection and transmission coefficients depend on the incident and refraction angle and on the intrinsic impedances of the two materials. The reflection angle does not appear in these relations since we have made use of Snell's law which indicates that the reflected and incident angle are the same. Also, from Eqs. (3.56) and (3.57) we have

$$\sin\theta_t = \sqrt{\frac{\mu_1\varepsilon_1}{\mu_2\varepsilon_2}}\,\sin\theta_i \tag{3.68}$$

Now, the reflection and transmission coefficients can be written in terms of the incident angle and material properties alone. These are (from Eqs. (3.64) and (3.65))

$$\Gamma = \frac{\eta_2\cos\theta_i - \eta_1\cos\theta_t}{\eta_2\cos\theta_i + \eta_1\cos\theta_t} \tag{3.69}$$

$$T = \frac{2\cos\theta_i}{\eta_2\cos\theta_i + \eta_1\cos\theta_t} \tag{3.70}$$

To see the general behavior of the reflection coefficient, consider the plot in Figure 3.15 where the reflection coefficient for water ($\varepsilon_r=80$) is plotted against the angle of incidence for the horizontally polarized wave. For simplicity, it is assumed that $\varepsilon_1=1$, $\mu_1=\mu_2=\mu_0$.

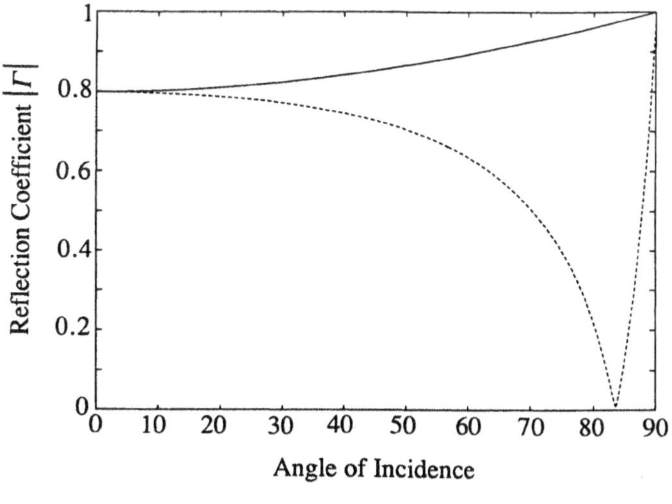

Figure 3.15. Absolute value of the reflection coefficient versus angle of incidence for parallel polarization (dotted line) and perpendicular polarization (solid line).

We note that at an angle of incidence equal to 83.75 degrees, the reflection coefficient is zero. This angle of incidence is called Brewster's angle and it occurs when the numerator in Eq. (3.69) is zero. For a parallel polarized wave, Brewster's angle is found from the condition

$$\eta_2 \cos\theta_t = \eta_1 \cos\theta_i$$

or

$$\cos\theta_i = \frac{\eta_2}{\eta_1}\cos\theta_t \tag{3.71}$$

From Eq. (3.68) we have for the refraction angle (Snell's law)

$$\cos\theta_t = \sqrt{1-\sin^2\theta_t} = \sqrt{1-\frac{\mu_1\varepsilon_1}{\mu_2\varepsilon_2}\sin^2\theta_i} \tag{3.72}$$

Substituting Eq. (3.72) in Eq. (3.71) for $\cos\theta_t$ and writing η_2 and η_1 in terms of permeabilities and permittivities, we get for the incident angle that satisfies these conditions

$$\sin\theta_i = \sqrt{\frac{\varepsilon_2}{\mu_1}\frac{(\mu_2\varepsilon_1 - \mu_1\varepsilon_2)}{(\varepsilon_1^2 - \varepsilon_2^2)}}$$

Or, we can write the angle itself as

$$\theta_b = \sin^{-1}\left[\sqrt{\frac{\varepsilon_2}{\mu_1}\frac{(\mu_2\varepsilon_1 - \mu_1\varepsilon_2)}{(\varepsilon_1^2 - \varepsilon_2^2)}}\right] \tag{3.73}$$

where the notation θ_b was used to distinguish this angle from other incident angles that do not satisfy the conditions above.

This angle is called Brewster's angle or polarizing angle since any wave will become linearly polarized at this angle, regardless of its polarization.

If $\mu_1=\mu_2$ (as would be the case for nonmagnetic materials), Brewster's angle can be written as

$$\theta_b = \sin^{-1}\left[\sqrt{\frac{\varepsilon_2}{\varepsilon_2+\varepsilon_1}}\right] \tag{3.74}$$

For perpendicular polarization, the condition for zero reflection coefficient is

$$\eta_2 \cos\theta_i = \eta_1 \cos\theta_t$$

Following the same steps as for horizontal polarization, the incident angle required for zero reflection coefficient is

$$\theta_b = \sin^{-1}\left[\sqrt{\frac{\mu_2(\varepsilon_1\mu_2 - \varepsilon_2\mu_1)}{\varepsilon_1(\mu_2^2 - \mu_1^2)}}\right] \tag{3.75}$$

However, for two dielectrics with identical permeabilities the condition for no reflection cannot be satisfied. This can be seen from Eq. (3.75). If μ_1 and μ_2 are not the same, the condition can be satisfied and a Brewster angle exists. For materials with identical permittivities, the Brewster angle is given by

$$\theta_b = \sin^{-1}\left[\sqrt{\frac{\mu_2}{\mu_2 + \mu_1}}\right] \tag{3.76}$$

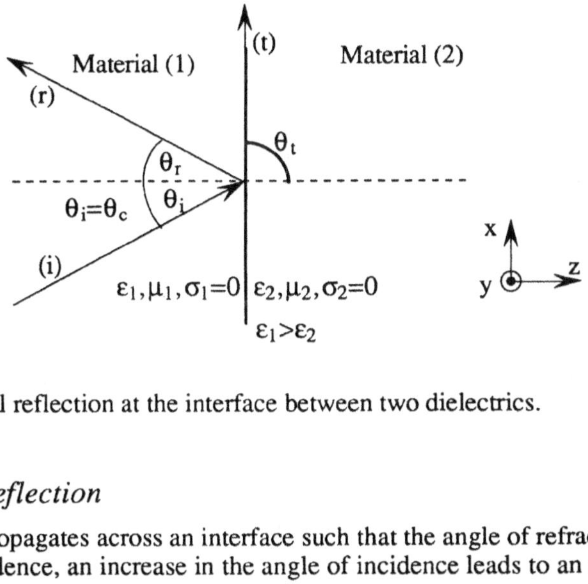

Figure 3.16. Total reflection at the interface between two dielectrics.

3.4.1. Total Reflection

If the wave propagates across an interface such that the angle of refraction is larger than the angle of incidence, an increase in the angle of incidence leads to an angle at which the refracted wave propagates at 90° to the normal (see Figure 3.16). This angle is called a critical angle. Any increase in the angle of incidence results in total reflection of the incident wave. This condition can occur, for example, in lossless dielectrics if $\varepsilon_1 > \varepsilon_2$. The phenomenon exists in either perpendicular or parallel polarization. In terms of the reflection coefficient, total reflection occurs when the reflection coefficient is equal to unity. To find the required conditions for total reflection we take the reflection coefficient for perpendicular polarization and set its magnitude to *1*

$$|\Gamma| = \frac{|\eta_2\cos\theta_i - \eta_1\cos\theta_t|}{|\eta_2\cos\theta_i + \eta_1\cos\theta_t|} = 1 \tag{3.77}$$

The only way this condition can be satisfied is if the reflection coefficient is a complex number. From Eq. (3.72) we can write

$$\cos\theta_t = \sqrt{1 - \frac{\mu_1\varepsilon_1}{\mu_2\varepsilon_2}\sin^2\theta_i} = j\sqrt{\frac{\mu_1\varepsilon_1}{\mu_2\varepsilon_2}\sin^2\theta_i - 1} \tag{3.78}$$

from which the necessary condition for this equation to be satisfied is for the term under the square root sign to be positive:

$$\frac{\mu_1\varepsilon_1}{\mu_2\varepsilon_2}\sin^2\theta_i \geq 1$$

We define now the critical angle as

$$\sin^2\theta_c = \frac{\mu_2\varepsilon_2}{\mu_1\varepsilon_1} \qquad \text{or} \qquad \theta_c = \sin^{-1}\left(\sqrt{\frac{\mu_2\varepsilon_2}{\mu_1\varepsilon_1}}\right)$$

Therefore, total reflection occurs for $\theta_i \geq \theta_c$ or

$$\theta_i \geq \sin^{-1}\left(\sqrt{\frac{\mu_2\varepsilon_2}{\mu_1\varepsilon_1}}\right) \tag{3.79}$$

In addition we must require that $\mu_1\varepsilon_1 \geq \mu_2\varepsilon_2$ otherwise the argument of the inverse sine function is larger than one. In dielectric media for which the permeability is equal in both materials, the condition for total reflection is:

$$\theta_i \geq \sin^{-1}\left(\sqrt{\frac{\varepsilon_2}{\varepsilon_1}}\right) \tag{3.80}$$

and this can only be satisfied if $\varepsilon_1 \geq \varepsilon_2$, as mentioned at the beginning of this section.

In the case of parallel polarization we start with the condition

$$|\Gamma| = \frac{|-\eta_1\cos\theta_i + \eta_2\cos\theta_t|}{|\eta_2\cos\theta_t + \eta_1\cos\theta_i|} = 1$$

However, since the relation given in Eqs. (3.79) and (3.80) is independent of polarization, the condition required to satisfy this equation are exactly the same as for perpendicular polarization. Therefore, the conditions in Eqs. (3.78) through (3.80) apply to either polarization.

3.4.2. Surface Waves

In the discussion above we have completely neglected the transmission coefficients. By substituting the value for $cos\theta_t$ from Eq. (3.79) into the transmission coefficient for the perpendicular polarization we get:

$$T = \frac{2\eta_2 cos\theta_i}{\eta_2 cos\theta_i + \eta_1 j\sqrt{\frac{\mu_1\varepsilon_1}{\mu_2\varepsilon_2}sin^2\theta_i - 1}} \tag{3.81}$$

This is clearly not zero, for any angle. Moreover, looking at T at the critical angle (i.e. $\theta_t = 90°$), we have $\Gamma = 1$ (total reflection) and $T = 2$. This indicates that there is a field in material (2), in spite of the total reflection condition. This field must in fact exist otherwise the continuity of the electric (or magnetic) field at the interface is not satisfied. The transmitted electric field has been found in Eq. (3.62) as

$$\mathbf{E}_t(x,z) = \hat{\mathbf{y}} T E_{i1} e^{-j\gamma_2(x sin\theta_t + z cos\theta_t)} \tag{3.82}$$

where T is given by Eq. (3.81) and

$$sin\theta_t = \sqrt{\frac{\mu_1\varepsilon_1}{\mu_2\varepsilon_2}}\, sin\,\theta_i$$

$$cos\theta_t = \pm j\sqrt{\frac{\mu_1\varepsilon_1}{\mu_2\varepsilon_2}sin^2\theta_i - 1}$$

These two conditions can only exist if θ_t is complex (since for any angle larger than the critical angle, $sin\theta_t$ must be larger than one). If we substitute these (assuming $\theta_i > \theta_c$) into the transmitted field we obtain

$$\mathbf{E}_t(x,z) = \hat{\mathbf{y}} T E_{i1} e^{-\alpha z} e^{-j\beta x} \tag{3.83}$$

where

$$\alpha = \omega\sqrt{\mu_2\varepsilon_2}\sqrt{\frac{\mu_1\varepsilon_1}{\mu_2\varepsilon_2}sin^2\theta_i - 1} \tag{3.84}$$

$$\beta = \omega\sqrt{\mu_2\varepsilon_2}\sqrt{\frac{\mu_1\varepsilon_1}{\mu_2\varepsilon_2}}\, sin\,\theta_i \tag{3.85}$$

We note here that the field in medium *(2)* propagates in the x direction (along the surface) but is attenuated in the z direction. The propagation in the x direction is at a velocity lower than that theat would normally exist in medium *(2)* (from Eq. (3.85)) and is attenuated rapidly in the z direction. For this reason, this is called a surface wave. More fundamental is the fact that there is no power propagated into medium *(2)*. This can be seen by writing the magnetic field and then the Poynting vector. The magnetic field has only a z component (see Eq. (3.63)) and therefore the Poynting vector only has an x component. The relations between fields and angles are shown in Figure 3.17.

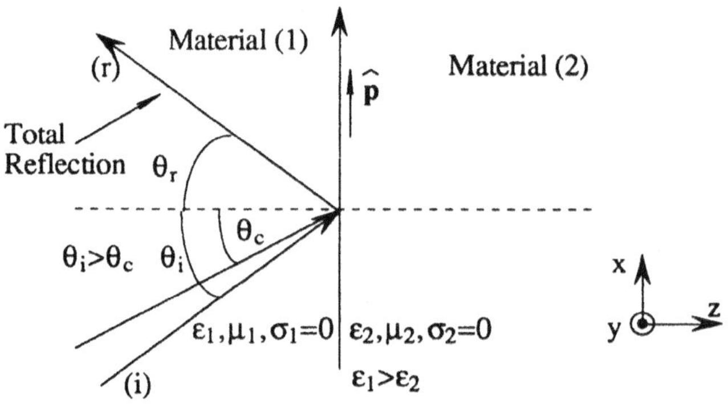

Figure 3.17. Total reflection occurs at angles of incidence larger than the critical angle.

3.5. Reflection and Transmission for Layered Materials at Normal Incidence

At multiple surfaces, such as between layered dielectrics, we expect both reflections and transmission at each surface. The treatment above, while certainly valid, is difficult to apply in terms of transmission and reflection coefficient. Multiple layers are treated by calculating the fields on each side of each layer and then applying the boundary conditions at each interface. From these conditions the fields on each side of each interface are calculated, completely specifying the problem. To illustrate the method in general terms, consider the two layered dielectric in Figure 3.18. For simplicity, we may assume that the layers are in free space although little will change if these were dielectrics or if one were a conductor. Also, the layered dielectrics should be viewed as general, lossy dielectrics. The general procedure is to write the electric and magnetic fields in each material. In material *(1)* we use the notation of incident and reflected waves while in materials *(2)* and *(3)* these are marked by *(+)* for the forward propagating wave and *(−)* for the backward propagating wave. In material *(4)* there is only a forward propagating waves

In material *(1)*

$$\mathbf{E}_1 = \hat{\mathbf{x}}\left[E_{io}e^{-j\beta_o z} + E_{ro}e^{j\beta_o z}\right] \tag{3.86}$$

$$\mathbf{H}_1 = \frac{\hat{\mathbf{y}}}{\eta_0}\left[E_{io}e^{-j\beta_o z} - E_{ro}e^{j\beta_o z}\right] \tag{3.87}$$

In material *(2)*

$$\mathbf{E}_2 = \hat{\mathbf{x}}\left[E_2^+e^{-\gamma_2 z} + E_2^-e^{\gamma_2 z}\right] \tag{3.88}$$

$$\mathbf{H}_2 = \frac{\hat{\mathbf{y}}}{\eta_2}\left[E_2^+e^{-\gamma_2 z} - E_2^-e^{\gamma_2 z}\right] \tag{3.89}$$

In material *(3)*

$$\mathbf{E}_3 = \hat{\mathbf{x}}\left[E_3^+e^{-\gamma_3 z} + E_3^-e^{\gamma_3 z}\right] \tag{3.90}$$

$$\mathbf{H}_3 = \frac{\hat{\mathbf{y}}}{\eta_3}\left[E_3^+e^{-\gamma_3 z} - E_3^-e^{\gamma_3 z}\right] \tag{3.91}$$

In material *(4)*

$$\mathbf{E}_4 = \hat{\mathbf{x}}E_4^+e^{-\gamma_4 z} \tag{3.92}$$

$$\mathbf{H}_4 = \frac{\hat{\mathbf{x}}}{\eta_4}E_4^+e^{-\gamma_4 z} \tag{3.93}$$

Now, using the dimensions in Figure 3.18 we equate the electric and magnetic fields on both sides of each interface.
At $z=0$,

$$E_1(0) = E_2(0)$$

$$H_1(0) = H_2(0)$$

at $z=d_1$

$$E_2(d_1) = E_3(d_1)$$

$$H_2(d_1) = H_3(d_1)$$

at $z=d_2$

$$E_3(d_2) = E_4(d_2)$$

$$H_3(d_2) = H_4(d_2)$$

These conditions form 6 equations in the unknown fields E_4^+, E_3^+, E_3^-, E_2^-, E_2^+, and E_{r0} are easily calculated although, the expressions are rather cumbersome. The magnetic fields are then evaluated from Eqs (3.87), (3.89), (3,91) and (3.93) by substitution. The incident electric field E_{i0} is assumed to be known. This method can be extended to any number of materials.

We discuss next the electric fields at the interfaces of a dielectric slab, followed by those for a lossy dielectric and then for a conductor (or high loss dielectric). Finally, we also discuss the conditions for a conductor backed dielectric as all of these are important conditions for testing. Testing aspects on dielectric slabs and metal backed dielectrics will be discussed in chapter 6.

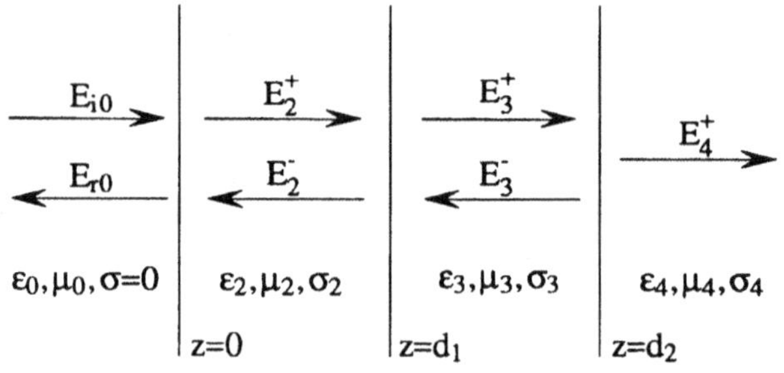

Figure 3.18. Reflection and transmission for multiple, lossy dielectric interfaces: normal incidence.

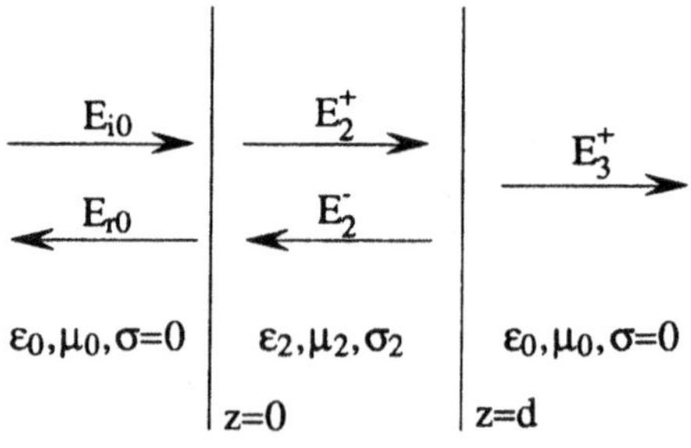

Figure 3.19. Reflection and transmission for a lossy dielectric slab in free space: normal incidence.

3.5.1. Reflection and Transmission for a Dielectric Slab at Normal Incidence

A dielectric slab in free space is shown in Figure 3.19 together with the electric and magnetic fields in each material. Since we assume here that all materials are lossless, the propagation constant is replaced by β and we can write

In material *(1)*

$$\mathbf{E}_1 = \hat{\mathbf{x}}\left[E_{i0}e^{-j\beta_0 z} + E_{r0}e^{j\beta_0 z}\right]$$

$$\mathbf{H}_1 = \frac{\hat{\mathbf{y}}}{\eta_0}\left[E_{i0}e^{-j\beta_0 z} - E_{r0}e^{j\beta_0 z}\right]$$

In material *(2)*

$$\mathbf{E}_2 = \hat{\mathbf{x}}\left[E_2^+e^{-j\beta_2 z} + E_2^-e^{j\beta_2 z}\right]$$

$$\mathbf{H}_2 = \frac{\hat{\mathbf{y}}}{\eta_2}\left[E_2^+e^{-j\beta_2 z} - E_2^-e^{j\beta_2 z}\right]$$

In material *(3)*

$$\mathbf{E}_3 = \hat{\mathbf{x}}E_3^+e^{-j\beta_0 z}$$

$$\mathbf{H}_3 = \frac{\hat{\mathbf{x}}}{\eta_0}E_3^+e^{-j\beta_0 z}$$

To find the fields, the boundary conditions at $z=0$ and $z=d$ must be satisfied. At $z=0$,

$$E_1(0) = E_2(0)$$

$$H_1(0) = H_2(0)$$

at $z=d$

$$E_2(d) = E_3(d)$$

$$H_2(d) = H_3(d)$$

From these, the four fields, E_2^+, E_2^-, E_3^+ and E_{r0} are calculated. These are

$$E_{r0} = -\frac{j\left[\eta_0^2 - \eta_2^2\right] \tan\beta_2 d}{2\eta_0\eta_2 + j\left[\eta_0^2 + \eta_2^2\right] \tan\beta_2 d} E_{i0} \tag{3.94}$$

$$E_2^+ = \frac{\eta_2\left[\eta_0 + \eta_2\right] e^{j\beta_2 d}}{2\eta_0\eta_2 \cos\beta_2 d + j\left[\eta_0^2 + \eta_2^2\right] \sin\beta_2 d} E_{i0} \tag{3.95}$$

$$E_2^- = \frac{\eta_2\left[\eta_0 - \eta_2\right] e^{-j\beta_2 d}}{2\eta_0\eta_2 \cos\beta_2 d + j\left[\eta_0^2 + \eta_2^2\right] \sin\beta_2 d} E_{i0} \tag{3.96}$$

$$E_3^+ = \frac{2\eta_0\eta_2\, \varepsilon^{j\beta_o d}}{2\eta_0\eta_2 \cos\beta_2 d + j\left[\eta_0^2 + \eta_2^2\right] \sin\beta_2 d} E_{i0} \tag{3.97}$$

Thus, we can define a general reflection coefficient for the slab as

$$\Gamma = \frac{E_{r0}}{E_{i0}}$$

This gives

$$\Gamma = -\frac{j\left[\eta_0^2 - \eta_2^2\right] \tan\beta_2 d}{2\eta_0\eta_2 + j\left[\eta_0^2 + \eta_2^2\right] \tan\beta_2 d} \tag{3.98}$$

and the transmission coefficient through the slab is defined as

$$T = \frac{E_3^+}{E_{i0}}$$

or

$$T = \frac{2\eta_0\eta_2\, e^{j\beta_o d}}{2\eta_0\eta_2 \cos\beta_2 d + j\left[\eta_0^2 + \eta_2^2\right] \sin\beta_2 d} \tag{3.99}$$

where d is the thickness of the slab and

$$\beta_2 = \omega\sqrt{\mu_2\varepsilon_2}$$

$$\eta_0 = \sqrt{\frac{\mu_0}{\varepsilon_0}} \qquad \eta_2 = \sqrt{\frac{\mu_2}{\varepsilon_2}}$$

The expressions for reflection and transmission coefficients will become useful in measuring the thickness of the slab.

3.5.2. Reflection and Transmission for a Low Loss Dielectric Slab at Normal Incidence

Obviously, a more complex solution is obtained if the dielectric is lossy. In this case, the attenuation in the slab must also be taken into account. Assuming a low loss dielectric, with attenuation constant α, we can write the equations above as:
In material *(1)*

$$\mathbf{E}_1 = \hat{\mathbf{x}}\left[E_{i0}e^{-j\beta_0 z} + E_{r0}e^{j\beta_0 z}\right]$$

$$\mathbf{H}_1 = \frac{\hat{\mathbf{y}}}{\eta_0}\left[E_{i0}e^{-j\beta_0 z} - E_{r0}e^{j\beta_0 z}\right]$$

In material *(2)*

$$\mathbf{E}_2 = \hat{\mathbf{x}}\left[E_2^+e^{-\alpha_2 z}e^{-j\beta_2 z} + E_2^-e^{\alpha_2 z}e^{j\beta_2 z}\right]$$

$$\mathbf{H}_2 = \hat{\mathbf{x}}\left[\frac{E_2^+}{\eta_2}e^{-\alpha_2 z}e^{-j\beta_2 z} + \frac{E_2^-}{\eta_2}e^{\alpha_2 z}e^{j\beta_2 z}\right]$$

In material *(3)*

$$\mathbf{E}_3 = \hat{\mathbf{x}}E_3^+e^{-j\beta_0 z}$$

$$\mathbf{H}_3 = \frac{\hat{\mathbf{x}}}{\eta_0}E_3^+e^{-j\beta_0 z}$$

The boundary conditions are as before:
At $z=0$,

$$E_1(0) = E_2(0) \qquad H_1(0) = H_2(0)$$

and at $z=d$,

$$E_2(d) = E_3(d) \qquad H_2(d) = H_3(d)$$

From these, the four fields, E_2^+, E_2^-, E_3^+ and E_{r0} are calculated as previously, but with the obvious changes due to attenuation. Now we can write

$$E_{r0} = -\frac{j\left[\eta_0^2 - \eta_2^2\right] tan(\beta_2 - j\alpha_2)d}{2\eta_0\eta_2 + j\left[\eta_0^2 + \eta_2^2\right] tan(\beta_2 - j\alpha_2)d} E_{i0}$$

$$E_2^+ = \frac{\eta_2\left[\eta_0 + \eta_2\right] e^{\alpha_2 d} e^{j\beta_2 d}}{2\eta_0\eta_2 cos(\beta_2 - j\alpha_2)d + j\left[\eta_0^2 + \eta_2^2\right] sin\beta_2 - j\alpha_2)d} E_{i0}$$

$$E_2^- = \frac{\eta_2\left[\eta_0 - \eta_2\right] e^{\alpha_2 d} e^{-j\beta_2 d}}{2\eta_0\eta_2 cos(\beta_2 - j\alpha_2)d + j\left[\eta_0^2 + \eta_2^2\right] sin(\beta_2 - j\alpha_2)d} E_{i0}$$

$$E_3^+ = \frac{2\eta_0\eta_2 e^{\alpha_2 d} e^{j\beta_0 d}}{2\eta_0\eta_2 cos(\beta_2 - j\alpha_2)d + j\left[\eta_0^2 + \eta_2^2\right] sin(\beta_2 - j\alpha_2)d} E_{i0}$$

Thus, the reflection and transmission coefficients of the slab are

$$\Gamma = -\frac{j\left[\eta_0^2 - \eta_2^2\right] tan(\beta_2 - j\alpha_2)d}{2\eta_0\eta_2 + j\left[\eta_0^2 + \eta_2^2\right] tan(\beta_2 - j\alpha_2)d} \qquad (3.100)$$

$$T = \frac{2\eta_0\eta_2 e^{\alpha_2 d} e^{j\beta_0 d}}{2\eta_0\eta_2 cos(\beta_2 - j\alpha_2)d + j\left[\eta_0^2 + \eta_2^2\right] sin(\beta_2 - j\alpha_2)d} \qquad (3.101)$$

where, as previously, d is the thickness of the slab and

$$\gamma_2 = \alpha + j\beta$$

$$\beta_2 \approx \omega\sqrt{\mu_2\varepsilon_2}$$

$$\alpha_2 \approx \frac{\sigma_2}{2}\sqrt{\frac{\mu_2}{\varepsilon_2}}$$

$$\eta_2 \approx \sqrt{\frac{\mu_2}{\varepsilon_2}}\left(1 + j\frac{\sigma}{2\omega\varepsilon}\right)$$

These are obviously more complex expressions and, unlike the lossless case, both Γ and T are complex. Nevertheless, their derivation follows identical steps.

3.5.3. Reflection and Transmission for a High Loss Dielectric Slab at Normal Incidence

In a highly lossy material, the solution is similar. We start with the same steps as for the low loss dielectric. The equations in the three materials and boundary conditions are the same. For a high loss material, we use the expressions of attenuation and propagation constant for a good conductor. These are

$$\alpha_2 = \beta_2 = \sqrt{\pi f \mu_2 \sigma_2}$$

and the intrinsic impedance of the slab is

$$\eta_2 = (1+j)\frac{\alpha_2}{\sigma_2}$$

Thus, the various fields are

$$E_{r0} = -\left[1 + j\frac{\eta_2}{\eta_0}\frac{1 + j\tan\beta_2 d \tanh\alpha_2 d}{\tan\beta_2 d - j\tanh\alpha_2 d}\right]E_{i0}$$

$$E_2^+ = -j\frac{\eta_2}{\eta_0}\frac{e^{\alpha_2 d}e^{j\beta_2 d}}{\sin(\beta_2 - j\alpha_2)d}E_{i0}$$

$$E_2^+ = -j\frac{\eta_2}{\eta_0}\frac{e^{-\alpha_2 d}e^{-j\beta_2 d}}{\sin(\beta_2 - j\alpha_2)d}E_{i0}$$

$$E_3^+ = -j\frac{\eta_2}{\eta_0}\frac{2e^{j\beta_0 d}}{\sin(\beta_2 - j\alpha_2)d}E_{i0}$$

and the reflection and transmission coefficients are

$$\Gamma = -1 - j\frac{\eta_2}{\eta_0}\frac{1 + j\tan\beta_2 d\tanh\alpha_2 d}{\tan\beta_2 d - j\tanh\alpha_2 d} \tag{3.102}$$

$$T = -j\frac{\eta_2}{\eta_0}\frac{2e^{j\beta_2 d}}{\sin(\beta_2 - j\alpha_2)d} \tag{3.103}$$

It should be noted here that the transmission coefficient is rather low since η_2 is small compared to η_0. The reflection coefficient approaches -1 as we would expect from a good conductor.

3.5.4. Reflection and Transmission for a Lossless Dielectric Slab Backed by a Perfect Conductor: Normal Incidence

For a dielectric backed by a conductor, the only difference is tthat E_3 is zero. In this case, the solution starts as for the lossless dielectric case, but we modify the boundary conditions to account for the vanishing fields in the conductor:
In material *(1)*

$$\mathbf{E}_1 = \hat{\mathbf{x}}\left[E_{i0}e^{-j\beta_0 z} + E_{r0}e^{j\beta_0 z}\right]$$

$$\mathbf{H}_1 = \frac{\hat{\mathbf{y}}}{\eta_0}\left[E_{i0}e^{-j\beta_0 z} - E_{r0}e^{j\beta_0 z}\right]$$

In material *(2)*

$$\mathbf{E}_2 = \hat{\mathbf{x}}\left[E_2^+e^{-j\beta_2 z} + E_2^-e^{j\beta_2 z}\right]$$

$$\mathbf{H}_2 = \frac{\hat{\mathbf{y}}}{\eta_2}\left[E_2^+e^{-j\beta_2 z} - E_2^-e^{j\beta_2 z}\right]$$

To find the fields we satisfy the boundary conditions at $z=0$ and $z=d$. At $z=0$,

$$E_1(0) = E_2(0)$$

$$H_1(0) = H_2(0)$$

at $z=d$

$$E_2^-(d) = -E_2^+(d)$$

From these, the fields, E_2^+, E_2^- and E_{r0} are calculated. Thus, from the conditions above

$$E_2^+ = \frac{2\eta_2}{(\eta_o + \eta_2) + (\eta_o - \eta_2)e^{-j\beta_2 d}} E_{i0}$$

$$E_{r0} = -\frac{\eta_o - j\eta_2 tan\beta_2 d}{\eta_o + jh_2 tan\beta_2 d} E_{i0}$$

and the reflection coefficient is

$$\Gamma = -\frac{\eta_o - j\eta_2 tan\beta_2 d}{\eta_o + j\eta_2 tan\beta_2 d} \tag{3.104}$$

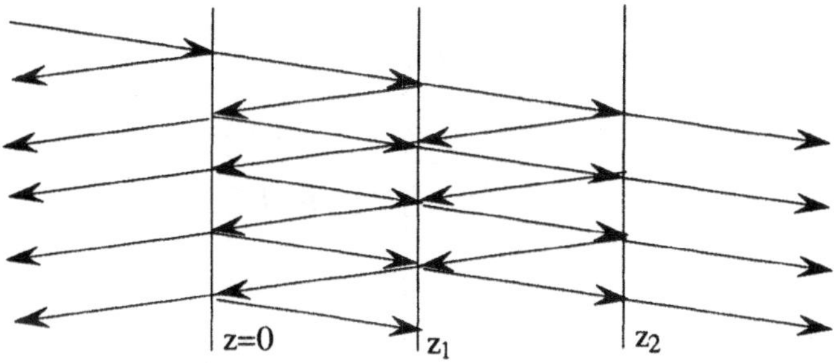

Figure 3.20. Multiple reflections at interfaces between materials.

3.6. Reflection and Transmission for Layered Dielectrics: Oblique Incidence

Reflection and transmission at multiple interfaces can also be calculated for oblique incidence in a manner similar to that done for a single interface. The simplest way to analyze this type of condition is to use the reflection and transmission coefficients at each interface. The basic relations are shown in Figure 3.20 where two dielectric layers are shown. The dielectrics can be lossless or lossy or may even be conducting. Viewing the incident wave as V_i, a reflected wave at the first interface will exist while the transmitted wave can be viewed as the incident wave on interface number 2. Again, a reflected and transmitted wave will exist on this interface. These waves bounce back and forth as shown in Figure 3.20. Calculation of fields consists essentially of calculating the various coefficients (calculated as for a single interface) and evaluation of the fields after any number of reflections. To get an exact solution, an infinite number of reflections are needed, and the fields anywhere in space are then calculated as superposition of all waves that exist in any material. Once the sums of the reflected waves, and transmitted waves in each material are known, a reflection coefficient and a transmission coefficient are calculated for the layered material. The same problem can be solved by other means, such as by use of impedance calculations. This can be done for any number of layers. However, it is possible to perform the same calculation in a much simpler manner by use of transmission line methods. To do so, we look at a dielectric layer as an equivalent transmission line and calculate the fields in terms of equivalent transmission line parameters.

Before we do so, it is useful to look at the transmission line parameters for a lossless transmission line of length d as shown in Figure 3.21 where the line impedance at one end is Z_l (we view this as the load) and at the other, the input impedance to the segment. We wrote the input impedance for this line in Eq. (2.37) as

$$Z_{in} = Z_0 \frac{[Z_l + jZ_0 \tan\beta d]}{[Z_0 + jZ_l \tan\beta d]} \tag{3.105}$$

For convenience we write $\tan\beta d = \sin\beta d / \cos\beta d$ and get

$$Z_{in} = Z_0 \frac{[Z_l \cos\beta d + jZ_0 \sin\beta d]}{[Z_0 \cos\beta d + jZ_l \sin\beta d]} \tag{3.106}$$

By definition,

$$Z_{in} = \frac{V_{in}}{I_{in}} \qquad and \qquad Z_l = \frac{V_l}{I_l}$$

Thus, we can write

$$Z_{in} = \frac{V_{in}}{I_{in}} = Z_0 \frac{\left[\frac{V_l}{I_l}\cos\beta d + jZ_0 \sin\beta d\right]}{\left[Z_0 \cos\beta d + j\frac{V_l}{I_l}\sin\beta d\right]} = \frac{\frac{Z_0}{I_l}\left(V_l \cos\beta d + jZ_0 I_l \sin\beta d\right)}{\frac{Z_0}{I_l}\left(I_l \cos\beta d + j\frac{V_l}{Z_0}\sin\beta d\right)}$$

or,

$$\frac{V_{in}}{I_{in}} = \frac{V_l \cos\beta d + jZ_0 I_l \sin\beta d}{I_l \cos\beta d + j\frac{V_l}{Z_0}\sin\beta d} \tag{3.107}$$

The voltage and current at the left side of the line can be written as

$$V_{in} = V_l \cos\beta d + jZ_0 I_l \sin\beta d \tag{3.108}$$

$$I_{in} = I_l cos\beta d + j\frac{V_l}{Z_0}sin\beta d \qquad (3.109)$$

Thus, we have an expression linking voltage and current at the two ends of the line.

The equivalence between the dielectric layer and the transmission line is defined based on the fact that the tangential component of the electric field at the layer interfaces is equivalent to the voltage at the ends of the transmission line and the tangential component of the magnetic field is equivalent to the current. The equivalence between the line and dielectric layer is shown in Figure 3.21.

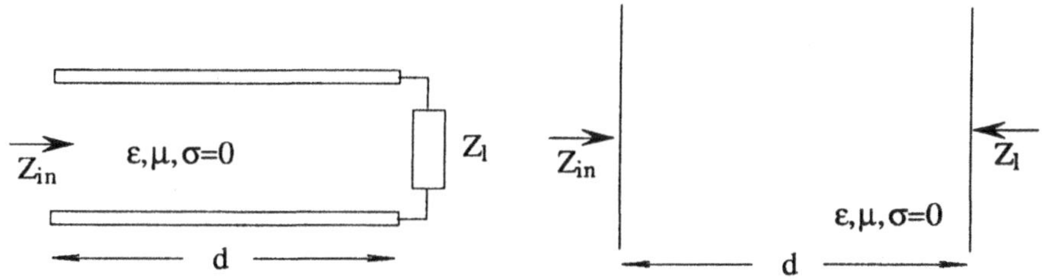

Figure 3.21. Equivalence between a lossless dielectric slab and a lossless transmission line.

3.6.1. Oblique Incidence on N Dielectric Layers: Perpendicular Polarization

Now we can consider an N layer material as shown in Figure 3.22. Using first the first layer we can write

$$E_{in1} = E_{l1} cos\beta_1 d_1 + jZ_1 H_{l1} sin\beta_1 d_1 \qquad (3.110)$$

$$H_{in1} = H_{l1} cos\beta_1 d_1 + j\frac{E_{l1}}{Z_1}sin\beta_1 d_1 \qquad (3.111)$$

where:

$$\beta_1 = \omega\sqrt{\mu_1\varepsilon_1}cos\theta_1$$

$$Z_1 = \frac{\sqrt{\dfrac{\mu_1}{\varepsilon_1}}}{cos\theta_1}$$

and θ_1 is the angle of the wave in material *(1)*. This angle (which is actually the angle of the transmitted wave across the interface between free space and material *(1)*) can be written in terms of the incident angle θ_i using Snell's law

$$n_0 sin\theta_i = n_1 sin\theta_1$$

where n_1 and n_1 are the indices of refraction of the two materials. Writing the indices of refraction in terms of ε and μ, and separating $sin\theta_i$ we get

$$sin\theta_1 = \frac{n_0}{n_1} sin\theta_i = \frac{1}{\sqrt{\varepsilon_{r1}\mu_{r1}}} sin\theta_i$$

or

$$cos\theta_1 = \sqrt{1 - sin^2\theta_1} = \sqrt{1 - \frac{1}{\varepsilon_{r1}\mu_{r1}} sin^2\theta_i}$$

Substituting these expressions in Eqs. (3.110) and (3.111) we get

$$E_{in1} = E_{t1}A_1 + H_{t1}B_1$$

$$H_{in1} = H_{t1}D_1 + E_{t1}C_1$$

or, in matrix form

$$\begin{bmatrix} E_{in1} \\ H_{in1} \end{bmatrix} = \begin{bmatrix} A_1 & B_1 \\ C_1 & D_1 \end{bmatrix} \begin{bmatrix} E_{t1} \\ H_{t1} \end{bmatrix} \qquad (3.112)$$

where

$$A_1 = D_1 = cos\omega\sqrt{\mu_1\varepsilon_1}\sqrt{1 - \frac{1}{\varepsilon_{r1}\mu_{r1}} sin^2\theta_i}\, d_1 \qquad (3.113)$$

$$B_1 = j\frac{\sqrt{\frac{\mu_1}{\varepsilon_1}}}{\sqrt{1 - \frac{1}{\varepsilon_{r1}\mu_{r1}} sin^2\theta_i}} sin\omega\sqrt{\mu_1\varepsilon_1}\sqrt{1 - \frac{1}{\varepsilon_{r1}\mu_{r1}} sin^2\theta_i}\, d_1 \qquad (3.114)$$

$$C_1 = j\frac{\sqrt{\frac{\varepsilon_1}{\mu_1}}}{\sqrt{1-\frac{1}{\varepsilon_{r1}\mu_{r1}}\sin^2\theta_i}}\sin\omega\sqrt{\mu_1\varepsilon_1}\sqrt{1-\frac{1}{\varepsilon_{r1}\mu_{r1}}\sin^2\theta_i}\,d_1 \tag{3.115}$$

This process is repeated for each of the layers by simply changing the material index, but, now E_{l1}, H_{l1}, are the incident fields to material *(2)*. Thus, for material *(2)* we get

$$\begin{bmatrix} E_{in2} \\ H_{in2} \end{bmatrix} = \begin{bmatrix} A_2 & B_2 \\ C_2 & D_2 \end{bmatrix}\begin{bmatrix} E_{l2} \\ H_{l2} \end{bmatrix} \tag{3.116}$$

with A_1 through A_4 obtained from Eqs. (3.113) through (3.114) by changing the index *(1)* by index *(2)*. Thus, for *N* layers, it is possible to write the following relation

$$\begin{bmatrix} E_{i0} \\ H_{i0} \end{bmatrix} = \begin{bmatrix} A_1 & B_1 \\ C_1 & D_1 \end{bmatrix}\begin{bmatrix} A_2 & B_2 \\ C_2 & D_2 \end{bmatrix}\cdots\begin{bmatrix} A_N & B_N \\ C_N & D_N \end{bmatrix}\begin{bmatrix} E_t \\ H_t \end{bmatrix} = \begin{bmatrix} A_T & B_T \\ C_T & D_T \end{bmatrix}\begin{bmatrix} E_t \\ H_t \end{bmatrix} \tag{3.117}$$

From this, the reflection and transmission coefficients for the *N* layer material are defined from the basic relations

$$E_{i0} = E_0(1+\Gamma_t)$$

$$H_{i0} = \frac{E_0}{Z_0}(1-\Gamma_t)$$

$$E_t = E_0 T_t$$

$$H_t = \frac{E_0}{Z_t}T_t$$

Substituting in Eq. (3.117) and solving for Γ_t and T_t we get

$$\Gamma_t = \frac{A_T + \frac{B_T}{Z_t} - Z_1\left(C_T + \frac{D_T}{Z_t}\right)}{A_T + \frac{B_T}{Z_t} + Z_1\left(C_T + \frac{D_T}{Z_t}\right)} \tag{3.118}$$

$$T_t = \frac{2}{A_T + \frac{B_T}{Z_t} + Z_1\left(C_T + \frac{D_T}{Z_t}\right)} \tag{3.119}$$

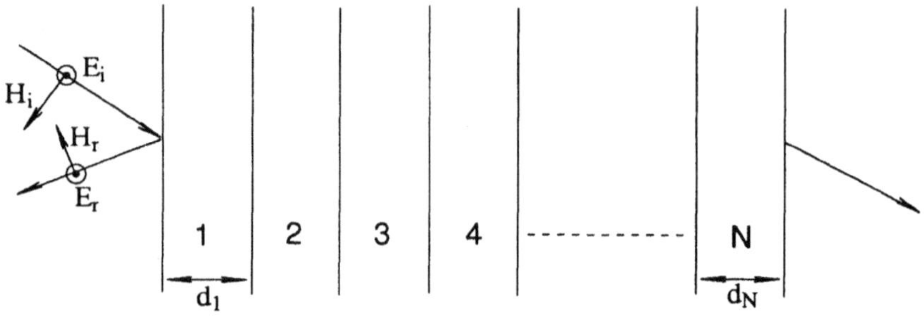

Figure 3.22. Reflection and transmission for an N layer dielectric: oblique incidence with perpendicular polarization.

For a slab of thickness d, in free space, as shown in Figure 3.23, the reflection and transmission coefficients are

$$\Gamma_t = \frac{A_I + \dfrac{B_I}{Z_0} - Z_I\left(C_I + \dfrac{D_I}{Z_0}\right)}{A_I + \dfrac{B_I}{Z_0} + Z_I\left(C_I + \dfrac{D_I}{Z_0}\right)} \tag{3.120}$$

$$T_t = \frac{2}{A_I + \dfrac{B_I}{Z_0} + Z_I\left(C_I + \dfrac{D_I}{Z_t}\right)} \tag{3.121}$$

where the various terms are given in Eqs. (3.113) through (3.115). Z_0 is the wave impedance of free space.

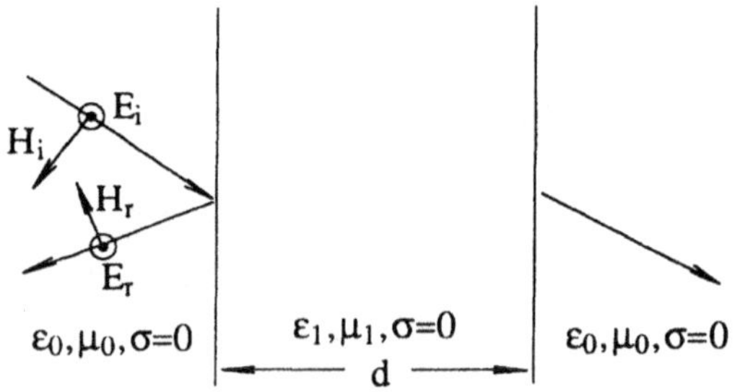

Figure 3.23. Reflection and transmission for a lossless dielectric slab: oblique incidence with perpendicular polarization.

3.6.2. Oblique Incidence on N Dielectric Layers: Parallel Polarization

A parallel polarized, obliquely incident wave on an N layer stack is shown in Figure 3.24. We can use an identical procedure as above except that the tangential component of the electric field now is equal to $E_x cos\theta_i$ in each layer. Also, the wave impedance is equal to

$$Z_l = \sqrt{\frac{\mu_l}{\varepsilon_l}}\, cos\theta_l$$

Thus, the coefficients A_l, D_l in Eq. (3.113) remain the same, but B_l, C_l in Eqs. (3.114) and (3.115) become

$$B_l = j\sqrt{\frac{\mu_l}{\varepsilon_l}}\sqrt{1 - \frac{1}{\varepsilon_{rl}\mu_{rl}}sin^2\theta_i}\; sin\omega\sqrt{\mu_l\varepsilon_l}\sqrt{1 - \frac{1}{\varepsilon_{rl}\mu_{rl}}sin^2\theta_i}\; d_l$$

$$C_l = j\sqrt{\frac{\varepsilon_l}{\mu_l}}\sqrt{1 - \frac{1}{\varepsilon_{rl}\mu_{rl}}sin^2\theta_i}\; sin\omega\sqrt{\mu_l\varepsilon_l}\sqrt{1 - \frac{1}{\varepsilon_{rl}\mu_{rl}}sin^2\theta_i}\; d_l$$

Therefore, the forms of the reflection and transmission coefficients remain unchanged as in Eqs. (3.120) and (3.121) but their values change.

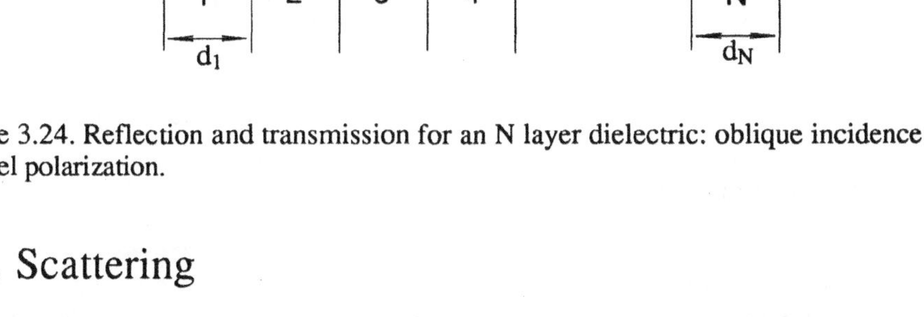

Figure 3.24. Reflection and transmission for an N layer dielectric: oblique incidence with parallel polarization.

3.7. Scattering

The foregoing sections dealt with reflection, transmission, and refraction at interfaces between materials. We found it useful to assume planar interfaces and, perhaps, materials that extend to infinity. This was actually necessary so that we may assume plane waves. It was also possible to say that these interfaces scattered the field due to differences in material properties. However, scattering normally refers to a similar effect produced by finite bodies at the boundaries of which the field is reflected and refracted (although, in most cases, scatterers will be viewed as conducting bodies, nonconducting bodies also scatter the field). Normally a scatterer is small (say, compared to wavelength) but does not have to be so. For testing purposes, this distinction is normally adequate. The scattering mechanism is shown schematically in Figure 3.25.

Scattering can take place from any object or any discontinuity in an object. However, the mechanism for scattering, and therefore the scattered field depends on a variety of parameters including frequency, material properties, size of scatterer, and the presence of other scatterers.

The calculation of scattered fields is rather difficult and requires specialized techniques. We will not attempt to do so here but, instead, limit ourselves to a number of well defined scattering parameters. These will allow discussion of basic concepts without the need for complex mathematics and will lead to the ideas of radar or radar-like applications which we will discuss again in chapter 6. The following assumptions are made throughout this discussion:

- The scatterer is in the far zone (Fraunhofer zone), meaning that the scatterer is far from the source or, in more exact terms $\beta a >> 1$ where a is the distance between source and scatterer and β is the phase constant of the material through which waves propagate.
- The scatterer is small compared to wavelength. This means that the scatterer's largest dimension (e.g. diameter of a sphere) is small compared to wavelength, say, $d<0.1\lambda$. This type of scattering is called Rayleigh scattering and is typically used for radar testing.

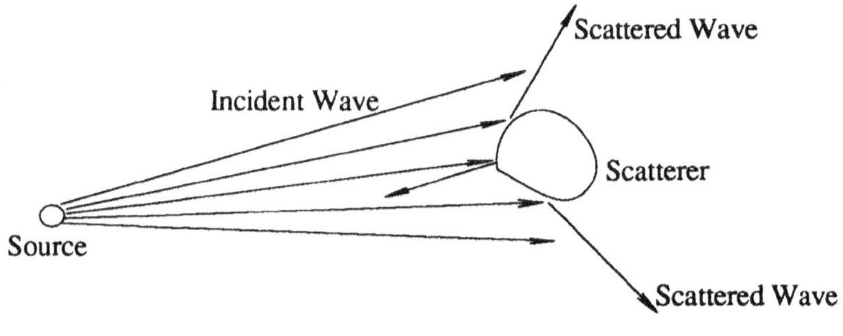

Figure 3.25. Scattering from an object. A point source is shown.

We note here that there are other types of scattering. Scattering due to objects that are large compared to wavelength (say $d>10\lambda$) is refered to as optical scattering while in cases where the dimension is of the same order of magnitude as the wavelength the scattering is referred to as resonant scattering, or Mie scattering.

The scattering configuration used here is shown in Figures 3.26a and 3.26b. Figure 3.26a refers to monostatic scattering, that is, the receiving and transmitting antenna are at the same location. This is typical of pulsed radar applications or, in the more general sense to a send-receive testing configuration. Figure 3.26b is a bistatic configuration in which the receiver and transmitter are at different locations.

The measure of detectability of an object is normally described in terms of backscatter cross section or radar cross section. The backscatter cross section is defined as the equivalent area that would scatter (or receive) the same amount of incident power to produce the same scattered power density at the receiver, if the object were an isotropic scatterer (that is, if the object scatters uniformly in all directions). Thus the radar cross section can be viewed as an equivalent object size. The larger the radar cross section, the larger the signal received.

The radar cross section of an object in the Rayleigh domain is given by

$$\sigma = 4\pi d^2 \frac{P_s}{P_i} \tag{3.122}$$

where:

σ = backscatter (radar) cross section $[m^2]$
d = distance between source and scatterer $[m]$
P_i = time averaged power density at the scatterer $[Watt]$
P_s = time averaged scattered power density at the receiver $[Watt]$.

Thus, it is obvious that the larger the scattered field, the larger the radar cross section.

For measurements, the radar cross section is normally not the quantity we are observing but rather the received power in relation to the transmitted power. To do so we consider the two antennas, a transmitting and a receiving antenna as in Figure 3.28b. If the area of the antennas are A_r and A_t, and the transmitted power is P_t, the power received by the receiving antenna is

$$P_r = P_t \frac{A_t A_r}{d^2 \lambda^2} \tag{3.123}$$

This is known as Friis' transmission formula (λ is the wavelength). Assuming the antennas are at the same location (as in Figure 3.26a), and they are of the same size (that is, the same antenna is used for transmission and reception, as is normally done in many radar applications) this can be written in terms of the scattering cross section as

$$P_r = P_t \frac{\sigma}{4\pi} \frac{A_t^2}{(d^2 \lambda)^2} = P_t \frac{\sigma}{4\pi} \frac{A_t^2}{d^4 \lambda^2} \tag{3.124}$$

which is known as the radar equation. The dependence on d^4 is typical and it is obvious that the power received is rather small, especially at large wavelengths. Thus, any radar system, including those used for *NDT* must operate at rather high frequencies, and often at high transmission powers to be able to detect targets with small radar cross sections.

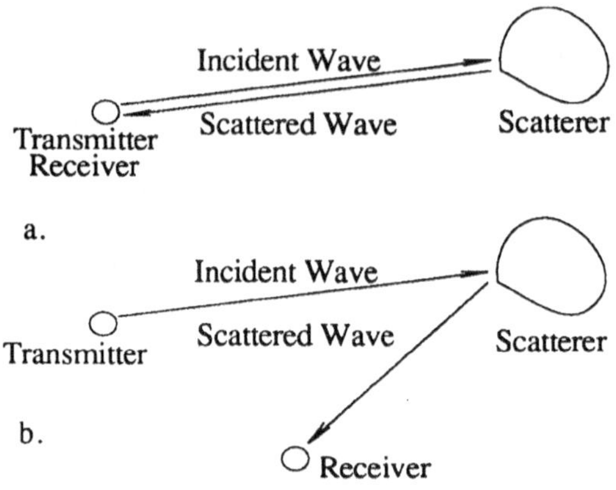

Figure 3.26. Two types of scattering. a. Monostatic scattering and b. Bistatic scattering.

3.8. Bibliography

Topics relating to polarization, transmission, reflection, and refraction of plane waves are standard topics in electromagnetic field theory. Most, but not all the topics covered here can be found in many textbooks. Examples are references [1] through [3]. The treatment in this chapter follows those in Plonus [1] and Cheng [2]. Transmission and reflection with two lossy materials is normally not included in textbooks but can be found in Balanis [3] and (for normal incidence) in Rao [4].

The general problem of transmission and reflection for layered materials is normally not discussed in textbooks. However, more advanced books treat this problem since it is important for design of layered dielectrics such as radomes. The method used here follows the discussion in Ishimaru [5]. Another method often used for this purpose is based on impedance transformation across interfaces. This can be found in references [3] and [5] through [7]. Treatment of reflection and transmission at oblique incidence in lossy dielectrics can be found in Balanis [3] and in [9,10].

A more detailed discussion on the subject can be found in Wait [8]. He discusses propagation of waves through the earth which is modeled as a layered material.

Scattering is an extensive topic and can be found in many books. We mention here those of Balanis [3], Ishimaru [5], and Collin [7]. The subject of radar cross section in particular has received attention and many articles on the subject can be found in journals treating microwaves and propagation aspects of electromagnetics.

[1] M.A. Plonus, "Applied Electromagnetics", McGraw-Hill, Inc., 1978.

[2] D.K. Cheng, "Field and Wave Electromagnetics", 2nd ed., Addison-Wesley, Reading, Ma., 1989.
[3] C.A. Balanis, "Advanced Engineering Electromagnetics", John Wiley & Sons, New York, 1989.
[4] N.N. Rao, "Elements in Engineering Electromagnetics", 3rd ed., Prentice- Hall, Inc., Englewood Cliffs, N.J., 1991.
[5] A. Ishimaru, "Electromagnetic Wave Propagation, Radiation, and Scattering", Prentice Hall, Inc., Englewood Cliffs, N.J., 1991.
[6] D.T. Paris and F.K. Hurd, "Basic Electromagnetic Theory", McGraw-Hill, New York, 1969.
[7] R.E. Collin, "Foundations for Microwave Engineering", McGraw-Hill, Inc., New York, 1992.
[8] J.R. Wait, "Electromagnetic Waves in Stratified Media", 2nd Edition, Pergamon Press, Oxford, 1970.
[9] J.J. Holmes and C.A. Balanis, "Refraction of a uniform plane wave incident on a plane boundary between two lossy media", IEEE Transactions on Antennas and Propagation, 1978.
[10] R.D. Radcliff and C.A. Balanis, "Modified propagation constants for nonuniform plane wave transmission through conducting media", IEEE Transactions on Geoscience and Remote Sensing, 1982.

Chapter 4. Microwave Measurement Techniques

4.1. Introduction

Measurements in the microwave range have the same purpose as at other frequencies but are different in nature. The quantities measured are often different and the methods employed tend to be more complex. As a rule, the basic lumped parameter measurements of voltage, current and lumped impedance cannot be used in their simple, low frequency form. This means, for example, that measurement of power cannot be done by simply measuring voltage and current. On the other hand, there are many parameters, like attenuation or reflection coefficient that are normally not used in the low frequency domain. The measurement environment at microwave frequencies is also unique. Some parameters cannot be measured directly and require specialized techniques. Power, for example, is always measured indirectly by use of thermal effects or even mechanical means.

The choice of measurement techniques presented here is selective and only those methods and variations that are applicable to nondestructive testing are mentioned. This is partly to keep the focus on *NDT* and partly because of the vast nature of the subject. As examples we will not discuss mechanical methods of measuring power because of their applicability to high power measurement, and calorimetric methods because they are used as standards and can seldom be used for field measurements. Similarly, many measurements in the microwave region are geared towards measurements in waveguides and transmission lines. For example, most attenuation measurements and standards assume that the device in which attenuation is measured is connected in line with some type of transmission line. For testing, attenuation in the bulk of a material is more important as, for example in a sample between a transmitting and a receiving antenna.

Finally, it should be noted that methods of measuring material properties are excluded from this chapter. This will be discussed in chapter 8 at some length. The reason we do not introduce them here is because measurement of material properties either relies on one of the general methods of measurements, or constitutes a specialized application which is best introduced separately.

4.2. Power Measurements

Measurement of power is fundamental in microwaves as it is at other frequencies. In addition to indicating the obvious power properties of microwaves it is also used extensively for measurement of other properties such as attenuation and absorption. The measurement of absolute power is often used to establish the output from a device (such as a horn antenna) while relative power measurements are used to determine attenuation, losses, and a variety of other properties. Most of the methods discussed below will deal with measurement of relatively low power. The distinction is not really important other than for *NDT*, it is natural to opt for low power. However, low in this context means low in comparison to the absorption capabilities of the material under test. This means, for example, that moisture in a material does not heat up to the point where the test might not be a true nondestructive test. On the other hand, the power requirements from the source might actually be high. An example is testing of thick concrete pavements or, perhaps, testing of materials in a highly absorptive environment such as underground objects. In cases such as these, because of high attenuation, the source power needs to be high to transfer enough power into the test sample for meaningful measurements. We will not specifically define low and high power as these are relative terms. We merely point out that most *NDT* applications require relatively low power, in many cases well below *1W*. The range of perhaps 10^{-2} *[Watt]* and 10^{2} *[Watt]* covers most applications.

4.2.1. Definition of Power

Power is defined through the Poynting theorem. However, for practical measurements we define peak and average power as:

- Peak power: The maximum value of power during a pulse.
- Average power: Average power over a given time length, often many pulses.

Peak and average power are shown schematically in Figure 4.1. Some sources also define pulse power as the average power taken over a single pulse. Pulse power is calculated or measured by integrating over the pulse. Average power as defined here is calculated or measured from a relatively large number of pulses.

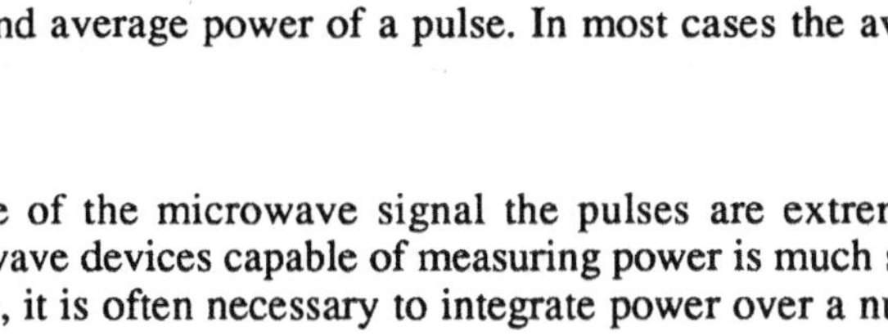

Figure 4.1. Peak and average power of a pulse. In most cases the average is taken over many pulses.

Since by nature of the microwave signal the pulses are extremely short, and the response of microwave devices capable of measuring power is much slower than say, the duration of a pulse, it is often necessary to integrate power over a number of pulses and calculate the peak power by dividing the average power by the duty cycle (that is frequency multiplied by the pulse width in time. Thus, one way of finding the peak power is to measure the average power and calculate

$$P_{peak} = \frac{P_{av}}{f_p t_p}$$

where f_p is the frequency of the waveform and t_p is the pulse width. This, of course, cannot be done for nonrepetitive pulses. Also, this particular formula is exact only for rectangular pulses but can be used for other pulses as well as an approximate expression.

4.2.2. Methods of Measurement

In most cases, measurements are made on the basis of absorption of power by an absorbing device and measurement of the response of the device. Either all the power is absorbed or part of the power is sampled. Absorption of power changes the temperature of a device, its resistance, or some other measurable property. This property is then measured, essentially by non-microwave means. There are many variations that can be used but this principle is widely followed. There are also methods of direct measurement of power as well as direct conversion methods such as the thermoelectric effect which converts power directly into a potential difference. Some of these methods are discussed next.

4.2.3. Thermo-Resistive Methods

In a thermo-resistive (or bolometric) measurement, power is absorbed in a thermistor, a baretter or a thin-film resistor. The absorbed power raises the temperature of the device. Its resistance, which either increases (as for baretters) or decreases (for thermistors) is then used as power indicator. A basic measuring circuit is shown in Figure 4.2. The indication can be measured directly on the meter or, the bridge may be balanced and the *dc* power required to balance the bridge is measured. Normally the latter method is used.

Thermistors are often used as sensors in power meters since they are simple, rugged, inexpensive, and relatively insensitive to overload. They require compensating circuitry to compensate for temperature variations. A second thermistor is often used, most often close to the measuring thermistor but isolated from the microwave power to maintain a reference reading and compensate for common-mode temperature variations. Thermistors, housed in standard mounts are available. These mounts allow connection of thermistors to waveguides for measurement of power in the waveguide. Because thermistors have relatively large negative temperature coefficients, they often require matching to avoid reflections. Attenuators are also sometimes employed to extend the useful range of measurement in the high power range. With proper design, matching and compensation, thermistors make excellent, accurate power sensors as is evident from their use as power sensing devices in commercial power meters. The response of thermistors is relatively slow (of the order of one second) therefore they should be viewed as average power sensors.

Baretters, because they employ a short, thin metallic wire, have a much faster response than thermistors and are somewhat more sensitive at low power. Their thermal coefficient is positive and is lower than that of thermistors. In all other respects these devices are similar to thermistors. Because of the small volume of material involved, baretters are quite sensitive to overload. As with thermistors, baretters are mounted in special mounts, or holders for installation in waveguides.

Thin film resistors are made by deposition of the resistive material on a dielectric such as glass or quartz. Their operation is again similar to those of the thermistor or baretter but, because of their planar nature and substrate tend to be slower in response. However, it is easier to match them to particular impedances, such as those of waveguides.

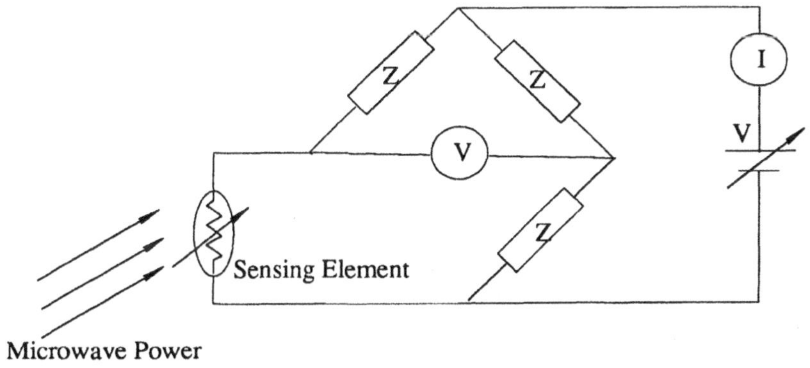

Figure 4.2. A thermo-resistive power measurement configuration.

4.2.4. Thermocouple Measurements

Thermocouples are used in two distinct types of measurements. In some applications, thermocouples or thermopiles are used to measure the temperature rise of the primary interaction material. This is the case, for example, in calorimeters. We will not discuss this application here primarily because it is not an appropriate measurement for field application. A second mode, one often used in power measurements is the direct absorption of power by the thermocouple junction. Although a variety of thermocouples are used, monolithic silicon thermocouples are particularly useful. They are compatible with other semiconductor devices and incorporate both the cold and hot junction on one, small substrate. This facilitates compensation for common mode temperature variations. Measurement of the thermoelectric effect is essentially one of measuring the potential on the thermocouple. As with thermistors, standard mounts for various applications exist. Thermocouples are used in power meters and, other than the method of measuring their effect, they are very similar to thermo-resistive devices, including in their response, power handling capabilities and stability.

Figure 4.3 shows a basic thermocouple measuring circuit. Initially, both junctions are at the same temperature and the reading of the device is zero. As microwave power is absorbed in junction *A* (hot junction) while junction *B* is shielded from the measured power, the temperature difference generates a potential difference, proportional to the power absorbed. Calibration of the device is normally done using simple *dc* or low frequency power substitution. In addition to the basic circuit, a variety of methods for signal conditioning, attenuation and amplification are sometimes used to extend measurement ranges.

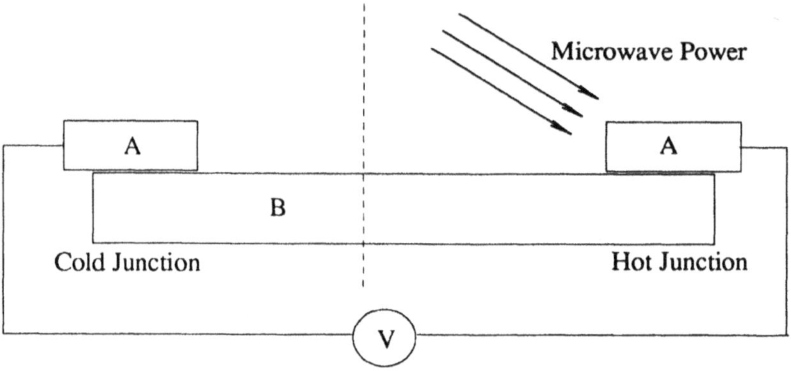

Figure 4.3. The thermocouple in a measuring circuit. The cold junction is isolated from the input power.

4.2.5. Measurement of Power Density

Sometimes the total power is of less interest than the power density. This is the case for example when a small sample is illuminated by an antenna and only a small fraction of the total power is linked to the sample. Similarly, in radiation exposure measurements, only the power density is normally specified. The main difference between measurement of power and power density is that in power density measurements, only a small fraction of the power is sampled. One obvious method of measuring flux is to use a small loop. The loop operates as an antenna and has a relatively narrow bandwidth. Dipole antennas of various forms can be used for this purpose. Perhaps the most common method for flux measurement is the use of an array of thermocouples on a substrate. The individual thermocouples are fabricated on a substrate with alternating hot and cold junctions, at intervals smaller than a quarter wavelength such that the net result is a thermopile with a number of thermocouples connected in series. The main advantage of this device is that it is nonresonant (unlike the small loop) and therefore broadband. For applications at very high frequencies, it is not normally possible to use loops for measurement but thermopiles, fabricated by integrated circuit means can always be made small enough to be below the required quarter wavelength intervals. The use of power substitution for measurement simplifies calibration and measurement. A thermopile array is shown schematically in Figure 4.4. A practical array is built with alternating hot and cold junctions on a substrate, with individual elements insulated from each other.

H	C	H	C		
C	H	C			
H	C				
C					
H					
C					

H=Hot, C=Cold

Figure 4.4. A thermopile array for measurement of power density. Junctions alternate and cold junctions are isolated from the microwave source.

4.3. Frequency Measurements

Frequency is one of the most fundamental parameters one often needs to measure. It is almost always required as part of the microwave measurement setup. In some cases, we must measure it because the frequency of the source is either not known exactly, or is not stable enough to assume it is known. In other instances, frequency is the measured quantity, as, for example, in frequency domain radar measurements. Sometimes the frequency is of interest because some other parameter is frequency dependent. Attenuation and impedance are two such parameters. However, often, frequency must be measured accurately as an indicator of the test. As an example, the dielectric constant of a material may be measured in a cavity resonator. In this case, the resonant frequency shift is a direct measure of the dielectric constant. Unless an accurate absolute measurement of frequency can be made, this type of method cannot be used.

There are three basic methods of measuring frequency. These are: electronic frequency counting, heterodyne or comparison methods and electromechanical wavelength measurements. It is also possible to measure frequency using standing wave techniques by measuring the distance between two peaks or two minima in the standing wave pattern (using a slotted line) but this is an inconvenient method, normally only applicable to waveguides, and almost always in a laboratory environment.

4.3.1. Frequency Counting

The main advantage of this method is that it is essentially electronic (other than the obvious need for a coupling mechanism). As with any frequency counter, the number of pulses over a fixed period are counted, giving a direct reading in, say, *MHz*. A schematic circuit is shown in Figure 4.5. The microwave field is sampled and a frequency divider is used in any combination required for the appropriate output. Digital, decade counters are used extensively because they are simple and inexpensive. The only limitation on this straightforward method is the frequency response of digital circuitry. Counters of this type exist with an upper limit of about *3 GHz*, while standard universal counters measure frequency to about *500 MHz*. Although perhaps not high enough for many microwave measurements, these are simple, accurate and relatively inexpensive instruments. Unlike other types of measurements (such as the wavemeters discussed below) the power needed for sampling is insignificant, and the sampling mechanism does not load the microwave source.

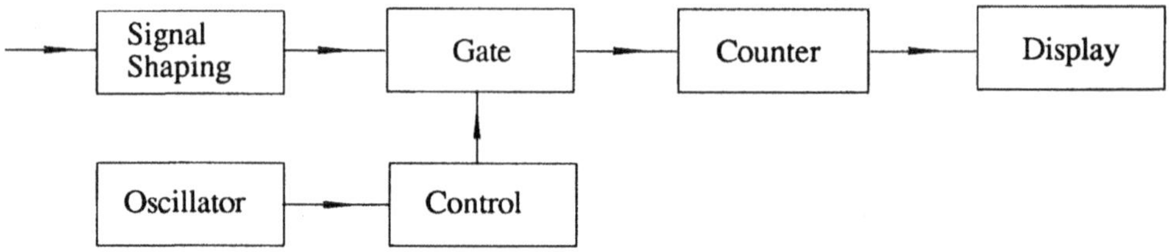

Figure 4.5. The basic frequency counter.

4.3.2. Heterodyne Frequency Measurement

Measurement at higher frequencies is possible if the frequency is first converted to a lower range, at which counting can be done using the direct method described above. The principle used is the same as in receivers: the input signal at high frequency is mixed with a convenient local oscillator signal at a fixed frequency. The signal out of the mixer contains two parts: a low frequency part that is the difference between the incoming signal frequency and the oscillator frequency and a high frequency signal that is the sum of the two. By appropriate filtering, the high frequency part is filtered out and only the low frequency component is measured. The low frequency is counted and from knowledge of the reference signal, the input frequency is deduced. The basic diagram of this process is shown in Figure 4.6. Unfortunately this very simple process is seldom satisfactory. For a measurement over a wide range, the local oscillator must have variable, accurately known frequency. Also, if the difference between the two frequencies must be low (up to a few hundred megahertz at most), a very stable, microwave oscillator is needed. Practical frequency counters based on heterodyning use a fixed frequency local reference provided by an oven stabilized crystal oscillator. This allows a very stable reference frequency. This signal is then entered into a harmonic generator which generates a large number of harmonics of the original frequency. The harmonics are connected to the mixer and the difference is adjusted to an IF (intermediate frequency) value. This frequency is counted and by computation the output frequency is determined and then displayed. In practical measuring equipment, either the input signal is scanned at fixed frequency intervals, or the output from the harmonic generator is filtered for a given range, to provide a frequency difference in the range of at most a few hundred *MHz* which can then be counted directly. A simple schematic of this type of frequency meter is shown in Figure 4.7. Often the functions of harmonic frequency generation, selective filtering and local oscillator are controlled by a microprocessor.

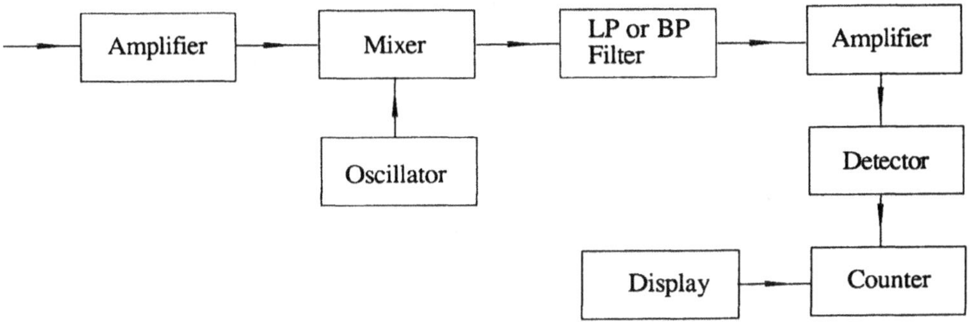

Figure 4.6. Basic heterodyne frequency meter.

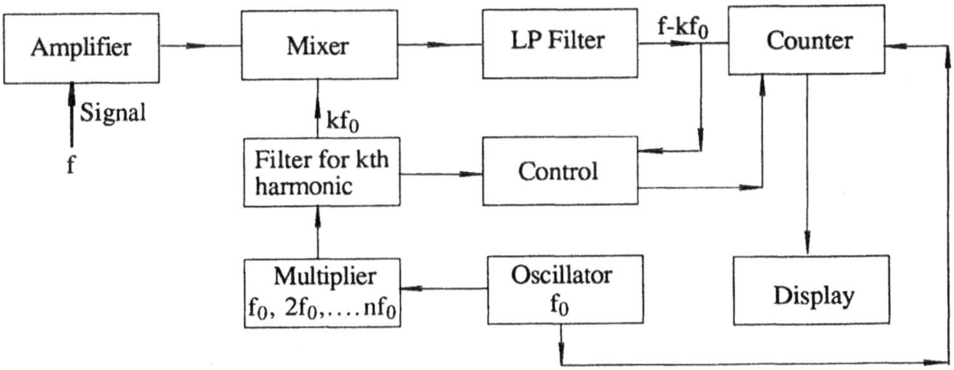

Figure 4.7. A practical heterodyne frequency meter.

4.3.3. Wavemeters

Mechanical wavemeters were in the past the main method of measuring frequency and, because they are simple and inexpensive, are still used. Their accuracy is, however, lower than that of electronic frequency meters and, because they operate "off the signal" they tend to load the input. Wavemeters are based on direct measurement of the length of a cavity resonator. There are three types of wavemeters: transmission, reaction and absorption wavemeters. Among these, the transmission wavemeter is most often used, primarily because it is connected in line with waveguides or transmission lines, simplifying connections, and does not load the waveguide other than at resonance.

In a transmission wavemeter, power flows through the cavity which, at all frequencies except resonance acts as a short circuit. At resonance, there is a certain amount of energy absorbed by the cavity. This can be read as a dip in the output power. The output power is then used as an indication of resonance. The resonant frequency of the cavity is adjusted mechanically and the frequency read directly off the adjusting screw.

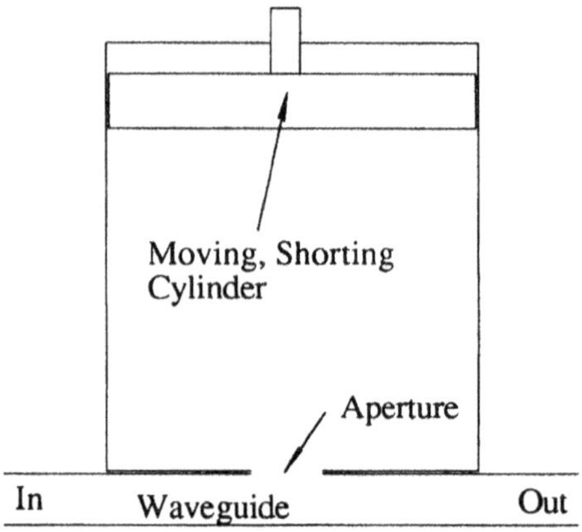

Figure 4.8a. Cylindrical wavemeter (reaction type) for measurements in waveguides.

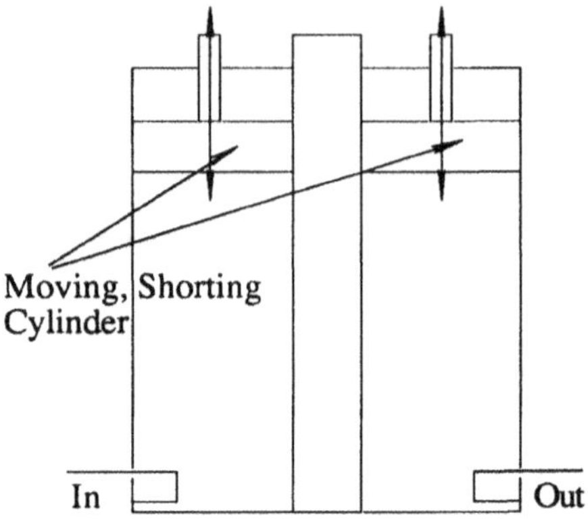

Figure 4.8b. Coaxial wavemeter (transmission type) for use with transmission lines.

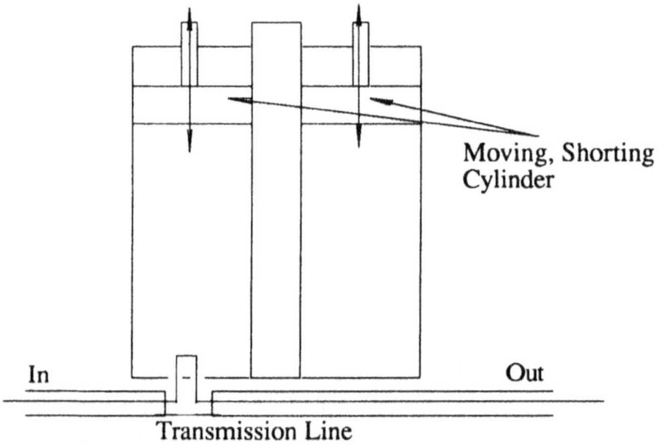

Figure 4.9. A reflection type wavemeter.

Two types of wavemeters are shown in Figure 4.8. One is a cylindrical cavity, the other a coaxial cavity. Coaxial wavemeters are more often used, primarily because they have a higher bandwidth (larger frequency band between consecutive modes). The cavity is normally a quarter wavelength cavity with the plunger being adjusted. Half wavelength cavities are also used with one of the cylindrical walls being adjusted. Coupling to the cavity is by loops, that excite the magnetic field inside the cavity. In a transmission mode, the coaxial line is terminated in the cavity and another line couples energy out (as in Figure 4.8a). Reaction type waveguides also exist, in which the transmission line is not interrupted but the cavity is coupled to the line by exposing the inner conductor in the cavity. In effect, the inner conductor serves as a coupling loop to couple energy into the cavity. This is shown schematically in Figure 4.9. Absorption type wavemeters also exist but are rarely used because they require more energy from the measuring environment. In all cases, the measurement is completed by adjusting the cavity to resonance at which point a dip in the signal is observed due to resonance.

Waveguide wavemeters are made of a cylindrical waveguide with an adjustable shorting wall, normally operating in the TE_{111} mode. Coupling to waveguides is by aperture. In all cases, the frequency range of a wavemeter is that between two consecutive modes and the frequency is read directly off the adjusting screw. Figure 4.8b shows a waveguide wavemeter of this type.

4.4. Q Measurements

The need to measure the Q-factor of microwave cavities stems from the fact that the Q-factor is a very sensitive measure of losses in the cavity and therefore of the contents of the cavity. The definition of Q was given in chapter 2, Eq. (2.133). Here we merely indicate that it is a ratio of the stored energy (time average energy stored in the cavity) to dissipated energy in one period of the signal (multiplied by 2π). The simplest method of measuring the quality factor of a cavity is to measure the bandwidth of the cavity from its half-power points. If the half power points are known, the Q-factor is given by:

$$Q = \frac{\omega_0}{\omega_2 - \omega_1}$$

where ω_0 is the resonant frequency, and ω_1 and ω_2 are the frequencies at the half-power points as shown in Figure 4.10. In principle, all that is required is scanning of the frequency and measuring the power in the cavity. However, this is not normally practical. Instead, either the transmission through the cavity may be measured (Figure 4.11a) or, for a single port cavity, the detector may be connected across the port in a shunt mode (Figure 4.11b). In the first case, the signal is passed through the cavity and the output, or transmitted signal detected. Since the Q-factor is a ratio of powers, the actual amplitude is not important but often, the signal is also amplified. The signal one expects is similar to the one shown in Figure 4.10. Alternatively, for single port cavities, the detector may be connected across the port and measures the energy transferred into the cavity. In this case, the cavity passes through a minimum at resonance and the power indicated by the detector describes a curve that is the inverse of the curve in Figure 4.10 but the measurement in all other respects is the same. The functions necessary for Q measurements are easily achieved using a spectrum analyzer since both the scan and measurement functions are available.

The Q-factor can also be measured from the properties of the cavity as viewed from the input, that is: the power reflected from the cavity can be measured through a standing wave measurement as shown in Figure 4.12. Since the cavity presents a pure resistance at resonance, the standing wave ratio at resonance is the ratio between the cavity impedance and the line impedance:

$$VSWR = \frac{Z_c}{Z_0}$$

if the cavity is overcoupled (if the cavity impedance Z_c is higher than Z_0) or

$$VSWR = \frac{Z_0}{Z_c}$$

if the cavity is undercoupled (if the cavity impedance Z_c is lower than Z_0).

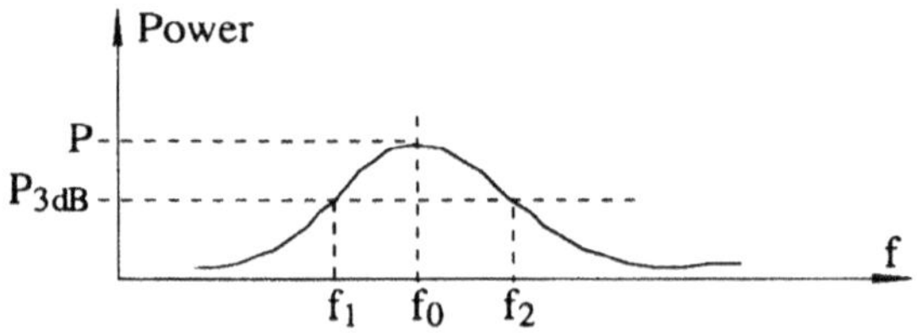

Figure 4.10. *3 dB* points and corresponding frequencies used to determine the quality factor of a cavity resonator.

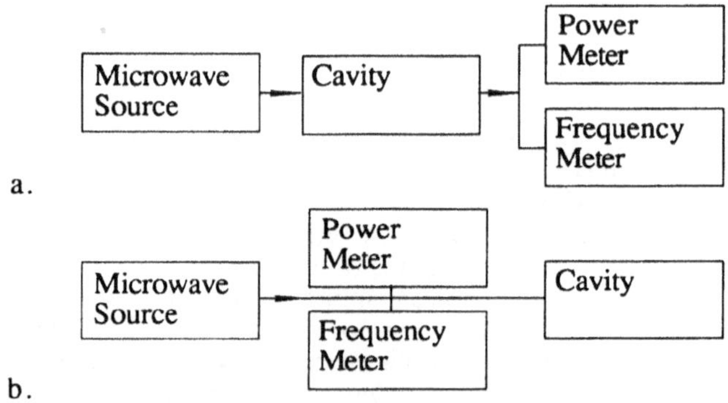

Figure 4.11. Two basic methods used to measure quality factor. a. Power transmitted through the cavity shows a peak at resonance. b. Power measured at the cavity port shows a dip at resonance.

The measurement proceeds by detecting resonance, then moving the detector to minimum *VSWR*. This indicates power being absorbed by the cavity and therefore the reading of the standing wave meter increases. Off resonance is indicated by a decrease in the standing wave meter reading. The measurement starts by tuning the cavity to resonance. The standing wave meter reading represents power at resonance. Detuning the cavity and measuring the frequency at which the standing wave meter reads *3dB* below that at resonance on each side of the resonant frequency gives the two *3dB* points. These frequencies are f_1 and f_2. The Q-factor is then the ratio between f at resonance and f_2-f_1. This represents the unloaded Q-factor, or Q_0. If the loaded cavity Q-factor is required, this is easily calculated from the unloaded Q-factor and the *VSWR* as

$$Q_l = \frac{Q_0}{1+VSWR}$$

and, based on the definition above, *VSWR* is always greater than *1*. This gives a loaded cavity Q smaller than Q_0.

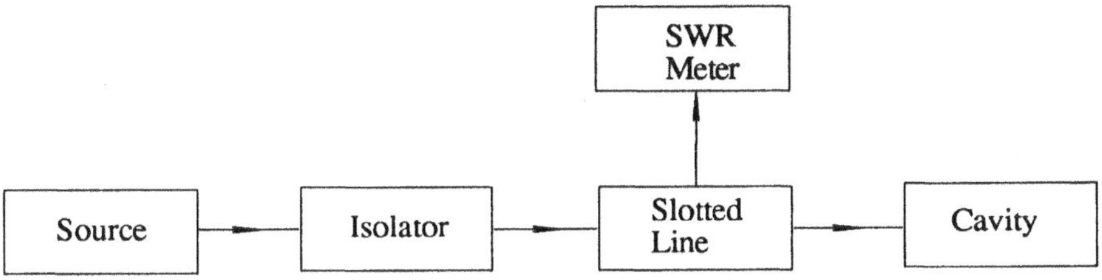

Figure 4.12. Standing wave method of measuring half bandwidth frequencies for Q-factor measurements.

4.5. Attenuation Measurements

Attenuation measurements are essentially comparative power measurements. The power going into the sample and the power out of the sample are measured. The difference is the power absorbed. In the microwave range there are two types of attenuation measurements that are of importance. The first relates to measurements of attenuation in a waveguide or transmission line. The equivalent testing environment would be a sample inserted in a waveguide. However, in most cases, this is a method for measuring the attenuation of a device that is matched to the input line and to the load. Under these conditions (see Figure 4.13). attenuation is defined as

$$L = 10 \log \frac{P_i}{P_o}$$

where P_i is the power measured without the material insertion while P_o is the power measured with the material inserted, after attenuation. This is also called the insertion loss since it can be characterized by measuring the power with and without the attenuator. In this sense, we are measuring the loss inserted by the attenuator under matched conditions, in an otherwise lossless system.

There are a number of other methods for measurement of attenuation in waveguides, including substitution methods (*rf*, *if* and *dc* methods) as well as measurements based on the measurement of the reflection coefficient. These will not be discussed here as they only apply to measurements in waveguides or transmission lines and, therefore, are used mostly for characterization of attenuators. This type of attenuation measurement is very important for characterization of devices but, for nondestructive evaluation, attenuation in free space rather than in a waveguide must be used.

For this we must use the basic definition of attenuation as in chapter 1. For a microwave test in free space, Figure 4.14 may be used. In this case, a wave is transmitted through the sample. Some of the energy is reflected at the surface, the rest transmitted and attenuated through the sample. If we can assume that there is no dispersion (that is: without the sample, all energy from horn *(1)* is transmitted to horn *(2)*), then we can measure the power on both sides of the sample and deduce the attenuation through the sample. This measurement is accurate if the reflections at the surface are small or can be taken into account. If not, then this can only be used as an estimate measurement of the attenuation. Since this is often used for testing,

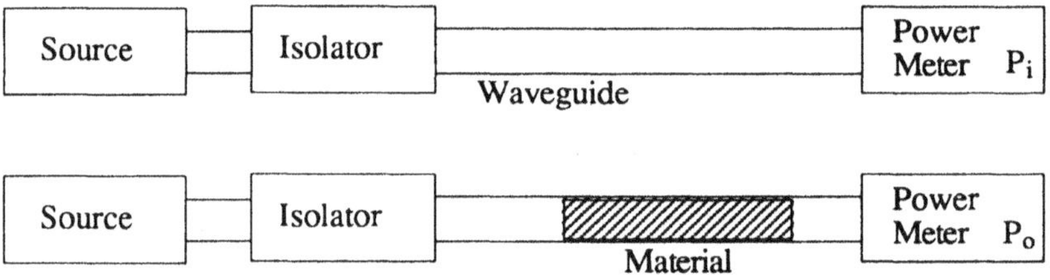

Figure 4.13. Attenuation measurement by comparative measurements of power in waveguides.

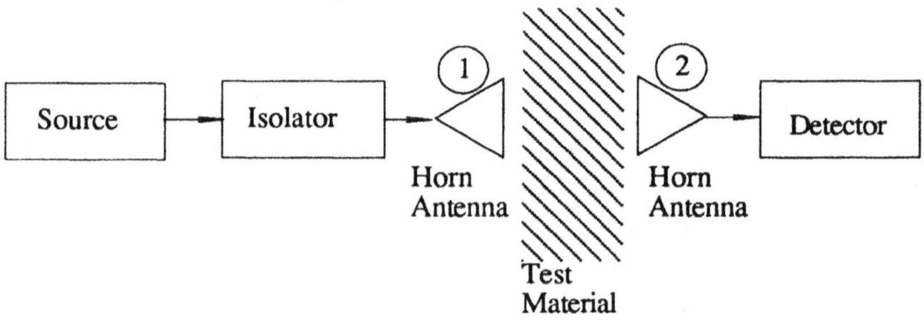

Figure 4.14. Attenuation measurement in free space.

4.6. Standing Wave Ratio and Reflection Coefficient Measurements

Standing waves are a very good indication of mismatch between load and line (in transmission lines), or, in a more general sense, between any two materials. Thus, if a wave propagates from one material to another, there will be a reflected wave and the interference between the forward and backward propagating waves results in a propagating wave and a standing wave. As the reflection coefficient increases, the standing wave component increases until, in the limit, we may have only standing waves. Thus, standing wave ratio, and standing wave pattern are basic measurements and are used extensively in testing. The most fundamental method of measuring standing waves is the slotted waveguide or the slotted transmission line. This is simply a line or waveguide with a thin slot through which a probe is inserted. This probe measures either the electric field or the magnetic field (voltage or current in transmission lines). The probe is attached to a sliding mechanism which allows the probe to be moved in a certain range. The detector is attached to the probe and slides with it. Figure 4.15 shows a basic microwave *SWR* measurement arrangement. The probe is moved and the reading meter shows the magnitude of the field. The ratio between maximum and minimum of the meter reading is the standing wave ratio (see chapter 2 for definitions). Figure 4.16 shows a typical standing wave pattern, with the standing wave ratio shown. This measurement is based on the definition

$$VSWR = \frac{E_{max}}{E_{min}} = \frac{V_{max}}{V_{min}}$$

The reading of E_{max}, E_{min} are taken directly from the meter and the *VSWR* is calculated. Some devices have an indicator in terms of *VSWR*, allowing direct measurement.

From the definition of standing wave ratio, the same measurement can also be used to calculate the reflection coefficient since, by definition,

$$VSWR = \frac{1 + |\Gamma|}{1 - |\Gamma|}$$

and, rewriting this, we get

$$|\Gamma| = \frac{VSWR - 1}{VSWR + 1}$$

We note also that, from the definitions in chapter 2, the distance between any two maxima on the line is equal to one half wavelength. Thus, the same method can be used to measure the frequency of the wave. This method of frequency (or wavelength) measurement is not normally considered as a basic method because it cannot be used for frequency measurement if the reflection coefficient is zero. However, because it is simple, it is often used. When theis is used, the normal method is to measure the distance between as many minima (or maxima) as possible to reduce errors.

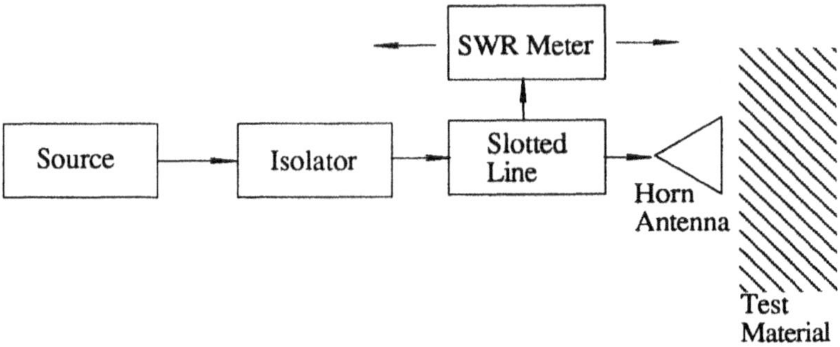

Figure 4.15. Standing wave measurements.

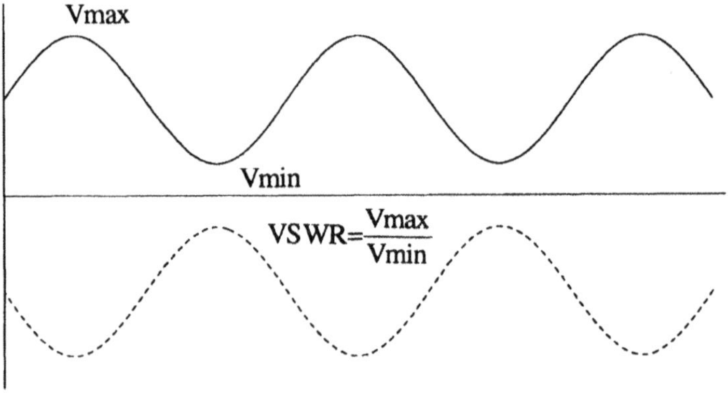

Figure 4.16. Standing wave pattern.

There are a number of other methods for measurement of standing wave ratio and reflection coefficient, including resonant methods, shorted waveguides and transmission lines but we will only mention here one more method based on measurement of the reflection coefficient. Assuming there are no losses, the incident and reflected waves can be measured directly be separating them through directional couplers. The basic principle is shown in Figure 4.17 in which two directional couplers are used. We will call this a bi-directional coupler. One leg of the coupler measures the amplitude of the incident wave, the second, of the reflected wave. These waves are detected and the ratio between the two amplitudes is equal to the magnitude of the reflection coefficient. From $|\Gamma|$, the standing wave ratio is calculated as above. This method, often referred to as reflectometry, is fundamental in microwave measurements, and we will use this or one of its many variations for testing arrangements in chapters 6 through 8.

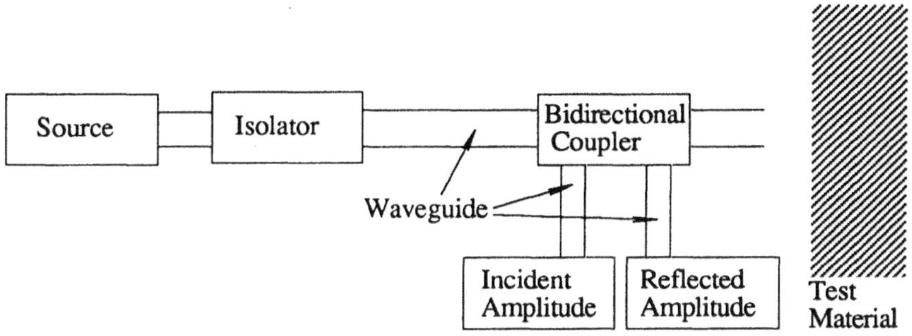

Figure 4.17. Measurement of reflection coefficient.

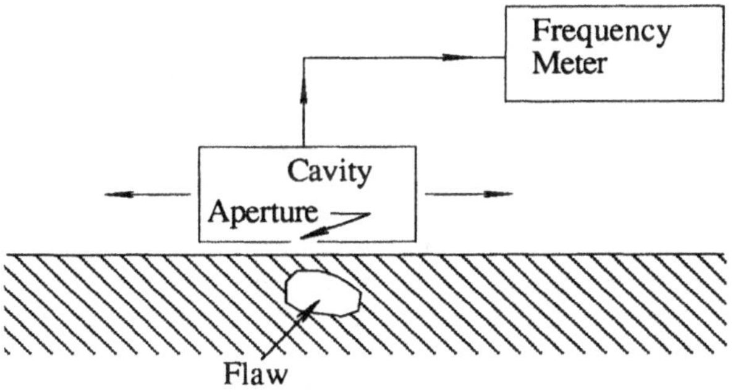

Figure 4.18. Use of small apertures to measure small discontinuities.

4.7. Microwave Microscopy

Microscopy, as with any other type of sources indicates a method of increasing resolution of a system so that smaller effects can be detected. In microwaves, this takes added significance because, even at the highest frequencies feasible, the wavelength is often longer than the artifact we wish to measure. As an example, at *30 GHz*, the wavelength is *10 mm*. Even at the upper range of the microwave region, it is probably not feasible to measure artifacts much smaller than 1 mm. If microwaves are to be used to test for cracks or small surface variations, a means of increasing the resolution must be used.

One principle that has been used for this purpose is a cavity perturbation technique (see bibliography). In principle, any perturbation in the cavity results in a shift in resonant frequency of the cavity. If the cavity has a high Q, this method can be very sensitive. Instead of including the test sample in the cavity, which for testing purposes is not always possible, an open cavity may be used or, the cavity may be coupled with the test sample through a small aperture. Figure 4.18 shows this principle. By doing so, the artifact, (in this case a small inclusion in a lossy or lossless dielectric) is illuminated by evanescent fields from the cavity (evanescent fields are fields below cutoff). Any change in the material will perturb the resonant frequency.

Depending on the sensitivity of the cavity to perturbations (i.e. its Q-factor), it is possible to detect very small perturbations in the cavity and, therefore very small defects. As a rule, the resolution is of the order of the size of the aperture. Since the aperture can be much smaller than the wavelength, the resolution is higher than what one would obtain with standard techniques. Thus the name microwave microscopy.

4.8. An example of Test Setup: The Balanced Microwave Interferometer

The following examples is given here to demonstrate some of the principles of setting up a test using many of the components required for most tests, including scanning of the test sample. The particular test chosen is the detection of edges of a lossless dielectric over a lossy dielectric. The edges are viewed here as defects but actual defects or discontinuities in the lossy dielectric could be measured just as easily. In addition, some of the aspects of data processing often needed in testing are discussed.

The basic circuit is shown in Figure 4.19. It consists of a single source used to illuminate the test sample and two receivers connected in a differential loop. Each receiving antenna is fed into a mixer. This provides a convenient method of amplification in the intermediate frequency range rather than in the microwave range. The two mixers are supplied by the same local oscillator and the path lengths are identical to ensure symmetry in the two propagation and amplification paths.

The differential amplifier amplifies the difference between the two signals received by the antenna. If the two antennas are symmetric about the source, the output of the differential amplifier is zero under all conditions that are symmetric. Any local change in the sample alters this balance and the output will deviate from zero. This can be caused either by a nonsymmetric location of a flaw or inclusion or by variations in material properties. The signal one should expect depends on the size and shape of the variation.

Operation of this type of arrangement consists of moving the antennas as a unit over the sample or, as would be more feasible in monitoring in a production environment, of moving the sample underneath the antenna cluster. The signal from the peak detector is recorded and then used for analysis of the material.

If detection is the purpose of the measurement, the signal can be used directly, without further processing.

Figure 4.20 shows a testing procedure where a piece of Plexiglass is located over a concrete slab. The measurement is achieved by first balancing the system without the Plexiglass and then moving the antennas over the sample. The signals obtained for two measurements are shown in Figure 4.21 taken at a distance of *26 cm* from the plate and Figure 4.22 taken at a distance of *65 cm* from the plate. These are digitized data plots of the signal as obtained from the peak detector. The distance shown is the scan distance, clearly showing the edges of the Plexiglass sample. The scans were taken by moving the whole antenna assembly over the plate. Alternatively, and in most cases the only possible method would be to move the sample.

The size of objects detectable by this method is of the order of the wavelength of the microwaves, in this case, *3 cm*

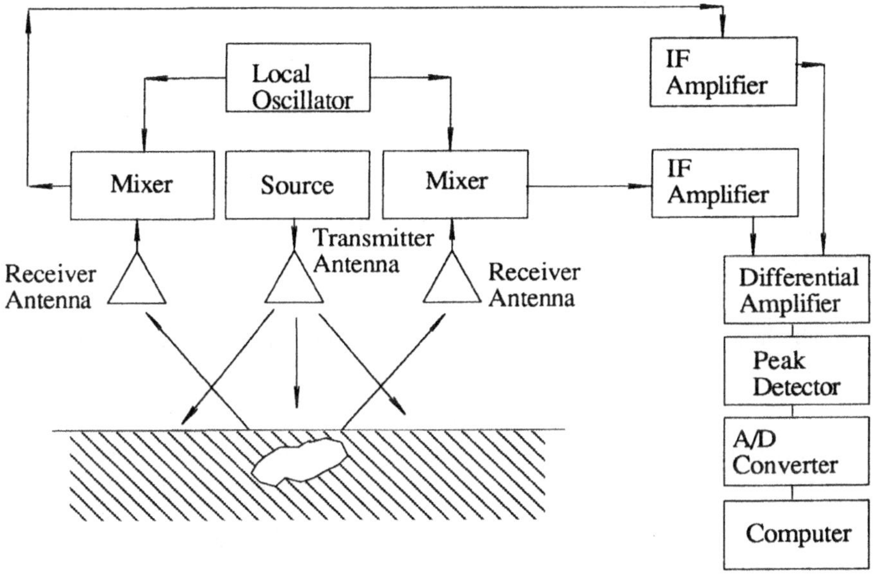

Figure 4.19. Test setup using a balanced microwave interferometer.

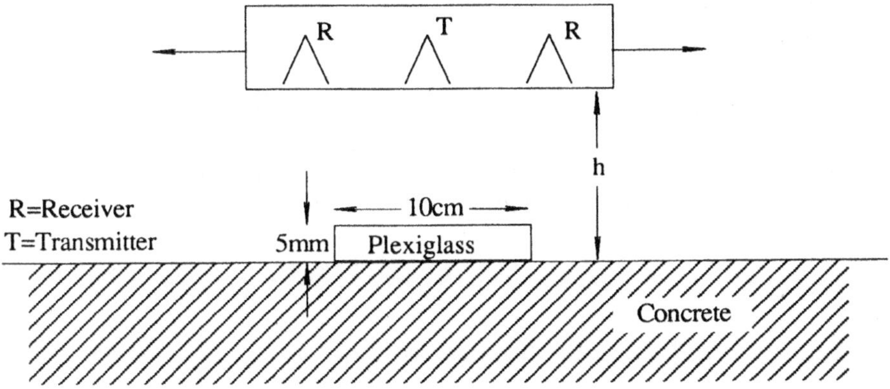

Figure 4.20. Method of scanning the test sample using the balanced interferometer and dimensions in test setup.

4.9. Bibliography

The subject of microwave measurements is so extensive that it is almost futile to cover the subject completely. The references given here are intended to complement the material that has not been covered in this chapter. The number of references cited has been kept to a minimum. While some sources discuss only specific measurements, there are sources dedicated exclusively to measurements. One of the classic books on the subject is Ginzton's book [1]. Notable, because of the extent of material, is the three volume book edited by Sucher and Fox [2]. Lavergheta [3] discusses a wide range of methods from a practical point of view with extensive discussion on equipment. Other sources are references [4] through [9].

Microwave microscopy is discussed by Husain and Ash [10]. General sources on measurements can also be found in Harvey [11] and Fink [12]

[1] E.L. Ginzton, "Microwave Measurements", McGraw Hill Book Company, Inc., NY, 1957.

[2] M. Sucher and J. Fox, Eds., Handbook of Microwave Measurements", Polytechnic Press, New York, 1963. (Vol. I, II, II).

[3] T. Laverghetta, "Handbook of Microwave Testing", Artech House, Boston,

[4] A.E. Bailey, "Microwave Measurements", IEE Electrical Measurement Series, Peter Peregrinus Ltd., London, 1985.

[5] H.E. Thomas, "Handbook of Microwave Techniques and Equipment", Prentice-Hall, Inc., Englewood Cliffs, NJ., 1972.

[6] P.I. Somlo and J.D. Hunter, "Microwave Impedance Measurements", IEE Electrical Measurement Series, Peter Peregrinus Ltd., London, 1985.

[7] F.L. Warner, "Microwave Attenuation Measurements", IEE Electrical Measurement Series, Peter Peregrinus Ltd., London, 1977.

[8]J.A. Lane, "Microwave Power Measurement", IEE Electrical Measurement Series, Peter Peregrinus Ltd., London, 1972.

[9] F.E. Gardiol, "Introduction to Microwaves", Artech House, 1984.

[10] Husain A. and E.A. Ash, "Microwave scanning microscopy for non-destructive testing", Proceedings of the European Microwave Conference, 1976, pp. 213-217.

[11] A.F. Harvey, "Microwave Engineering", Academic Press, London, 1963.

[12] D.G. Fink, ed., Electronics Engineer's Handbook, McGraw-Hill Book Company, New York, 1975.

Chapter 5.
Microwave Sources, Sensors, and Devices

5.1. Introduction

The generation of microwaves has been the main impediment to the development of microwave systems until the late 1930's, when the first practical microwave sources were developed. Naturally, these were essentially vacuum tubes but the principles involved had to be different than those used for low frequency vacuum tubes. The limitations of inter-element capacitance, for example, was one of the major problems in extending the operation of vacuum tubes to higher frequencies. The transit time of the electrons was of such orders of magnitude that it was not practical to operate beyond the lowest ranges of microwaves. The specially developed microwave tubes operate on velocity modulation of the electron beam and transition time does not play a role in these devices.

For most applications in *NDT*, low power sources are used. These are mostly semiconductor devices. In the higher power range, microwave tubes are still in use. In particular magnetrons and klystrons are still in wide use. Traveling wave tubes also exist but, because these operate at lower power, they have been largely replaced by semiconductor devices.

While there are a large number of microwave sources, the emphasis in this short overview is on those that can be used as sources of microwave energy useful for *NDT*. In particular low power solid state devices are discussed but, microwave tubes are also described. One reason for their inclusion is that these are widely available and in some cases, used exclusively. Some microwave tubes are used for other applications, accounting for their low prices. One such device is the magnetron which, because of its use in microwave ovens is widely available at very low costs. Magnetrons are particularly useful at higher powers but low power magnetrons also exist.

For obvious reasons, it is not possible to describe the whole spectrum of microwave sources. Representative devices within each type will be described to outline principles.

In addition to microwave sources and sensors, there is a large number of passive microwave devices that are very often used in conjunction with sources and sensors. In the testing environment, as in any microwave application, the required operation is obtained by a combination of active and passive circuits. As a rule, active circuits are used for generation, detection, modulation, amplification, etc., while passive devices are used for transmission of power, attenuation, splitting of power, as well as a variety of more complex functions such as isolation, directional coupling, bridge functions, etc.

It is obviously not possible to describe all passive devices in any detail. However, since many of these will be referred to in later sections on measurements, their functional properties will be described briefly in the sections that follow. Only the basic structures are discussed, leaving the details and other structures to the bibliography section.

5.2. Generation of Microwave Fields: Microwave Tubes

Historically, the first practical source was the magnetron, invented in 1921 and developed in the late 1930's for radar applications. This was followed by the klystron amplifier and then in the early 1960's by the traveling wave tube (*TWT*). After that, a large number of solid state devices were developed including diodes, transistors and field-effect transistors (*FET*).

5.2.1. The Magnetron and M Tubes

The magnetron is one of the so called crossed field tubes in which a *dc* magnetic field is applied perpendicular to the electric field used to accelerate the electron beam. For this reason, the electron beam takes curved paths rather than the familiar linear motion in other vacuum tubes.

In a normal tube, the electric field between anode and cathode accelerates the electrons emitted by the cathode. In a crossed field device the magnetic field also affects the electrons forcing them to move in circular paths of a radius that depends on their velocity. The essentials of the magnetron are shown in Figure 5.1. A cylindrical cathode emits electrons which are accelerated towards a concentric, cylindrical anode by the electric field between the two electrodes. A magnetic field is applied perpendicular to the plane as shown (therefore the crossed field name). The electrons are forced to move in a circle of radius r where

$$H^2 = \frac{8mV}{er} \qquad or \qquad r = \frac{8mV}{eH^2}$$

where V is the accelerating voltage (*dc*) and H is the magnetic field intensity. The magnetic field can be supplied by a permanent magnet or an electromagnet although permanent magnets are more common, especially at lower power levels.

If the electric and magnetic fields between the electrodes are properly applied and adjusted, the electrons may hit the anode (also called the interaction structure), graze the interaction structure surface, or return to the cathode. For an electron to reach the anode, its velocity must be smaller than the cyclotron velocity (i.e. the velocity needed to move in a constant radius circle). Under the conditions above this can only happen if the accelerating *dc* voltage is low enough (assuming H to be constant). If an *rf* field is applied to the anode, the interaction is more complex. The *rf* field either accelerates or decelerates the electrons. Those electrons entering the field during the negative half cycle will be slowed down, giving some of their energy to the *rf* field. The slower electrons now move in larger circles, eventually hitting the anode. The faster electrons, those entering the field during the positive half cycle, will be returned to the cathode since they are forced to move in tighter circles that do not reach the anode. The energy given up by the electrons increases the energy in the *rf* field. The question is where does the *rf* field come from. Because this device is an oscillator the field must be internal (i.e. an external *rf* field cannot be supplied). The answer is obvious: it is generated by the induced currents in the anode due to movement of the electrons. The frequency of the field is the cyclotron frequency.

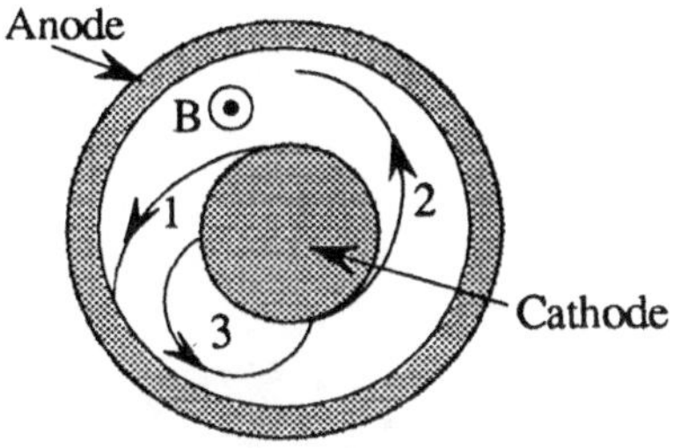

Figure 5.1. Principle of the magnetron.

This type of magnetron, while certainly practical (and appropriately called a cyclotron magnetron) is quite inefficient and can only be used at low power. The main reason for this inefficiency is the accelerated electrons that return to the cathode and generate heat in the process.

A more practical and more efficient device employs cavity resonators around the periphery of the interaction structure as shown in Figure 5.2.

The cyclotron frequency depends on the magnetic field

$$\omega_c = \frac{eB}{m}$$

The electrons make a complete revolution in time T

$$T = \frac{2\pi}{\omega_c} = \frac{2\pi m}{\varepsilon B}$$

Now, if there are N cavities, the phase shift in each cavity is:

$$\phi_n = \frac{2\pi m}{N}$$

where n is the nth mode of oscillation.

For proper operation the voltage in the structure is adjusted such that this phase adds up to a multiple of 2π. The normal mode of operation for a magnetron is the π mode in which the phase shift per cavity is equal to π. For the magnetron to operate, the electrons must be decelerated as they pass through each cavity. If they travel a distance L between adjacent cavities, their phase velocity must be

$$\beta = \frac{2\pi n}{N\Lambda}$$

β is also called the phase velocity of the slow wave (the wave that travels around the structure). This produces oscillations of frequency

$$\omega_c = \beta \frac{d\phi}{dt}$$

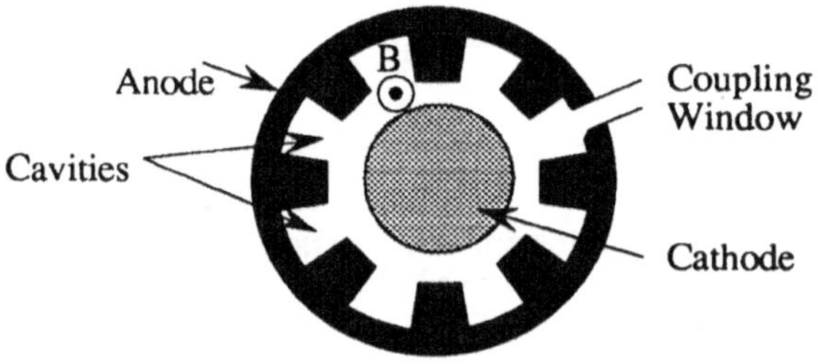

Figure 5.2. The multi-cavity magnetron.

There are many variations in the structure of magnetrons, including linear magnetrons, but the operation is essentially as described above.

The power generated by the magnetron is proportional to the number of cavity resonators, voltage across the resonator gap, frequency, and inversely proportional to the cavity quality factor. This means that the highest efficiency is obtained under high load conditions (low Q).

The power is coupled out of the magnetron by an aperture in one of the cavity resonator. A dielectric window is employed to allow transmission of power and keep the vacuum of the tube.

The magnetron is simple, efficient (up to *80%* efficiency, making it the most efficient of microwave sources). It is also inexpensive and used extensively in microwave ovens. Unlike many other microwave sources it is essentially an oscillator although crossed field amplifiers also exist. The magnetron operates well at high peak powers but they tend to be noisy and their frequency stability is not very good. However, they offer a simple, inexpensive source in many application. In addition, they are probably the smallest tubes for a given power output.

There are a number of variations of the basic magnetron design. These include the coaxial magnetron, the inverted coaxial magnetron (in which the anode and cathode are interchanged), the Ditter magnetron and linear magnetron. In addition there are special purpose magnetrons for use in radar (for example the frequency agile magnetron, employed for fast frequency scanning purposes)

Tuning of a magnetron to the correct frequency is done in one of several methods. The simplest is mechanical tuning of the cavities. By inserting mechanical elements in the cavity, either a capacitive or inductive effect can be employed to tune the cavity. This is identical to the tuning of any resonant circuit. However, in a cavity, an element at a location of high electric field causes a capacitive variation and an element at a location of high magnetic field causes an inductive variation. Tuning can also be achieved by varying the voltage across the electrodes. The tuning mechanism is part of the magnetron structure.

5.3. Microwave Liner-Beam Tubes (O-Tubes)

5.3.1. The Klystron

The klystron has been developed shortly after the magnetron and since the early 1940's has become perhaps the most important microwave device, especially where high power, stable, efficient operation are required. While it can reach peak powers of over *30 MW*, klystrons in the range of a few hundred *milliWatts* also exist. Efficiencies of *50* to *60%* are possible. The klystron is essentially a microwave amplifier but, by proper positive feedback can be operated as an oscillator. The reflex klystron is a special device that is designed to operate as an oscillator by internal feedback.

The basic klystron amplifier is shown in Figure 5.3. It consists of an electron gun, two resonant cavities and a focusing magnet.

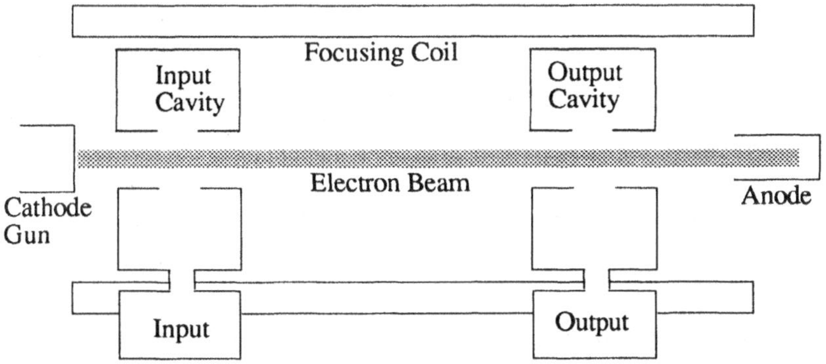

Figure 5.3. A two-cavity klystron.

The klystron amplifier is based on emission of electrons from the cathode as in any tube. The electrons are accelerated by the anode which is near the cathode as in an electron gun. The acceleration is high and from this point on the electrons travel at essentially this velocity. The beam is focused by a magnetic field. The field is in the same direction as the electric field and is generated by a coil or magnet, coaxial with the beam axis. Beyond the cathode assembly, which can accelerate the electrons to very high velocities, the electrons enter into a cavity known as the input cavity. The *rf* input signal to be amplified is entered into this cavity by means of a coaxial transmission line or a waveguide and a dielectric window. Electrons that enter into the input cavity during the positive half cycle are accelerated while those that enter during the negative half cycle are decelerated. The electrons exiting the cavity are at different velocities and tend to bunch up in some areas and deplete other areas. For this reason, the input cavity is also called a bunching cavity. The electron beam travels along a drift zone, allowing the fast electrons to catch up with the slower electrons and form the bunching effect. The electrons now enter the second cavity. The velocity variations of the electrons constitutes a velocity modulation. The electron density varies with time, which in turns generates currents in the cavity wall. These currents are the source of the microwave power. The electrons are decelerated in the output cavity, giving up their energy to the output cavity (catcher cavity). The output is transferred out by an aperture in the output cavity and a proper dielectric window. The slowed down electrons exit the cavity and are absorbed in a structure called beam collector.

More complex klystrons employ additional resonant cavities in the drift zone. These are self contained and are driven by the beam itself. Each cavity can add of the order of *20 dB* to the gain of the klystron. By use of multiple cavities *40* to *60 dB* gain is possible. Figure 5.4 shows a klystron with two cavities in the drift zone.

Klystrons use a solenoidal coil for focusing (Figure 5.3) at high powers or permanent magnets at lower power, although, some high power pulsed klystrons use permanent magnets for focusing. Tuning of the klystron to the required frequency is done by essentially mechanical means of changing the cavity resonant frequency. Each cavity is tuned independently.

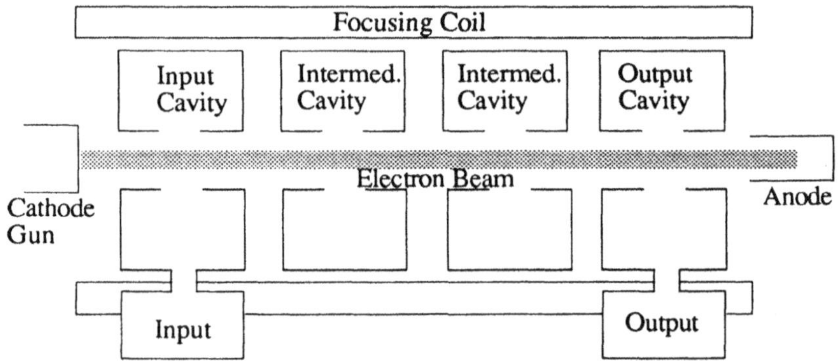

Figure 5.4. A klystron with two cavities in the drift zone, in addition to the input and output cavities.

To use the klystron as a microwave source or oscillator, requires the incorporation of a positive feedback path. This is done by coupling the output and input cavities and disposing with the input connection as shown in Figure 5.5.

An alternative is to couple all cavities, including the intermediate cavities into a single output cavity which can then be tuned mechanically as one structure instead of tuning each cavity separately. This is normally referred to as an extended interaction klystron.

A klystron designed specifically as an oscillator is the reflex klystron. This is shown in Figure 5.6.

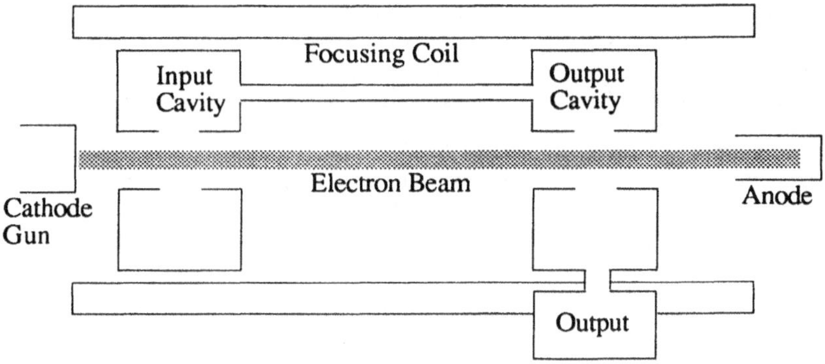

Figure 5.5. A klystron oscillator.

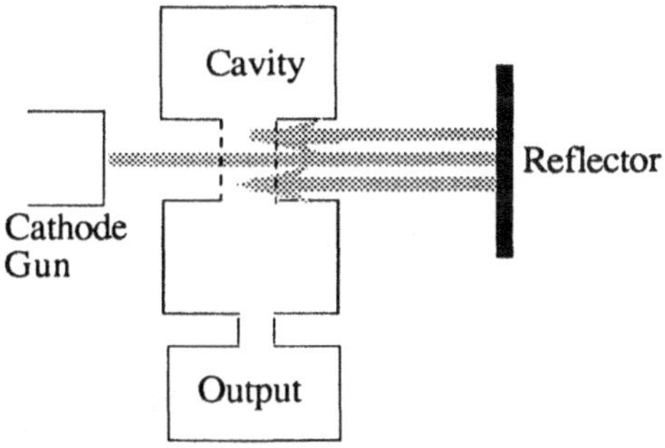

Figure 5.6. A reflex klystron oscillator.

The reflex klystron is a single cavity device (only an output cavity is used). The operation is identical to the klystron amplifier but part of the electrons are repelled back by the reflector and enter back into the cavity where the energy interchange takes place. This is normally a low power oscillator that can be modulated by connecting the modulating voltage to the cavity. The output power is coupled out by a transmission line (coaxial) or a waveguide.

Klystrons have a relatively narrow bandwidth as one would expect with tuned cavities.

5.3.2. The Traveling Wave Tube (TWT)

The basic structure of the *TWT* is similar to the klystron in most respects but the cavity resonators are absent. It is a nonresonant device and for that reason it tends to have a much wider bandwidth. Instead of the input, output and intermediate cavities, the input and output in a *TWT* are coupled together through a continuous interaction zone. One particularly useful method of coupling is by use of a helix. This is shown in Figure 5.7. The name *TWT* comes from the fact that a wave travels along the tube at the speed of the electron beam. The beam is modulated by the input signal creating the bunching effect as in the klystron. However, since the helix is continuous, those electrons that enter the interaction zone during the positive half cycle are continuously accelerated while those that enter in the negative half cycle are continuously decelerated. As the fast electrons catch up with the slow electrons the bunching effect occurs. The beam induces currents in the interaction structure, causing amplification.

An alternative structure to the helical *TWT* is shown in Figure 5.8. The helix is replaced by a series of cavities, all coupled together through apertures in their common walls. Because the metallic structure is more massive than for the helical *TWT*, the power capability of the coupled cavity *TWT* is higher but, because of the use of resonant structures, the bandwidth is lower.

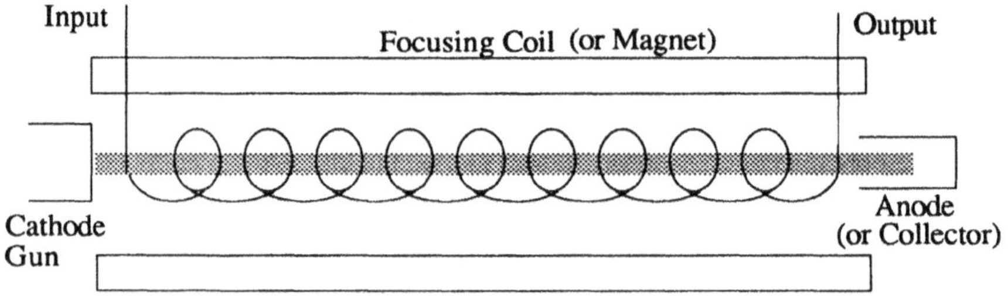

Figure 5.7. The helical Traveling Wave Tube.

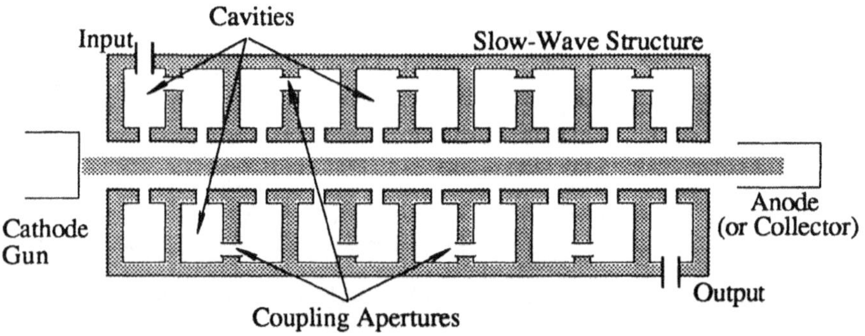

Figure 5.8. A coupled cavity *TWT*.

In general *TWT's* are used where high bandwidth is required. The tubes have low efficiency (lower than any other microwave tube) and can only operate at moderate power.

As with any other amplifier, an oscillator can be built by adding a proper positive feedback network. However, in the microwave region, this network is essentially a coupling from the output (almost always a cavity) to the input by use of an aperture or any other guiding structure.

5.4. Solid State Microwave Devices

Solid state devices have all but replaced microwave tubes for low power applications. As an example, most applications that previously used *TWT's* now use solid state devices. In fact, some solid state devices like the crystal diode preceded some microwave tubes. The emphasis in this short review is on active, semiconductor devices. These are divided into three groups. The first includes Transfer Electron Devices *(TED)*. These are diodes based on the bulk effect in semiconductors. A well known example is the *Gunn* diode. The second group is the Avalanche Transit Time Devices and includes the *IMPATT*, *TRAPATT* and *BARITT* diodes. The third group are the microwave transistors. While some of these devices were developed specifically for microwave applications, some are essentially developments of low frequency devices, whose high frequency response has been improved through processing. The *BJT*, *JFET*, *MESFET*, *HFET* and other transistors as well as the tunnel diode find applications in many microwave systems. These are standard transistor devices that have been improved through processing and dimension control. In the overview below, devices like the *BJT* or the *FET* will not be discussed. Only those devices or aspects of operation that are peculiar to the microwave range will be outlined here.

5.4.1. The Tunnel Diode

The tunnel diode is a *pn* junction and properly belongs to the same group as the various transistors. The tunnel diode employs heavily doped *p* an *n* materials. Under normal conditions, the flow of carriers across the junction creates a barrier voltage. This barrier voltage prevents flow of carriers because of the energy levels required. However, because of quantum effects there is a final probability of carriers "tunneling" across the barrier even though normal current is not possible. (The name tunnel diode comes from the fact that under classical considerations the carriers cannot pass "over" the barrier but can pass "under" it). The probability of tunneling occuring increases with a decrease in the barrier width (junction width), doping of the materials, and the forward biasing potential. Figure 5.9 shows a tunnel diode with its forward biasing voltage and the *I-V* characteristics.

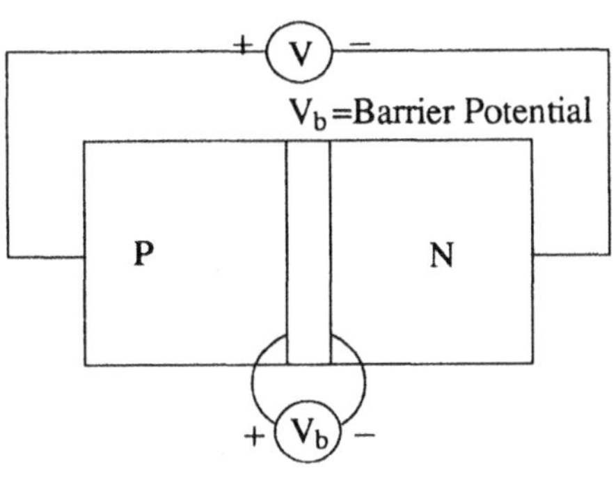

Figure 5.9a. Tunnel diode and biasing.

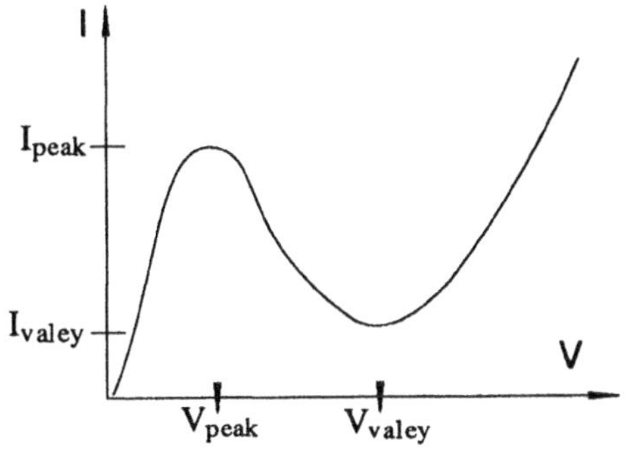

Figure 5.9b. Current-Voltage *(I-V)* characteristics of the Tunnel diode.

Initially, the only current in the diode is due to tunneling. This current peaks at a forward potential V_P, at which the maximum number of carriers can cross the barrier. Beyond this point the number of available vacancies on the other side of the barrier decreases and the current decreases to a minimum or valley at V_V. By further increasing the forward bias, the normal forward conduction of carriers takes over. The characteristic negative resistance of the diode (an increase in biasing voltage reduces the current through the diode) is the main characteristic of interest in the microwave region. It allows the diode to function as an amplifier or oscillator.

5.4.2. The Gunn Diode

The *Gunn* diode is made of a single piece of material (e.g. *n* type Gallium Arsenide *(GaAs)*, but other materials exhibit the same effect as for example Indium Phosphide *(InP)*), therefore there is no junction and the effect is that of a bulk effect. In fact, the diode can be properly seen as a resistor. The *Gunn* effect exists only in compound materials like *GaAs* but not in single materials like silicon. The effect is due to nonuniform electric fields within the material. Therefore, an electron moving from one electrode to another (Figure 5.10) will undergo acceleration or deceleration or, what more properly called change in mobility. This causes the negative resistance effect seen in Figure 5.10b. The effect of this velocity change is to generate electron domains which, after building up, they transit through the material and the process is repeated. This periodic transit of carriers is at a frequency that depends on the width of the material. The frequency of oscillation therefore depends on the width of the diode. In simple terms, this is the lower frequency at which a *Gunn* diode is capable of oscillation. The diode can also operate at higher frequencies by placing it in a tuned microwave cavity. Now the oscillations are determined by the cavity's resonant frequency. The higher frequency is explained by the fact that now the electrons are not allowed to "wait" until a domain has built up. More accurately, a domain that has formed, is dissipated (quenched) before it reaches the electrode by reversal of the *rf* polarity, if the frequency of the *rf* field is higher than the transit frequency. As the *rf* field switches polarity again, a new domain is formed. Thus, the formation of domains is controlled by the *rf* field and oscillations occur at the resonant frequency of the cavity. This frequency can be significantly higher than the natural frequency of the material.

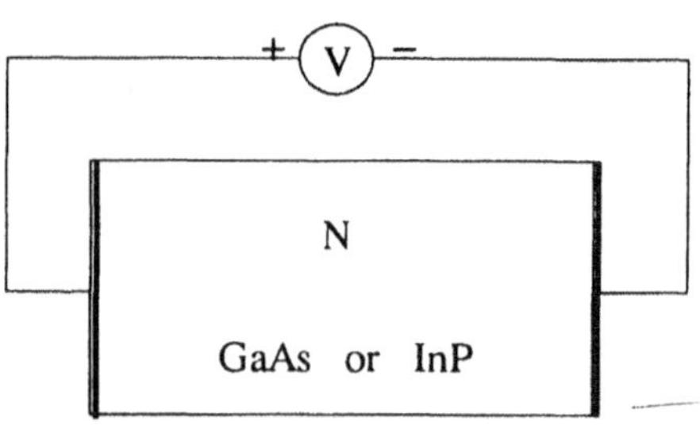

Figure 5.10a. *Gunn* diode and biasing.

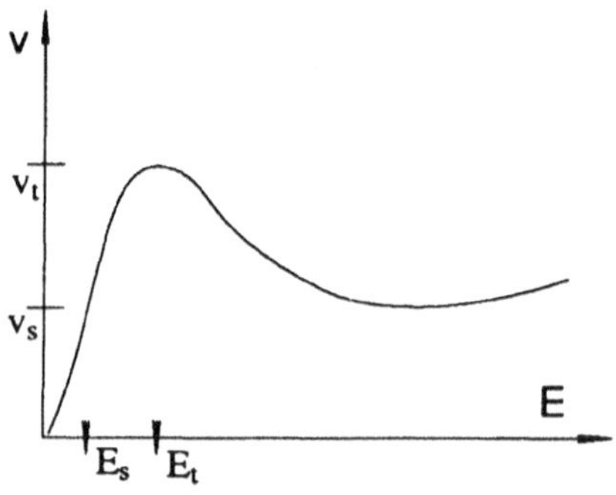

Figure 5.10b. *Gunn* diode carrier velocity versus electric field..

Another mode in which a *Gunn* diode can be operated is in the Limited Space Charge Accumulation *(LSA)* domain. in this mode, the diode is biased such that for most of the signal cycle, the diode cannot conduct and, therefore there is very little space-charge accumulation. This occurs at higher operating frequencies. This mode is quite useful especially for higher frequencies as it allows efficient operation, considerably beyond the transit frequency mode. While it is possible to increase the frequency by decreasing the size of the diode but this reduces power handling capabilities of the device. The operation in *LSA* mode is similar to that in the quenched mode but the biasing is much higher. *Gunn* diodes designed to operate in this mode are normally called *LSA* diodes. They are capable of handling more power because the frequency is independent of transition time and therefore the diode can be larger. An oscillator using this mode consists of a tuned cavity and a diode, with the cavity defining the frequency of oscillation.

The *Gunn* diode can also be used for amplification at frequencies near the transit frequency, by utilizing the negative resistance characteristics of the diode. The only requirement is that the frequency be low enough and the device impedance have a negative real component.

Gunn diodes can operate at power levels of about *1W* or lower *(CW)* with perhaps a few hundred *Watts* in pulsed mode and at frequencies of up to about *30 GHz*. The theoretical limit is much higher both in frequency and power. Generation of microwaves by these devices offers all the advantages of solid state devices but they also tend to be sensitive to temperature and load and tend to loose efficiency at higher frequencies. Compared to tubes they also tend to be noisy. However, they are used extensively for low power applications

5.4.3. The Avalanche Diode

Unlike the tunnel diode or the *Gunn* diode, the avalanche diode depends on a "controlled breakdown" across a reverse biased *pn* junction to provide the negative-resistance characteristic. Two types of avalanche diodes are used for microwave applications. One is the *IMP*act ionization *A*valanche *T*ransit-*T*ime diode *(IMPATT)* and the other is the *TRA*pped *P*lasma *A*valanche *T*riggered *T*ransit diode *(TRAPATT)*.

5.4.3a. The *IMPATT* Diode

The construction and *I-V* characteristic of an *IMPATT* diode are shown in Figure 5.11. This figure shows one type of diode, sometimes called an abrupt *pn* junction diode. The reverse bias of the *pn* junction prevents flow of carriers and the current through the diode is essentially zero. A depletion zone exists in the junction. As the reverse bias is increased, the avalanche voltage is reached. At this point, the attraction of the positive electrode on electrons in the *p* type material is high enough to force some of the carriers across the junction. Since the voltage is high, these electrons will move and collide with atoms in their path. An avalanche effect occurs whereby more electron-hole pairs are released moving in opposite directions and therefore creating an avalanche current through the junction. In the *IMPATT* diode, the biasing is such that it extends into the p^+ and n^+ (heavily doped *p* and *n* regions. The depletion zone consists of the *n* material. If an *IMPATT* diode is properly biased, and then an *rf* field applied to it, the *rf* field will force the diode into avalanche during part of the cycle, causing a current pulse through the diode. In the rest of the cycle, no avalanching occurs. This pulsed conduction occurs with a current that is 180° out of phase and therefore, the diode exhibits negative resistance characteristics. However, unlike other diodes this is a dynamic (or *ac*) negative resistance.

The operation of the diode as an oscillator depends on the diode being mounted in a resonant cavity, to oscillate at a resonant frequency defined by the diode characteristics and the cavity.

IMPATT diodes are very useful both as oscillators and as amplifiers. They can operate at frequencies throughout the microwave range and at powers of a few Watts. Efficiencies of *10%* to *30%* are possible. *CW* power of about *10W* at efficiencies of *30%* are possible, making these perhaps the highest output microwave solid state devices. They are however relatively noisy. *IMPATT* diodes are made of Silicon or *GaAs* but other materials can also be used.

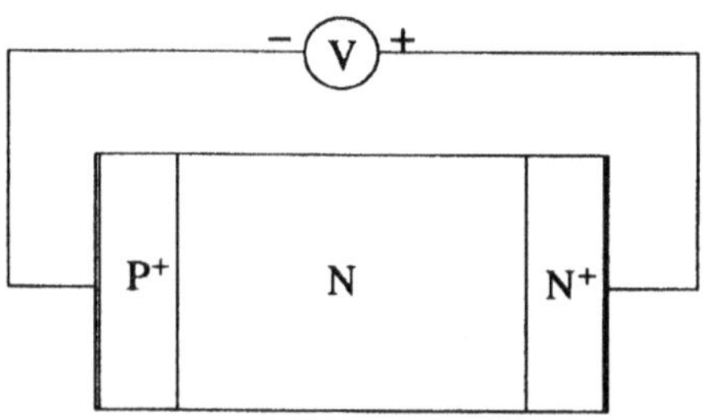

Figure 5.11a. Construction of the *IMPATT* diode.

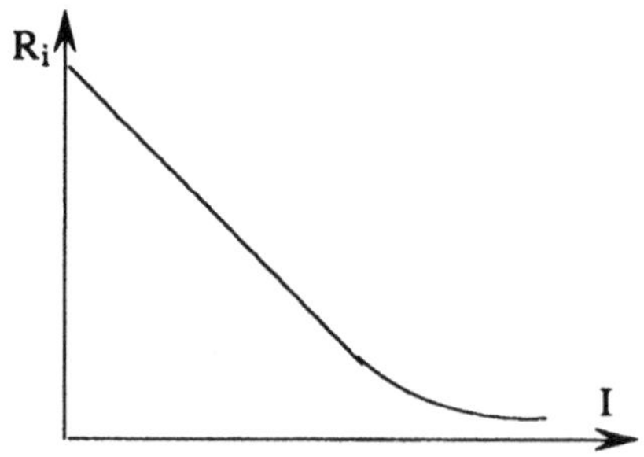

Figure 5.11b. Negative resistance characteristics of the *IMPATT* diode.

5.4.3b. The *TRAPATT* Diode

The *TRAPATT* diode is similar to the *IMPATT* diode in construction and biasing but the diode is biased to currents much higher than in *IMPATT* diodes. The high fields produce a plasma region in the depletion zone. The carriers in this zone become trapped because once avalanche occurs, the field intensity in the plasma zone decreases, increasing the transit time of the carriers. After the plasma has been removed, the process can be repeated. Because of reliance on transit time, *TRAPATT* diodes operate at lower frequencies than *IMPATT* diodes but they are capable of much larger peak power. Efficiency is also higher and can exceed *60%*. *TRAPATT* diodes are useful for pulsed operation at medium power (up to a few *kW*) and relatively low frequencies (a few hundred *MHz* to a few *GHz*).

5.4.4. The BARITT Diode

The *BARITT* (*BAR*rier *I*njected *Transit T*ime) diode is similar in structure to the *IMPATT* diode but the highly doped regions are replaced by Schotky barrier contacts as shown in Figure 5.12a. The main difference is that the operation is based on minority carriers injected across the Schotky barrier. The *I-V* (current-voltage) characteristic of a *BARITT* diode in Figure 5.12b. This shows a rapid increase in minority carriers in the forward biased diode until saturation of carrier formation is reached.

BARITT diodes operate in the low frequency, low power, continuous wave microwave region. They are less noisy than *IMPATT* diodes.

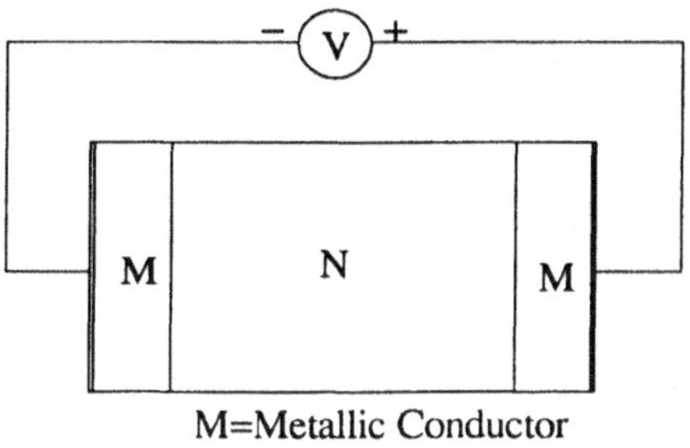

Figure 5.12a. Schematic of a *BARITT* diode.

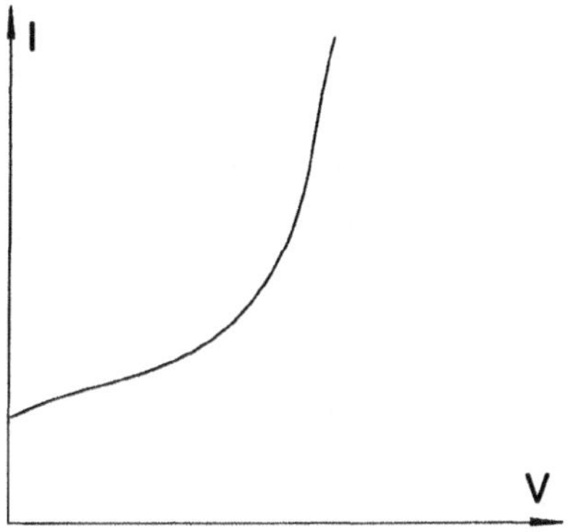

Figure 5.12b. *I-V* characteristics of the *BARITT* diode.

5.4.5. The PIN Diode

PIN diodes are built as *pn* diodes with an additional domain between the *p* and *n* materials made of intrinsic material, with little or no doping (Figure 5.13a). Thus, under normal conditions, the intrinsic layer acts as a very high resistance. When reversely biased, the resistance of the device is high. When forwardly biased, the carriers from both the *p* and *n* areas make the intrinsic layer conductive and the total resistance is reduced in proportion to the current through the diode. The resistance of the device changes over a wide range. The *I-V* characteristic of the diode is shown in Figure 5.13b. This characteristic is only applicable to high frequency operation. At low frequencies, the diode behaves similar to a rectifier. It can be used either as a variable resistor in variable attenuators or as a basic switching device. Other uses include phase shifters and modulators.

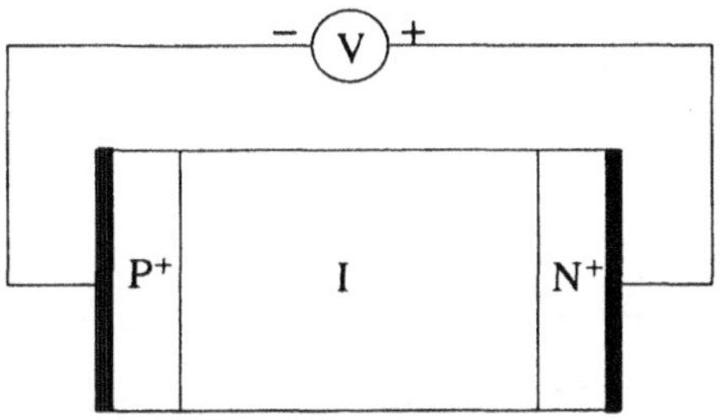

Figure 5.13a. The *PIN* diode.

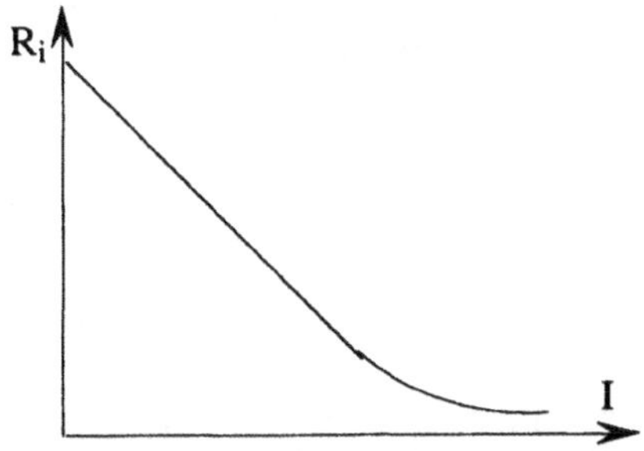

Figure 5.13b. *PIN* diode *I-V* characteristics.

5.5. Microwave Circuits

In principles, microwave circuits are not different than low frequency circuits. We can identify the same circuit relations and the same types of circuit elements. Thus an inductive element behaves as an inductor, but, its physical construction is quite different. Perhaps the easiest way to view the difference is to look at the cavity resonator. A cavity resonator is equivalent to an *LC* resonant circuit. A lossy cavity would then be equivalent to an *RLC* circuit. However, the cavity resonator is a shorted section of a waveguide. Physically it looks as a chamber of some defined shape with conducting walls. The inductive and capacitive effects are defined by the electric and magnetic fields within the cavity. Thus, for example, adjusting the inductance of a cavity resonator may mean modifying the magnetic field in the cavity. This of course is not different in principle than what we do at low frequencies but is very different in implementation.

In general terms, the role of current is taken by the magnetic field and that of voltage by the electric field. Circuit parameters cannot be considered to be lumped but rather, must be considered to be distributed. Conductors are replaced by transmission lines and waveguides. Reflections of waves from discontinuities are much more important than at low frequencies therefore, proper impedance matching in microwave circuits is absolutely essential since mismatch is not only a matter of power transfer. Reflected waves propagate back into sources and produce standing waves. In fact, matching impedances is often more important than matching power (that is, conjugate matching, in which maximum power transfer occurs).

Microwave circuits is a subject in itself, and a very extensive one. We will only mention a few basic, active circuits here, those that depend for their operation on diodes, transistors and tubes.

5.5.1. The Negative Resistance Oscillator

Most of the diodes discussed previously have a negative resistance characteristic. For operation as an oscillator, the device must operate in a nonlinear mode while exhibiting the negative resistance characteristics. This can be best seen from Figure 5.14. If the impedance of the diode is Z_d, then for oscillations to occur, with a nonzero current and an equivalent load impedance Z_L, the following must be satisfied

$$Z_d = -Z_l$$

or

$$R_d = -R_l \qquad and \qquad X_d = -X_l$$

Oscillations start by transients as the circuit is switched on and, as the impedance of the circuit changes (impedance of active devices is nonlinearly dependent on current or voltage) the condition above must be maintained.

Since the load is normally either constant (i.e. a transmission line) or is different than the diode characteristics, an impedance matching network is inserted in the oscillator circuit to maintain the impedance characteristics. This can take the form of a simple matching stub or a more elaborate network.

As an example, an oscillator based on an a *Gunn* diode is shown in Figure 5.15. The diode is placed in a coaxial cavity resonator. Tuning is achieved by moving the shorting plate on the right. The output is coupled out through an aperture into a waveguide.

Diode oscillators are single port devices. *IMPATT* and *Gunn* Diodes are very often used and are available as commercial devices. These often include a cavity resonator which can be adjusted (in a narrow band) to the required frequency and also ensure the operation of the device by increasing its stability. In general, *Gunn* diodes are used at lower power than *IMPATT* diodes as discussed in the sections on the various devices.

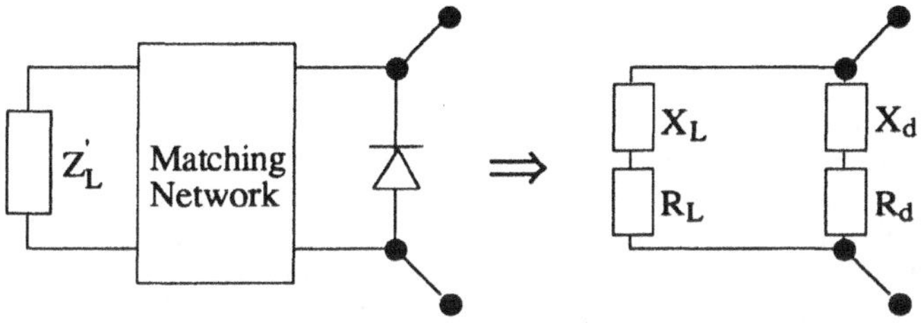

Figure 5.14. A negative resistance oscillator and its equivalent circuit.

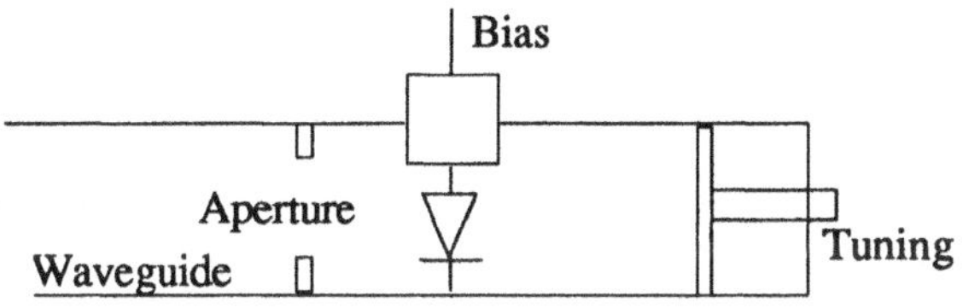

Figure 5.15. A diode oscillator. A *Gunn* diode oscillator is shown.

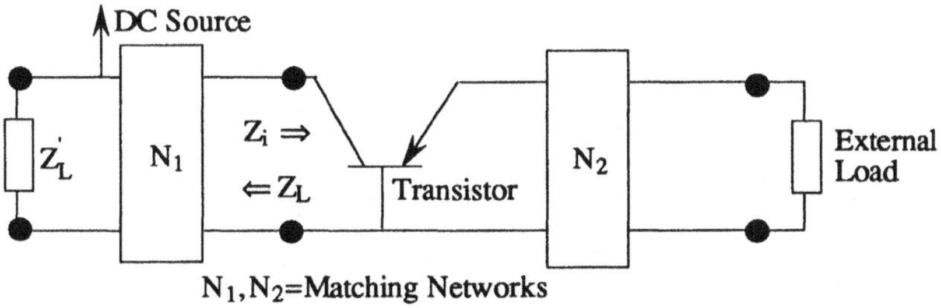

Figure 5.16. A basic transistor oscillator. The *BJT* transistor shown can be replaced by an *FET*.

5.5.2. *Transistor Oscillators*

Transistor oscillators operate on essentially the same principles as diodes. A transistor is a two port device one of which is terminated such that the internal impedance of the transistor satisfies the impedance conditions above. In effect, a stable device is taken into an unstable, nonlinear mode by use of the termination impedance (which, in effect creates positive feedback). An equivalent circuit for an oscillator based on an amplifier (i.e. a transistor, or a tube) is shown in Figure 5.16.

5.5.3. *Amplifiers*

Amplification in the microwave range is not fundamentally different than at lower frequency. The required elements are an active device and biasing circuitry. There are however two aspects that are unique in microwave amplification. The first has to do with matching of impedances. In general, the active device will have impedances that are different than the input or output lines. This of course happens at any frequency but is more important at higher frequencies. The second is the active device itself. There are many devices that can be used for amplification including tubes (i.e. crossed field amplifiers, klystron amplifier or *TWT* amplifiers), a large and increasing number of *BJT* transistors, and *FET*s. However, many of the diodes described above can also be used as amplifiers, and often are, by proper circuit design. A basic transistor amplifier is shown in Figure 5.17, where N_1, N_2 represent matching networks.

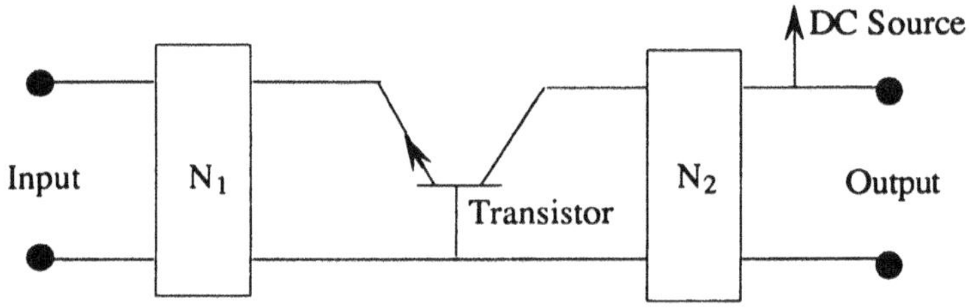

Figure 5.17. The basic transistor amplifier.

5.6. Coupling of Microwaves Sources to the Test Sample

Coupling of energy to a test sample can be done in a variety of forms, some simple, some quite complicated. For example, a test sample can be placed in a resonant cavity and some effect can then be measured. The sample is now part of the resonant structure and as such, will affect it. Alternatively, the sample may be placed in the path of a beam. The sample now reflects and transmits parts of the input energy. In still other forms, the sample may be placed in a waveguide, may be a waveguide itself or may be an absorber of microwave energy. In effect, any form of interaction between a sample and microwave energy can be utilized for measurement and testing. Coupling is achieved through use of any of the sources described above and a means of guiding the energy to the sample. The various types of antennas are perhaps the simplest methods of illuminating a sample if the sample is in free space. This can be done in the near field (that is, within a wavelength from the antenna) or in the far field. The antenna may also be used for sensing as is done with radar systems or, a separate antenna may be used for this purpose.

5.7. Microwave Probes and Sensors

Microwave sensing is obviously a very important subject in any application. In the *NDT* context, sensing means a method of measuring parameters of interest in the evaluation of materials. Probes and sensors fill this need. We will discuss here a few of the more important sensors and probes used for microwave measurements. In some cases, the distinction between sensors is clear. A thermistor is clearly a sensing device. In other cases, it is not. For example, an aperture in a microwave cavity can be viewed as a means of coupling energy into the cavity or as a source of microwaves or, as a sensor for measuring the frequency of a wave.

5.7.1. The Thermocouple

A thermocouple is formed by two different materials, with two junction. One junction is held at a reference temperature (normally called the cold junction) and the other is exposed to the measured quantity which, to be effective, should change the temperature of the hot junction. The thermocouple produces a potential difference which can be measured directly and which is directly proportional to the measured quantity. The thermocouples is used primarily for measurement of power. The microwave power is absorbed in the hot junction, elevating its temperature. The resulting potential difference is calibrated to read in *Watts* or *dBm*. For accurate measurements, monolithic silicon thermocouples are commonly used. The two junctions are produced on the same chip, so that the ambient temperature of the junctions is the same. Figure 5.18 shows the basic construction of the thermocouple. Conductor *A* is a good conductor while *B* is composed of a contact (i.e. gold or aluminum) and a material with high thermal potential such as a tantalum compound. The semiconductor between them is a diffused region. The junction between conductor *A* and silicon is the cold junction. During measurement, the cold junction is shielded from the measured microwave power.

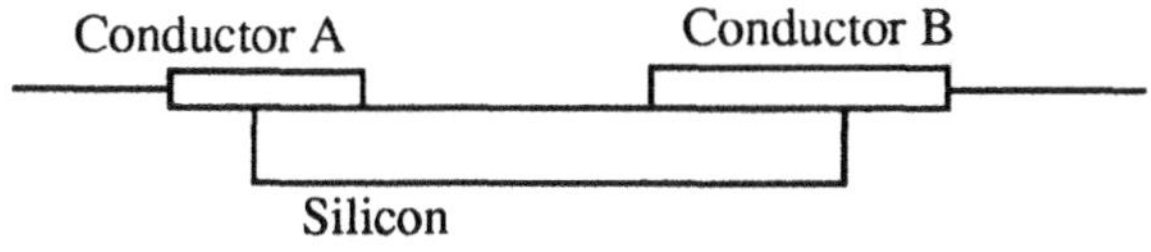

Figure 5.18. A semiconductor thermocouple.

5.7.2. The Thermistor

The thermistor is a negative temperature coefficient resistor. It is made of semiconductors with large temperature coefficients, normally as a mixture of metal oxides. A semiconductor thermistor is shown in Figure 5.19. A bead of the semiconductor material is formed around the two leads and is encapsulated in a protective material (glass). The thermistors for microwave applications are very small, unlike thermistors for other purposes. The size reduces the response time and increases accuracy. Like the thermocouple, the thermistor is a power sensing device but is based on changes in resistance. The resistance, connected in a circuit is then used to measure current or voltage which is proportional to the power absorbed in the bead. If properly chosen, and if the temperature rise is not excessive, thermistors can have a linear temperature dependence. Measurements are normally done in a bridge circuit as in Figure 4.2 with additional thermistors as temperature compensators. With proper design, the measuring and compensating thermistors are kept at identical temperatures and the reading of the power meter is zero unless the temperature of the measuring sensor changes. With automatic balancing circuitry, a very accurate device is obtained.

A variation of the thermistor is the baretter. It employs a fine metallic wire as the microwave absorber and, unlike the thermistor, its resistance rises with temperature. The baretter has a faster response time and can be used as integrator for measurement of peak power.

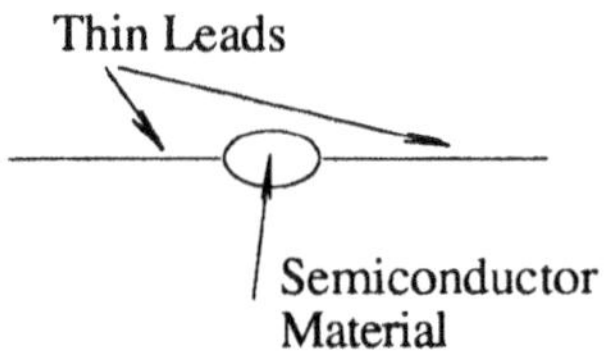

Figure 5.19a. A bead thermistor used for microwave applications. The bead is very small and the conductors are thin.

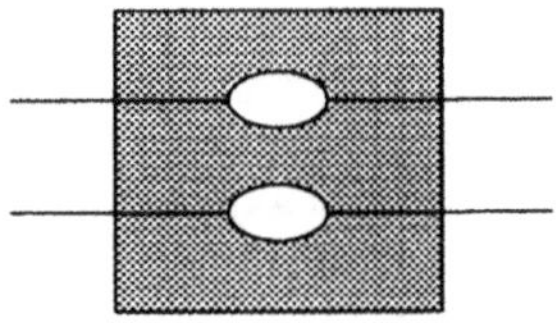

Figure 5.19b. The use of compensating thermistors. The two thermistors are mounted on the same substrate; one is used for measurements, the second for compensation.

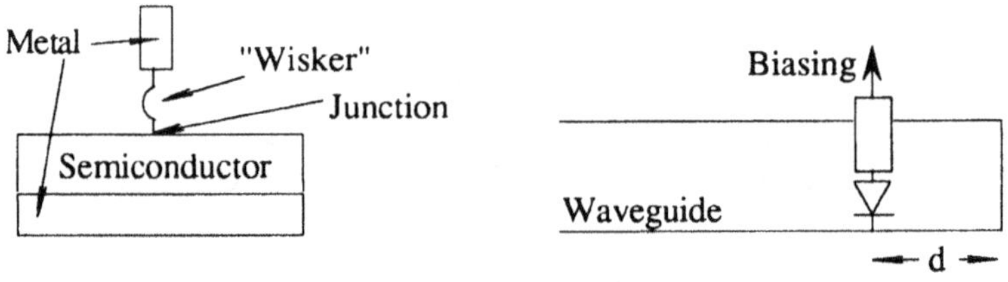

Figure 5.20. Metal contact diode and mounting of the diode in a waveguide.

5.7.3. Diode Detectors

Perhaps the most commonly used of all sensors is the diode or crystal detector. Most detectors now use the Schotky barrier diode or metal point contact diode but various solid state detectors have been used from the very early times of microwaves. These were various forms of the metal contact or "Whisker" diodes. Diodes in this mode serve as rectifiers, and are often used for detection of signals, as well as phase. They can be used as quantitative sensors because their output is proportional to the square of the input voltage and, therefore, to power. Although some power meters use calibrated diodes as the sensing element for power measurements, the diode is more often used as an indicator of power rather than as a measuring device. A metal contact diode and a diode mounted in a waveguide are shown in Figure 5.20. The distance *d* is a quarter wavelength to ensure the diode is at a point of maximum electric field.

5.8. Waveguide Probes

Probes in waveguides are designed to couple energy into and out of the waveguides in a manner similar to that used for coupling into and out of cavity resonators. However these have other uses as, for example, measurement of standing waves. These probes are often used either directly, as for measurement of standing wave patterns, or indirectly, such as to couple into or out of a device. There are three types of probes, as shown schematically in Figure 3.21. The first is a capacitive or electric field probe. A transmission line (e.g. a coaxial line) is terminated with the central wire extending into the waveguide. If this is located at a point where the wire is parallel to the electric field, energy will be coupled into or out of the transmission line. The second type is an inductive or magnetic field probe. The probe is in the form of a small loop or part of a loop. The loop generates a magnetic field perpendicular to the plane of the loop and couples or detects fields of the same type. The third type of sensing mechanism is the aperture. An aperture connecting two structures allows energy to pass between the two if the fields of the two structures (either electric or magnetic) are parallel to each other.

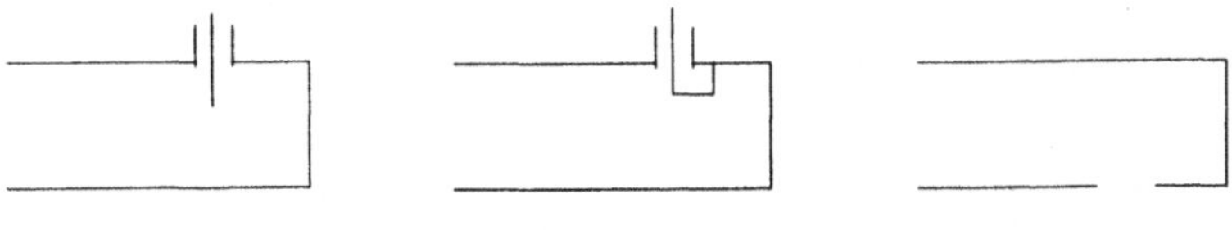

Figure 5.21. Waveguide probes. a. Capacitive probe, b. Inductive probe, c. Aperture probe.

5.9. Antennas

The most general sensing element is the antenna. In fact, any of the sensors described above and all others can be viewed as an antenna of one type or another. Antennas are used both as sources, to couple energy into the testing environment and as sensors, to receive signals. Some of the antennas used in microwaves are not applicable to *NDT*. For example, large parabolic antennas or reflectors are not normally used because of their size or shape, although they are very common in communication and radar applications. Others, particularly the simple, small antennas are often used in *NDT* work.

5.9.1. The Horn Antenna

One of the simplest and most useful antennas is the horn antenna. These are essentially extensions of waveguides, with flaring of the sides of the antenna into a horn shape. The antenna can be flared in the H-field direction, the E-field direction or both. The horn antenna shown in Figure 3.22 is flared in the E-field direction. Similarly, a cylindrical waveguide can be extended into a conical horn antenna. These antennas are very simple, relatively small, and are easily matched to waveguides. Their characteristics are very good and in addition to being used in their own right, they are often used as feeds to larger reflectors. Some horn antennas have specialized construction to enhance specific characteristics. An example is the corrugated horn which is designed for more uniform field and low cross polarization, and is often used as a feed for reflector antennas.

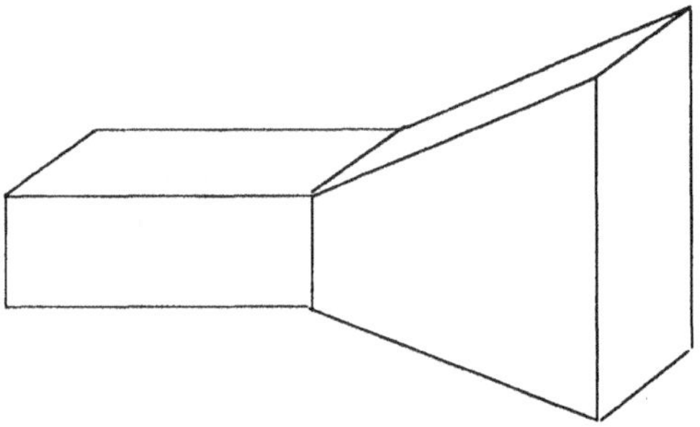

Figure 5.22. A waveguide horn antenna. The narrow end is connecter to a waveguide.

5.9.2. Microstrip or Patch Antennas

The patch antenna is made by etching a defined shape (rectangular, circular, ring, etc.) on a printed circuit board type of material. The connection to the patch can be as simple as a coaxial transmission line through the dielectric backing as shown in Figure 5.23. There are a variety of patches and feed techniques depending on the required field. Patch antennas tend to be low gain antennas and very narrow in bandwidth but, because of their simplicity and the fact that they can be made to conform to various shapes are used extensively, especially in the high *VHF*, *UHF*, and low microwave domains.

5.9.3. Spiral Antennas

Spiral antennas are similar to patch antennas and are also very simple and convenient for measurements. These are normally made of metal strips on a dielectric backing material by etching techniques. Under normal conditions, the antenna is bi-directional. If a more directional antenna is required, the spiral may be backed by an open cavity resonator. Spiral antennas are particularly convenient when the feed transmission line is a balanced two conductor transmission line. This type of antenna uses two conductors for the spirals, either terminated at the ends or open-ended, depending on the requirements of the system. Radiation from spiral antennas tends to be circularly polarized.

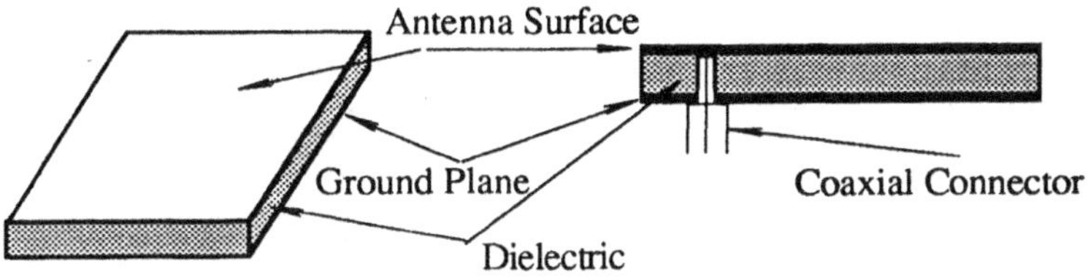

Figure 5.23. A rectangular patch antenna.

5.9.4. Slot Antennas

The slot antenna is essentially an aperture in a waveguide or cavity resonator and is not different than an aperture used for coupling two waveguides or a waveguide and a resonator. By properly cutting slots in the surface of a waveguide, (see Figure 5.24) the modes that generate fields parallel to the slot are radiated. This type of antenna is one of the easiest to implement and, like the horn antenna, requires minimal effort in matching and transmission of power to the slot. There are a variety of parameters that affect the slot antenna performance including waveguide modes, wall thickness, orientation and size of slot, etc., but, for many applications, this is a simple method of coupling energy into the testing environment.

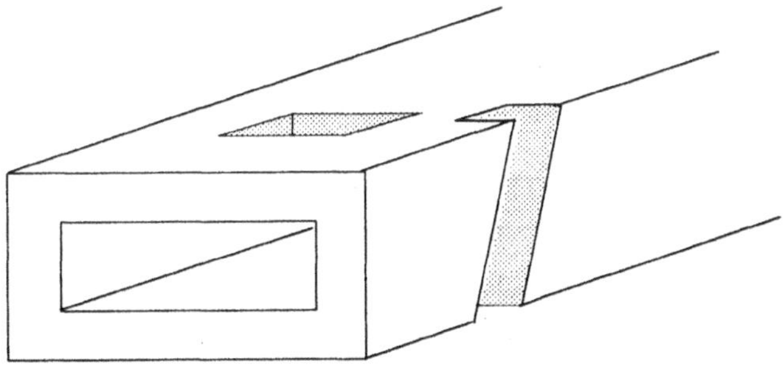

Figure 5.24. Slots in a waveguide used as radiating structures or antennas.

5.10. Open Waveguides as Sources and Probes

If an aperture or a slot in a waveguide can be used to couple energy in or out of the guide, then an open waveguide is merely a special case of an aperture. Unless properly terminated to avoid radiation of power, an open waveguide will radiate and can be used as a source. Also, as with any antenna, it can also be used as a receiver or sensor in its opposite mode. Open waveguides are often used as feeds mostly in their dominant mode, and mostly for reflector antennas. However they can be easily used as sources for testing. In a way, this would be equivalent to inserting the sample into the waveguide but it is more convenient. The open waveguide can also be viewed as a special case of the horn antenna. An open waveguide radiates power directly related to the aperture size.

An alternative method is to open a radiating section through a slot on the side of a waveguide rather than opening the waveguide itself. This branching of the waveguide is shown in Figure 5.25. for one possible orientations of the branching. By proper sizing of the aperture of the branched guide, different radiating patterns can be obtained. As an example, the radiating aperture in Figure 5.25 is parallel to the main waveguide and produces an impedance in series with the impedance of the main waveguide. The length of the extension should be close to $\lambda/4$ where λ is the wavelength in the branch for proper matching of the branch with the main waveguide. This type of radiating slot allows more flexibility in coupling energy to the testing environment.

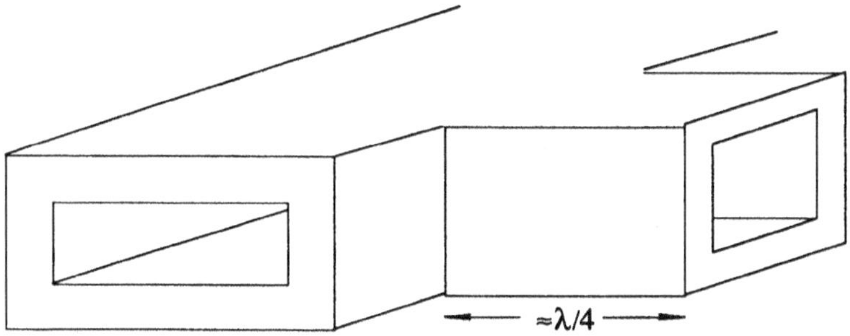

Figure 5.25. Waveguide branching used as a radiator. The length of the branch must be adjusted for matching with the main waveguide.

5.10.1. The Small Loop

A small loop can be viewed as a type of antenna but, we will view it here as a sampling device. The current in the loop is directly proportional to the magnetic flux through the loop. For this reason it is often used to measure flux. One example of its use in this mode is for measurement of radiation exposure. The small loop is useful in a variety of other application including for coupling in and out of cavities and waveguides and for exciting of Yitrium Iron Garnet (*YIG*) resonators.

5.11. Passive Microwave Devices.

Passive microwave devices include a variety of structure with specific functions. The most basic of these is the waveguide, intended to guide waves. This replaces the familiar transmission line used at lower frequencies. There are a variety of waveguides in use as well as guide sections such as bends, joints, junctions, shorts and couplers that need to be used in a microwave test setup, as well as microwave cavity resonators. In addition to the basic waveguide sections, other passive devices are used. These include attenuators, power dividers, directional couplers, mixers and the like as well as more complex devices such as the hybrid junctions, rotators, isolators and many others.

5.11.1. Waveguides, Waveguide Sections, and Cavity Resonators

The operation of waveguides and cavity resonators were described in detail in chapter 2. In general terms, a waveguide is any structure that can guide waves from one point to another. This may be as simple as a dielectric slab or a tube of some defined cross section. In chapter 2, the rectangular and circular cross-section waveguides were described. While these latter structures are the most common they are by no means the only structures. The choice of structure depends on application and on the type of modes that need to be propagated. We will only discuss here rectangular and circular waveguides in the most common modes.

A waveguide is characterized by the following:

1. Dimensions: these define the operating frequency range and the modes the waveguide can support. The cutoff frequency is also a function of dimensions.
2. Frequency range: the waveguide can only be operated within this range.
3. Modes: a waveguide may support more than one mode but often it is operated in a single mode, normally the dominant mode.
4. Cutoff frequency (or wavelength): defines the lowest frequency at which the waveguide can operate. Below cutoff there is no propagation of energy.

As an example a rectangular waveguide is given in Figure 5.26. This is a waveguide intended for operation at *10* to *15 GHz*, (or *3 cm* to *2 cm* wavelength) in TE_{10} mode. For the dimensions given the cutoff frequency is *7.87 GHz* (*3.81 cm*). A typical waveguide of this type is made of aluminum or brass with a polished finish but waveguides can be made of silver or silver coated brass. The purpose of the highly conducting smooth finish is to reduce losses in the waveguide. The normal mode used in rectangular waveguides is the TE_{10} mode. Unless specifically designed for other modes, only this mode is possible since the dimensions are such as to exclude other modes (the cutoff frequencies for other modes are above the limits of the waveguide range).

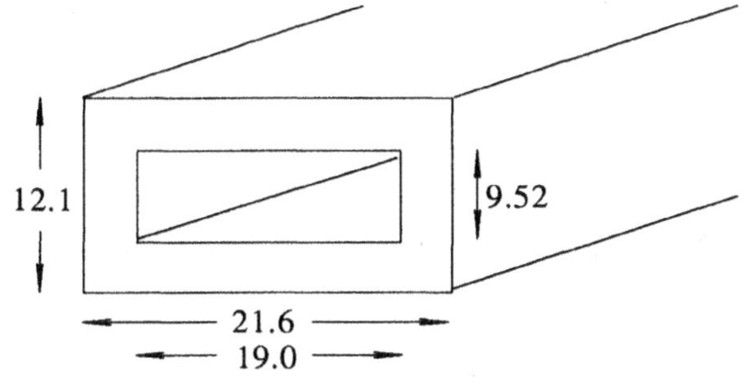

Figure 5.26. A rectangular waveguide intended for operation between *10* and *15 GHz*, with cutoff at *7.87 GHz* (TE_{10} mode).

A circular cross-section waveguide is shown in Figure 5.27. For the dimension given the waveguide operates in a TE_{11} mode (lowest mode), with a cutoff frequency of 2.88 GHz. As an example, it can also operate in the next higher mode which is a TM_{01} mode with cutoff at 3.76 GHz and so on. To ensure operation at the lowest mode, the waveguide must be operated between the two cutoff frequencies given above.

Circular waveguides are less commonly used that rectangular waveguides but there are applications where they are easier to use such as flexible guides and pivoted structures.

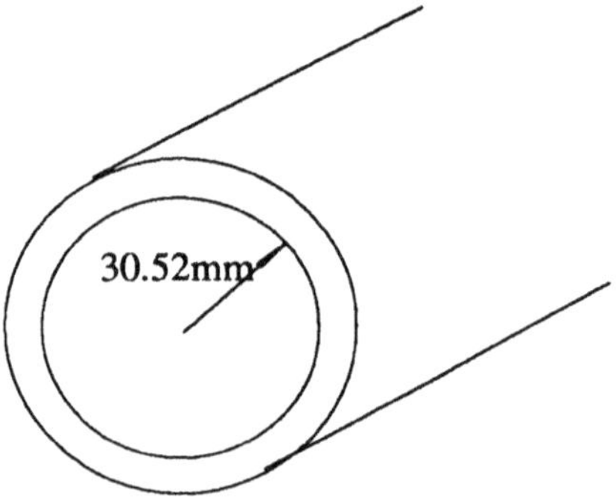

Figure 5.27. Circular waveguide with a cutoff at 2.88 GHz (TE_{11} mode).

Waveguide sections are connected to each other through flanges and, depending on application, may also include gaskets and matching sections. If a bent structure is required, we may use a variety of bends. These come in different shapes and sizes, some may be smooth, gradual bends, other may be twisted or sharp bends, depending on needs. Some bends are at $90°$, others at a variety of other angles, such that almost any shape can be accomplished. Two typical bends, one sharp, the other gradual are shown in Figure 5.28. The dimensions of the bends are designed to accomplish energy transfer through the bend without reflections or losses. The radius of the gradual bend (Figure 5.28a) must be larger than two wavelengths. The length of the reflecting surface in the sharp bend should be a quarter wavelength (for the lowest mode, in the waveguide). Bends in the narrow or wide dimension of the waveguide exist, allowing connection of waveguides in any configuration.

Figure 5.28. Two types of waveguide bends. a. Gradual bend and, b. Sharp bend.

Other type of connections include a variety of *T* junctions and *Y* junctions (Figure 5.29). These may be used to divide power between two guides or to accomplish other functions such as comparison and mixing of signals. In addition, joints to connect rectangular to circular guides and many other special purpose sections exist.

In addition to the waveguides discussed above, there are also special purpose waveguides such as ridged waveguides (used mainly to extend bandwidth) and elliptical waveguides which are much easier to bend without altering the dimensional properties (and therefore the electrical properties) of the waveguides.

Microwave resonant cavities are shorted waveguide sections as described in chapter 2. While in chapter 2 the cylindrical and rectangular cavity were discussed, many other cavities exist and most ar more complicated than the simple rectangular cavity. Although the method of operation remains essentially the same, the construction and coupling to the cavity varies. Common structures are the coaxial cylindrical cavity and the cylindrical ring as shown in Figure 5.30.

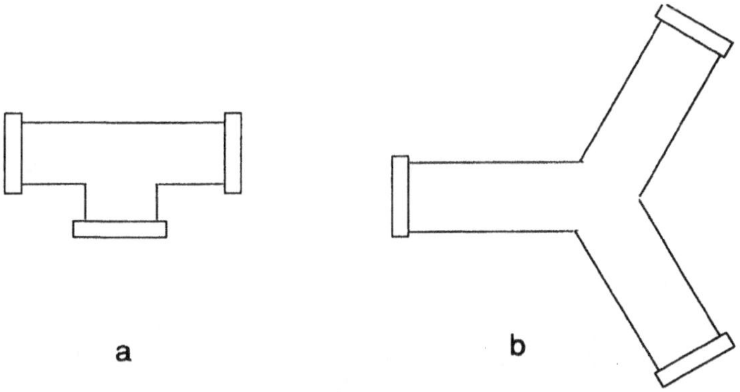

Figure 5.29. Two types of junctions; a. *T* junction, and b. *Y* (*120°*) junction.

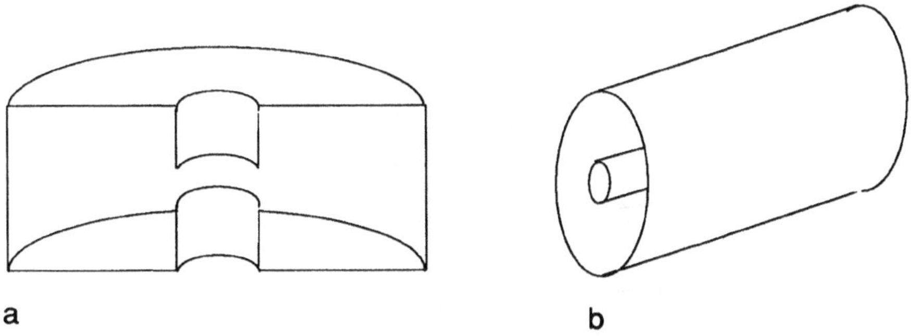

Figure 30. Two common cylindrical cavity resonators. a. Ring cavity resonator. b. Coaxial cavity resonator.

Although some cavity resonators are available as components, the cavity is more often used as part of a device such as the wavemeter (which typically uses a cylindrical cavity). A cavity can be tuned to a resonant frequency by sliding walls, distorting the walls, or tuning screws. Examples of tuned cavities are shown in Figure 5.31. Tuning of the cavity is equivalent to tuning of an *LC* circuit and can be done capacitively (by modifying the electric field) or inductively (by modifying the magnetic field).

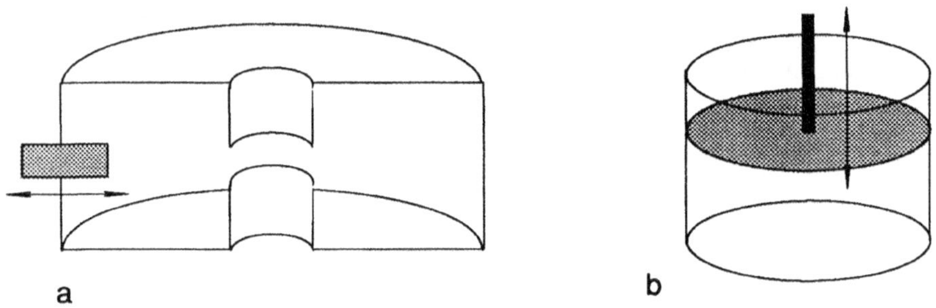

Figure 5.31. Tuning of cavities. a. Tuning by screw or slug. b. Tuning by moving, shorting wall.

5.11.2. The Magic T Hybrid Junction

A junction is used to connect waveguide sections so that one or both field components in the waveguide are coupled to another waveguide. There are junction that couple the *E* or *H* components and are therefore called *E* or *H* type junctions. A typical *T* junctions is shown in Figure 5.29a. A hybrid junction is a junction that combines the two modes in some given combination. One of the most useful devices is the magic *T* junction shown in Figure 6.32. This is the simplest method of achieving the function of the magic *T* junction.

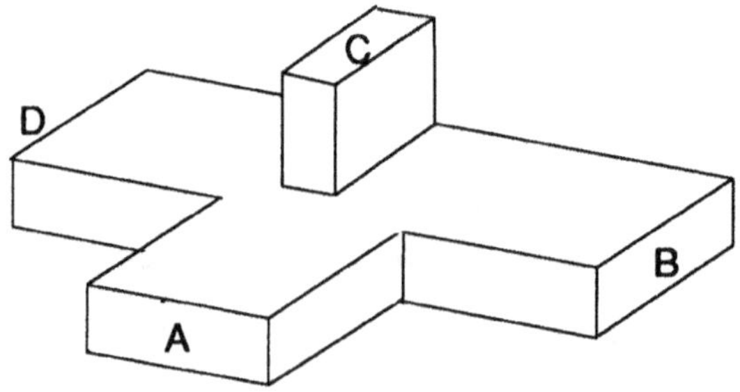

Figure 5.32. The magic *T* junction.

The properties of the junction are very important and can be summarized as follows:

a. A signal at port *C* appears at ports *B* and *D* at opposite phases.
b. A signal at port *A* appears at ports *B* and *D* at the same phase.
c. Port *A* and *C* are isolated from each other.

Because of these properties the magic *T* junction is useful in power dividers and balanced bridge circuits.

Extension to this type of hybrid junction is the four port (hybrid ring or rat-race) junctions as well as the six port junction. Also, junctions of various types may employ ferrite materials and act as circulators.

5.11.3. Directional Couplers

Directional couplers are sections of waveguides that can transfer power in one direction or in two directions (bi-directional couplers). The purpose is to allow measurement or coupling of a particular component of the wave. As an example, a directional coupler may be employed to measure the reflected wave without interfering with the incident wave. Similarly, a bi-directional coupler may be used to couple the incident wave into one section and the reflected wave into another for measurement purposes or for any other purpose. Although most couplers also include a probe (i.e for measurement), the idea of directional coupling simply implies that a means of transferring energy out of the wave guide, in a particular direction is available. This may be as simple as providing holes or apertures in the waveguide section such that only the required field is coupled out of the section. In most cases directivity is accomplished by using multiple apertures, each producing coupling at a given phase, such that the fields either cancel or add, depending on the direction of propagation in the waveguide. One configuration is the use of two holes, spaced $\lambda/4$ apart. A simple coupler of this type is shown in Figure 5.33. The coupler may also contain a matched load (to avoid reflections) and may be of a variety of types. Some couplers will couple the electric field, others the magnetic field. Circulators, isolators and duplexers are also types of directional couplers as are, in fact, the *N* port devices discussed above. In particular the magic *T* is a directional coupler.

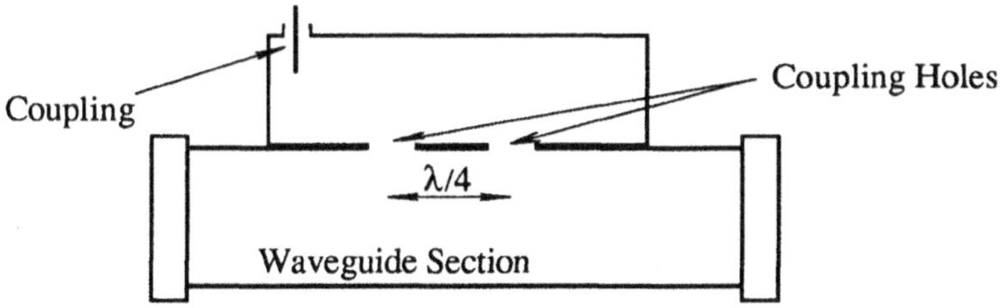

Figure 5.33. The basic structure of a directional coupler.

5.11.4. Isolators and Circulators

Isolating and circulating functions are obtained by use of ferrites and applied dc magnetic fields. The operation is based on the properties of ferrites and their precession frequencies. In the presence of an external magnetic field, the ferrite has a low loss in the forward direction and a very high loss in the opposite direction. For this reason, a backward propagating wave is absorbed in the ferrite while a forward propagating wave passes with very little attenuation. Because the backward propagating wave is absorbed, the power handling capabilities of isolators is limited by the power that can be absorbed in the ferrite. A simple isolator is shown in Figure 5.34.

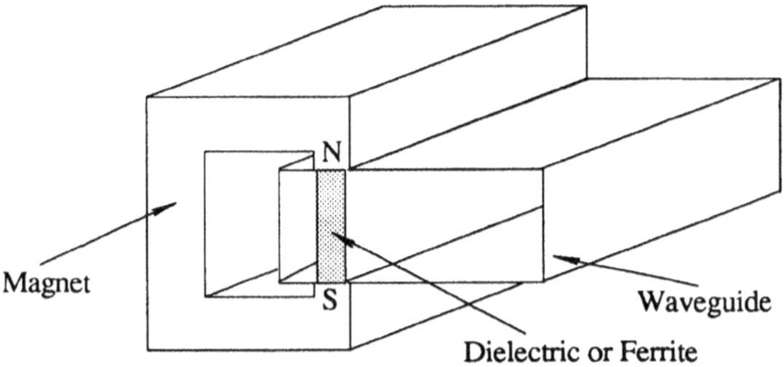

Figure 5.34. A simple isolator using a dielectric or ferrite card.

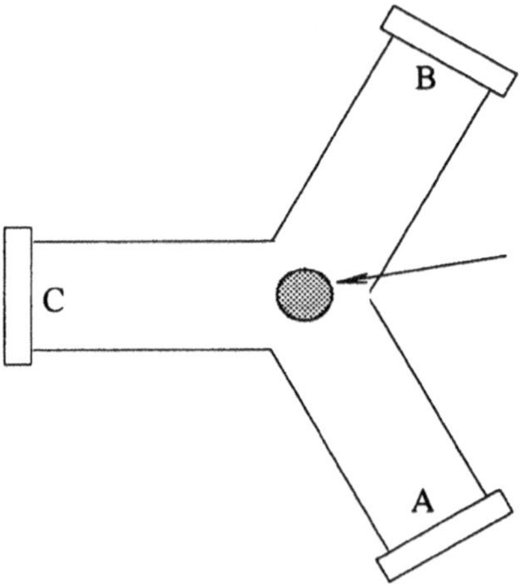

Figure 5.35. A *Y* junction circulator.

An isolator is defined by its forward attenuation and backward isolation. Typical values are *1 dB* loss or less in the forward direction and *30 dB* in the backward direction. Isolation is also dependent on frequency and, therefore, isolators are also specified by frequency range.

A circulator uses ferrite pieces in the vicinity of a port to force the fields into a particular direction around a ferrite piece. A circulator is shown in Figure 5.35. The circulation properties are accomplished by viewing the junction as a cavity resonator and the waveguides connected to the cavity as apertures. The presence of the ferrite shifts the field pattern such that one port is coupled with one of the other ports but not to the third port. If this shift corresponds to the location of port *B*, then energy from port A is coupled to port *B* but not to port *C*. Similarly, energy from port *B* can be coupled into port *C* and from port *C* to port *A*. Reverse circulation is also possible, depending on the design.

Any number of junctions may be present, provided the shift is adjusted for the separation between junctions.

5.11.5. Attenuators

Attenuators are sections or fixtures that allow monitored dissipation of energy so that the input energy can be reduced to the required level for measurement purposes or matching. In its simplest form, a dissipative element is simply inserted into a waveguide, taking care in design to avoid reflections due to this discontinuity. A simple example of a continuously adjustable attenuator is shown in Figure 5.36. This is the so called rotary vane attenuator. There are a number of other models of attenuators but in principle they all operate in the same fashion. Fixed attenuators also exist and one particularly useful fixed attenuator is the termination or load. These are designed either for absorption of energy in a guide or for impedance matching. They also exist in fixed and variable terminations, the latter used mostly for impedance matching.

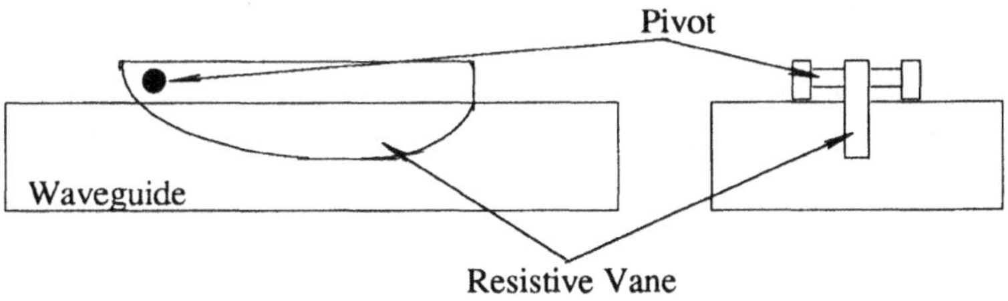

Figure 5.36. A waveguide attenuator.

5.12. Bibliography

Microwave sources are described in a large number of sources and, to a certain extent, can be found in almost all general texts on microwaves. In particular, Liao [1], [13], discusses both microwave tubes and semiconductor devices. A wealth of information on all aspects of microwave devices can be found in Fink [4] and Harvey [16]. A variety of microwave antennas are discussed by Silver and James [10]. Circuit aspects of microwaves can be found in [13-15].

[1] S. Y. Liao, Microwave Devices and Circuits, Third edition, Prentice Hall, Englewood Cliffs, NJ., 1990.
[2] W.S. Cheung and F.H. Levien, (eds.), Microwaves Made Simple: Principles and Applications, Artech House, MA., 1985.
[3] J.W. Lunden, "Microwave Amplifiers and Oscillators", in Electronics Engineering Handbook, D.G. Fink, Ed., McGraw Hill Book Company, NY, 1975, pp.13-108-13-127.
[4] D.G. Fink, Ed., Electronics Engineering Handbook, , McGraw Hill Book Company, NY, 1975.
[5] D.M. Pozar, "Microwave Engineering", Addison-Wesley, Reading, MA., 1990.
[6] M.J. Howes and D.V. Morgan, (Eds.), "Microwave Devices", John Wiley & Sons, London, 1976.
[7] F.E. Gardiol, "Introduction to Microwaves", Artech House, 1984.
[8] E.A. Wolf and R. Kaul, "Microwave Engineering and System Applications, John Wiley & Sons, New York, 1988.
[9] S.F. Adam, "Microwave Theory and Applications", Prentice Hall, Englewood Cliffs, NJ, 1969.
[10] S. Silver and H.M. James, "Microwave Antenna Theory and Design", IEE, London, 1984.
[11] T.S. Laverghetta, "Modern Microwave Measurements and Techniques", Artech House, Norwood, MA., 1988.
[12] G.D. Vendelin, A.M. Pavio and U.L. Rohde, "Microwave Circuit Design Using Linear and Nonlinear Techniques", John Wiley & Sons, New York, 1990.
[13] S. Y. Liao, "Microwave Solid-State Devices", Prentice Hall, Englewood Cliffs, NJ., 1985.
[14] I. Bahl and P. Bhartia, "Microwave Solid State Circuit Design", John Wiley & Sons, New York, 1988.
[15] W.A. Davis, "Microwave Semiconductor Circuit Design", Van Nostrand Reinhold Co., New York, 1984.
[16] A.F. Harvey, "Microwave Engineering", Academic Press, London, 1963.
[17] S.Y. Liao, "Microwave Electron Tube Devices", Prentice Hall, Englewood Cliffs, NJ., 1988.

Chapter 6. Dimensional Testing

6.1. Introduction: Testing with Microwaves

Testing with microwaves can be divided into four basic methods.

1. Reflection methods
2. Transmission methods
3. Scattering methods
4. Resonant methods

While the actual test, and the parameters that need to be determined vary, the basic principles involved remain the same. It is perhaps best that these measurements be defined in general terms before their application to specific tests.

6.1.1. Reflection Tests

This type of tests depends on the measurement of the reflection coefficient. A basic reflection test is shown in Figure 6.1. The incident wave is produced by a source and coupled through an antenna. Here we view the antenna in a generic manner to mean any mechanism of coupling energy to a test sample. The waves are reflected off the material either completely (as for conductors) or partially (as for dielectrics and lossy dielectrics). The reflected wave is received by an antenna (antenna B in this case). For simplicity, we have used two antennas: one for transmission, the other for reception. However, any transmitting antenna can also be used for reception. The separation of the two waves can be done by directional couplers or, in some cases, not done at all and measurements are performed on the total wave. There are many variations on this simple test but in all cases one can distinguish between an incident and a reflected wave. The processing module in Figure 6.1 identifies the two parts of the wave and produces the required display. This may be any of the effects on which reflection depends. In most cases, the changes in the dielectric constant or changes in losses (loss tangent) are correlated to a particular property. This may be as simple as thickness of material, variations in dielectric constant (for example delamination), moisture content, contamination, and a variety of other parameters. In fact, almost any material property can be identified with some change in the reflection coefficient.

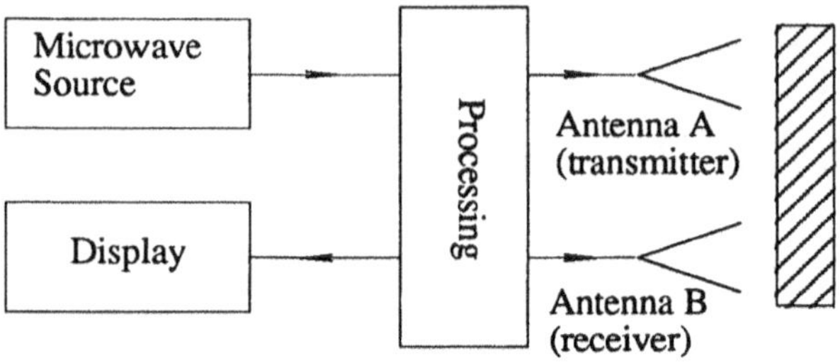

Figure 6.1. A basic reflection test.

6.1.2. Transmission Tests

Transmission tests rely on measurements on the transmitted wave. This is shown schematically in Figure 6.2. The transmission coefficient also depends on material properties as the reflection coefficient. However, transmission tests are sometimes easier to perform or to correlate to material conditions. Attenuation measurements are also transmission tests and are often used for evaluation of lossy dielectrics. The basic modules of the test are the transmitting and receiving antennas. Processing is similar to that done for reflection tests. There are some cases for which only one of the tests is practical. For tests on conductors (such as thickness gaging), only reflection tests are possible. Since the transmission is negligible, it is not possible to use it for testing. Similarly, if the reflection coefficient is very small, especially for low loss dielectrics, the transmission test might be more effective.

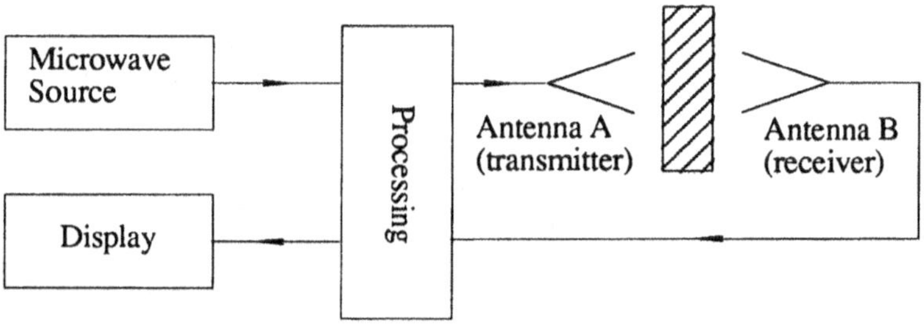

Figure 6.2. A basic transmission test.

6.1.3. Scattering Tests

Both reflection and transmission are properties associated with the bulk of the material. Scattering, as understood in the context of testing of material, on the other hand, is associated with local conditions in the material. While any variation in the material will affect both the reflection and transmission, we will view scattering as an indication of local effects such as flaws. This is in general a matter of convenience. We also include radar and similar effects with scattering. The two basic tests associated with scattering are shown in Figure 6.3a and 6.3b. In Figure 6.3a, the scattered field is picked up by antenna *C*. This antenna will pick little or nothing unless there is scattering within the test sample. Normally, only reflection and transmission will exist. Transmission is detected by antenna *B* while reflection can be detected by antenna *A*. This type of measurement can be used to detect scattering by foreign objects or sharp variations in material properties such as delaminations or inclusions. The second type is the basic radar system and is shown in Figure 6.3b. Here the pulsed radar method is shown although other radar methods can be used. In *NDT*, the pulses as scattered and picked up by the antenna are correlated with the location, size or properties of the material.

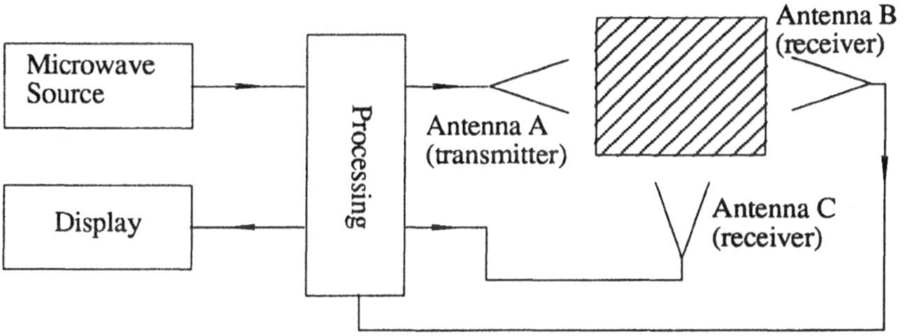

Figure 6.3a. Scattering test for inclusions or flaws in dielectrics.

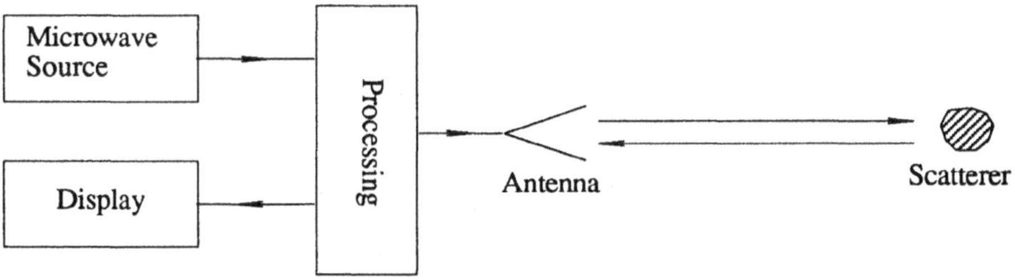

Figure 6.3b. Scattering measurement using a monostatic radar method.

6.1.4. Resonant Tests

In resonant methods of testing, the material is part of a resonant system and testing is based on variations of the resonant frequency or variations in the Q-factor of a resonant system. The resonant system is often a cavity resonator, a transmission line resonator or an open cavity. The basic test is shown in Figure 6.4. The figure shows a material in the cavity and the measurement equipment. Resonant frequency variations are due to changes in the dielectric constant (and, infrequently, to changes in permeability) while changes in the Q-factor are indicative of ohmic and dielectric losses in the cavity. Resonant methods are extremely sensitive and are best suited for testing of small samples with small variations in properties. With closed cavities, the test must be done on a small sample, inserted in the cavity. However, by using apertures, or windows in the cavity, or by using open resonators (both cavity and transmission lines) the material may be fed continuously and may be gas or liquid as well as solid. Again, as with the previous methods, the variations possible are many. Some of these will be discussed in tthis and the following two chapters while others are referred to in the bibliography sections of individual chapters.

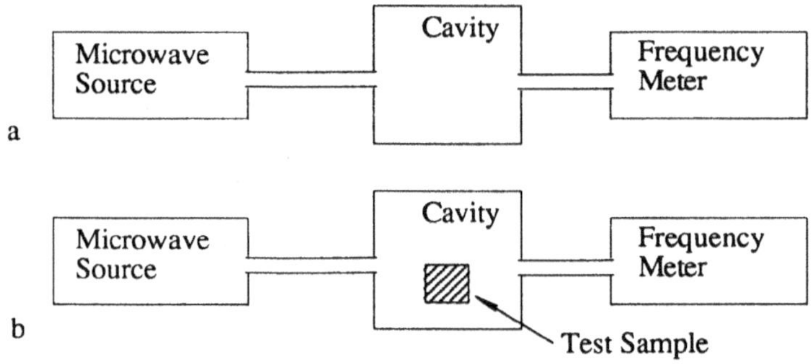

Figure 6.4. Resonant test. a. Measurement without material. b. Measurement with material present.

6.1.5. Testing Parameters

It would be tempting to try to list all parameters and effects that can be tested using microwaves. This, however would be futile since, as was pointed out in chapter 1, almost any material parameter affects the microwave beam. Anything that changes the dielectric constant, material permeability, or conductivity can, in principle, be detected and measured by use of microwave methods. The question is more often one of magnitude and discrimination between effects. We will look at testing in terms of three general groups of testing methods which, while not always clearly separated, serve as a guide in terms of applications and methods of testing. The three groups are

1. *Dimensional testing*, which includes thickness gaging, deals with dimensional variations of materials. Reflection, transmission and resonant methods can be used but, perhaps the most useful method is the reflection method.
2. *Testing for flaws*. Detection and evaluation of flaws can be done with any of the tests above, but we will emphasize here scattering methods, since scattering is normally used for this purpose. Both surface and volume defects are detectable.

3. *Monitoring material properties.* While the distinction is not always clear-cut, monitoring of material properties is based on measurements of specific effects that can be correlated with material properties. These may include water content in a material or degree of cure in polymers. All of the methods above are applicable with particular emphasis on resonant methods.

Perhaps the most important point in describing testing methods is that all measurements in the microwave domain are experimental methods that depend on many properties. It is only possible to test a parameter if all others are known or can be evaluated. This is not always possible while analytic expressions are of little use in most applications. Thus, most tests must rely on experimental data and calibration curves. These must be obtained for the system in use and for the type of materials used. Extreme care must be taken to positively identify the effect measured, otherwise, because of the large number of parameters involved, it is all too easy to miss or incorrectly identify a defect or to confuse a property with another. For example, a change in material permittivity due, say, to change in density of a material can be easily seen as an equivalent variation in material thickness, since the two effects cause the same electric thickness.

The use of microwave methods for dimensional testing is fundamental. Testing includes absolute thickness of materials, changes in thickness, thickness of coatings and layers of materials. Both thickness of conducting and dielectric materials can be measured accurately. By extension, delamination of layers in composite materials can also be viewed as a layer and therefore can be detected and measured.

A variety of measuring techniques can be used for this purpose. For metals and high conductivity materials, the strong reflections at the material surface is commonly used for testing. The amplitude or phase of the reflection is measured and the thickness of the material is inferred from these measurements either directly or by an appropriate calibration curve. To improve accuracy, differential measurement techniques are commonly used.

In dielectrics, transmission and attenuation measurements can also be used for this purpose, as well as standing wave and reflection methods.

Depending on the method used for testing, resolution of the order of a few microns in material dimensions is possible. This also applies to variations in size such as material thickness

Thickness and changes in material properties are often indistinguishable for many properties. A change in material properties will often look the same as a change in thickness. The effective "electrical thickness" is what is actually measured not the physical thickness. Of course, if material properties do not change, the electric thickness varies only with the physical thickness and the measurement is correct. If, however, material properties also vary, it is possible to incorrectly measure material thickness. We will normally assume that material properties are known for thickness measurements and that they do not vary. Similarly, when material properties are measured, the dimensions of the test sample are assumed to be constant.

6.2. Thickness Gaging

Measuring of material thickness is a common method of evaluation of materials and is an indispensable method for monitoring of production of many products. In the case of conducting materials, thickness means the actual dimension of the material. Because fields do not penetrate in highly conducting materials (the skin depth is of the order of a few microns), the interior of the conductor does not affect the measurement. In dielectrics, the properties of the material also affect thickness measurements.

Thickness gaging in conductors can be performed by two basic methods, both related to reflection measurements. First, the reflected wave or the reflection coefficient can be measured. Thickness can now be related to the amplitude or phase of the reflected wave (or the reflection coefficient). Alternatively, the standing wave pattern can be measured, from which the thickness of the material can be inferred. It is also possible to measure thickness by resonant means, from variations of resonant frequency of a cavity, provided that material handling can be performed satisfactorily.

6.2.1. Reflectometry at Conducting Interfaces

In conducting materials, the reflection coefficient is equal to -1 (see chapter 3). This means that at the surface of a conducting material the wave is totally reflected. Thus, a simple method of measuring thickness is to measure the reflected wave from the surface or surfaces of the conductor. Before discussing the measurements themselves it is useful to look at a wave, perpendicularly incident on a conducting surface as in Figure 6.5. This situation was discussed in chapter 3 (section 3.3.4) where the reflection and transmission coefficients were found to be

$$\Gamma = -1 \qquad and \qquad T = 0 \tag{6.1}$$

since the conductor behaves as a perfect conductor. From Eqs. (3.40) and (3.41), the electric and magnetic fields in free space (to the left of the conducting surface), at a distance d from the conducting surface are:

$$\mathbf{E}_l(z) = -\hat{\mathbf{x}} j2E_{il} \sin \omega\sqrt{\varepsilon_0\mu_0}d \tag{6.2}$$

$$\mathbf{H}_l(z) = \hat{\mathbf{y}} 2\frac{E_{il}}{\eta_0} \cos\omega\sqrt{\varepsilon_0\mu_0}d \tag{6.3}$$

Since $E_l(z)$ is the total field, we can either separate the field into an incident and reflected wave or use the total wave as in these equations. Now consider the situation in Figure 6.6 where the conductor changes in thickness by Δt. The equation for the total field is now

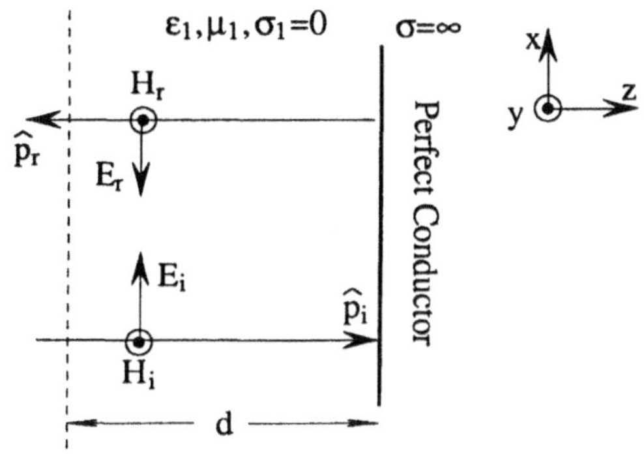

Figure 6.5. Reflection at an air/conductor interface: normal incidence.

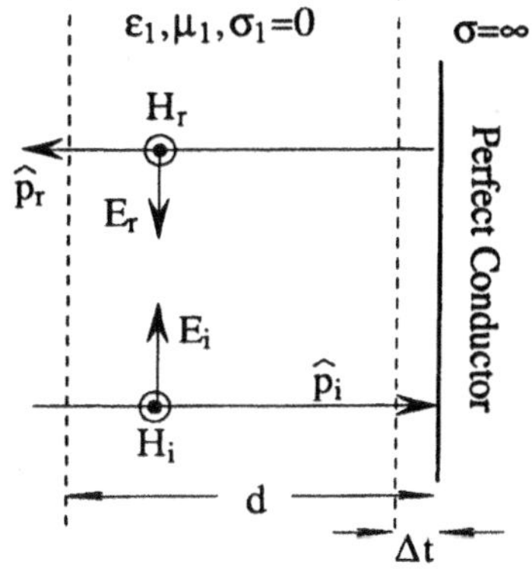

Figure 6.6. Reflection at an air/conductor interface showing the effect of a change in thickness.

$$\mathbf{E}_1'(z) = -\hat{\mathbf{x}} j2E_{i1}\sin\omega\sqrt{\varepsilon_0\mu_0}(d - \Delta t) \qquad (6.4)$$

Thus, the only term that changed is the phase of the wave or, more accurately, the phase of the generalized reflection coefficient (see Eq. (2.27)). The generalized reflection coefficient at a distance *d* from the conductor is

$$\Gamma_d = \Gamma_{d=0} e^{-j2\omega\sqrt{\varepsilon_0\mu_0}d} \qquad (6.5)$$

where $\Gamma_{d=0}=-1$

If the material thickness changes by Δd, the reflection coefficient at the same point is

$$\Gamma_{d-\Delta d} = \Gamma_{d=0} e^{-j2\omega\sqrt{\varepsilon_0\mu_0}(d-\Delta d)} \qquad (6.6)$$

and, again, this depends only on the change in material thickness.

If the electric field in this example can be measured, then this measurement can be directly interpreted in terms of the value of Δt.

In practice the incident and reflected electric fields are measured. From these, the generalized reflection coefficient at the measurement point is found and as for the electric field, the measurement is correlated with the change in thickness, based on the relations in Eqs (6.5) and (6.6). There are many variations of this basic method, some that compare the amplitudes of the reflection coefficient, some that compare the phase.

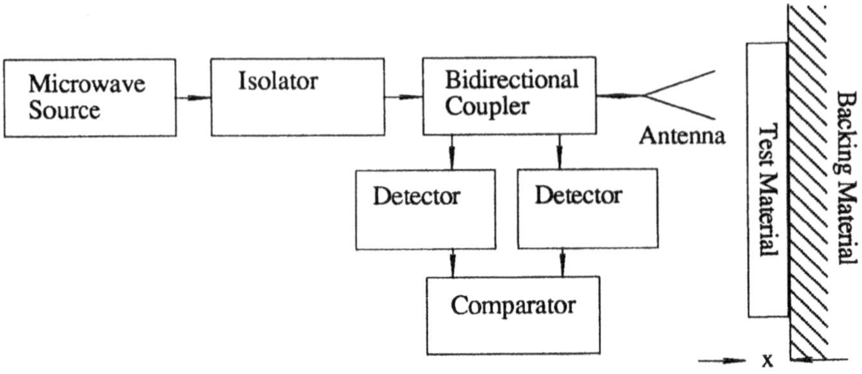

Figure 6.7a. Measurement of thickness of a conducting material using reflection from the conductor surface.

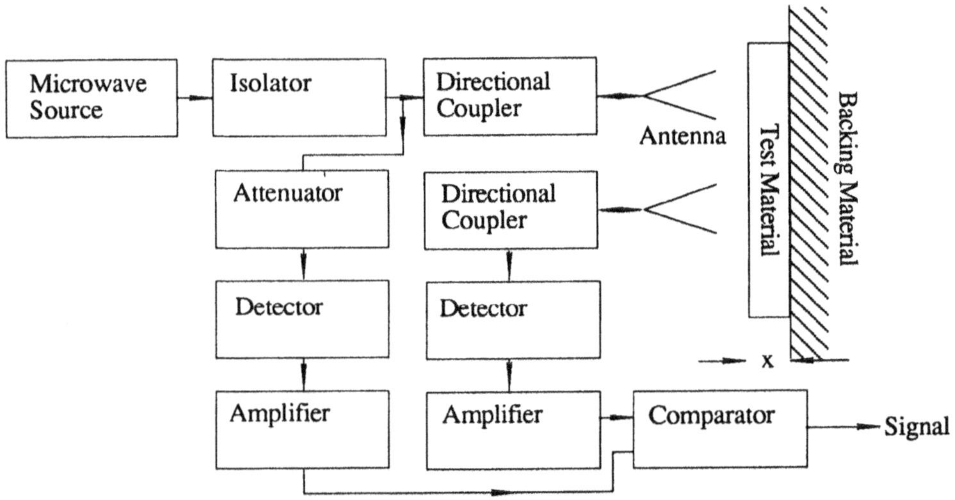

Figure 6.7b. Measurement of thickness of a conducting material using reflection from the conductor surface. Separate transmitting and receiving antennas are used.

The general method above can be applied directly if the test sample is flat and can be placed accurately against a backing material as shown in the configuration in Figure 6.7a and 6.7b. In this setup, a microwave source, such as a klystron or diode oscillator is connected through an isolator and a directional coupler to an open waveguide or a horn antenna. The directional coupler allows power to be transferred to the sample while the reflected power cannot return to the source. The reflected wave is transferred through another directional coupler to a detector. The signal is detected and amplified and compared to the detected signal from the incoming wave. By adjusting the attenuators to a null the change in thickness can be measured directly. For purpose of detection, the microwave signal is amplitude modulated with a low frequency signal which, after detection, is used for comparison. Normally, there would be no need for an amplifier since we assumed there are no losses either in free space or at the surface of a conductor. In practice, neither of these assumptions is accurate and the attenuator allows zeroing of the signal in the presence of losses while an amplifier may be necessary for measurements with low intensity sources. The test setup in Figure 6.7.b is similar except that a separate receiving antenna is used. The directional coupler in the receiving signal path can be omitted. Also, the power split between the transmission path and the comparison path in the incident signal is achieved by use of a power splitter between the isolator and the attenuator.

The various parameters in the measurements can be adjusted to a certain degree to produce the required test results. First, the frequency can be adjusted, or a different band may be used. This is not always possible and depends on availability of sources and other components. Normally, the frequency will be defined by the available waveguides and components such as waveguides. For example, measurements may be made in the *X* band (about *10 GHz*) with rectangular waveguides. Although, for conductors, any frequency will produce good results, higher frequencies allow higher resolution.

One of the most critical parameters is the distance between antenna (or waveguide) to the surface of the test sample. If this distance is short, the change in amplitude due to change in thickness is large. This is typical of materials in the near field of the antenna. At larger distances (as for example if the antenna cannot be placed in close proximity to the material) the change in amplitude may not be large enough for accurate measurement. In this case, the phase may be used instead, as we will see shortly.

The antenna used also affects the measurements. High gain antennas are not necessary in this case because of the strong reflection. Open waveguides work well in most cases.

The signal obtained may be linear, especially for small thickness variations. However, since it is relatively easy to generate calibration curves this is not normally a major consideration.

The main problem with this method is that the signal is proportional to the distance between antenna and sample as is obvious from Eqs. (6.4) through (6.6). Any movement of the sample will be read as a change in material thickness. Thus, this method is not suitable for measurements on a production line or other moving samples.

To cancel this sensitivity on location of the sample, two identical systems are used, one on each side of the sample as shown in Figure 6.8. In this configuration, a single source is used and the signal is split into two beams, travelling through identical paths to ensure identical phase. The coupling is the same as in Figure 6.7, and the signal from each amplifier is proportional to the distance between the respective antenna and the sample. The two signals are summed up and the summed signal is a direct measurement of the distance between the two antennas minus the thickness of the sample. Any increase in the output is due to an increase in material thickness and a decrease in signal indicates a thinner material. This test setup is a type of differential reflectometer and, therefore, only changes in material thickness are measured. Translation of the material does not affect the output as long as the thickness does not change, and the translation is not large. As an example, warping of the material will change the signal because the effective thickness changes. In this case we must also consider the fact that the reflection is reduced because of the angle the beam makes with the surface. Misaligned samples also affect the measurement. These situations are shown schematically in Figure 6.9.

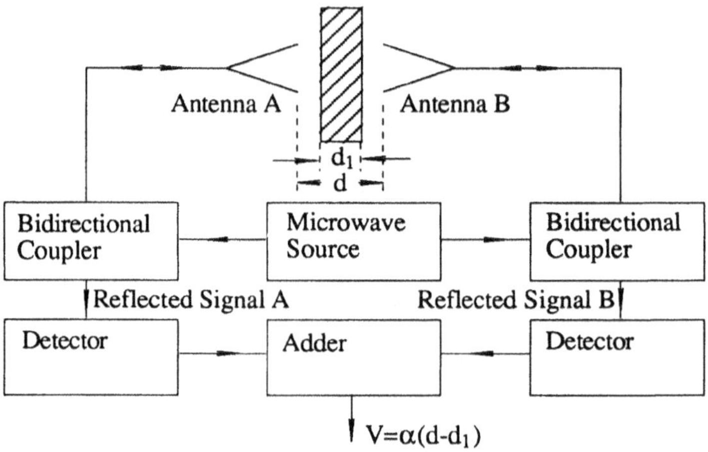

Figure 6.8. Measurement of thickness with insensitivity to sample location.

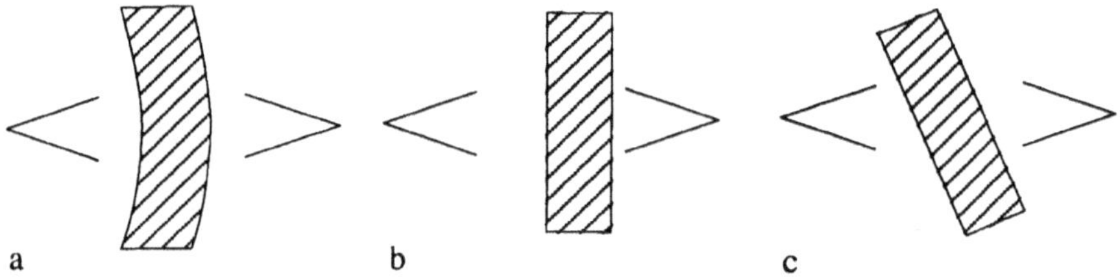

Figure 6.9. Misalignment of the test sample. a. Warping. b. Translation. c. Tilting.

A third possibility of dimensional monitoring is the use of a differential reflectometer on one side of the material as shown in Figure 6.10. In this configuration, the two transmitting horns are parallel to each other and a receiving horn is located midway between the transmitting horns. By simply shifting one transmitter by 180°, the normal signal in the receiver is zero because the signals reflected from the surface cancel each other at the receiver. If, however, the material is nonuniform, one of the horns illuminates it first, producing an indication in the receiver. Figure 6.11 shows the signal one expects to obtain from such a configuration. This method is particularly suitable for testing for thinning on continuous material such as rubber. Any thinning (for example due to stretching) will produce a signal at each edge of the thinned area. This only happens if the reflectometer is aligned as shown in Figure 6.12. If the thinning is transverse and the reflectometer is also transverse, there will be no indication in the receiver antenna.

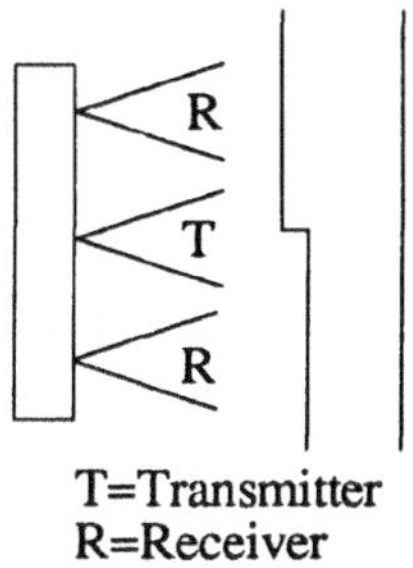

Figure 6.10. A differential reflectometer used to monitor material thickness.

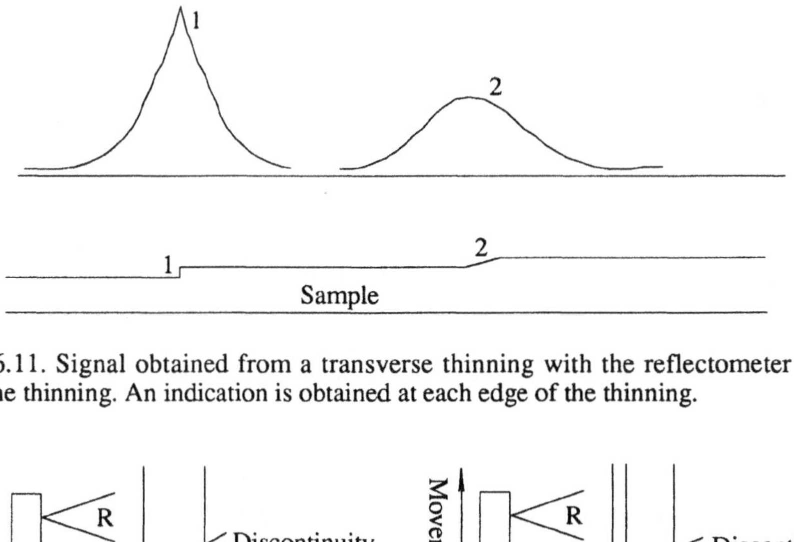

Figure 6.11. Signal obtained from a transverse thinning with the reflectometer aligned across the thinning. An indication is obtained at each edge of the thinning.

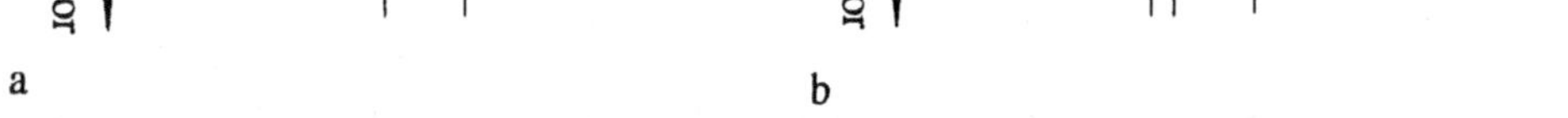

Figure 6.12a. Alignment of the reflectometer: nonzero indication.

Figure 6.12b. Alignment of the reflectometer: zero indication.

The considerations used above apply here as well since these two methods are very similar. Location of the horns above the sample is not critical but must be far enough to allow reflection into the receiver horn. On the other hand, shorter distance will allow larger reflections and better correlation with the signal and position of the thinning edge.

Another possibility often used for material thickness testing is to use a bridge detector to detect the variation in distance traveled by the waves. If this variation is due to change in thickness of a material, thickness can be easily and accurately measured. An appropriate configuration is shown in Figure 6.13. In this case, the detector is placed in one arm of a bridge (actually a magic T). To measure thickness, an attenuator and phase shifter are inserted in one arm and the material is placed at a given distance. The variable attenuator and variable phase shifter allow calibration of the bridge to null for nominal thickness. Any change in thickness will show up as an increase in the detector output.

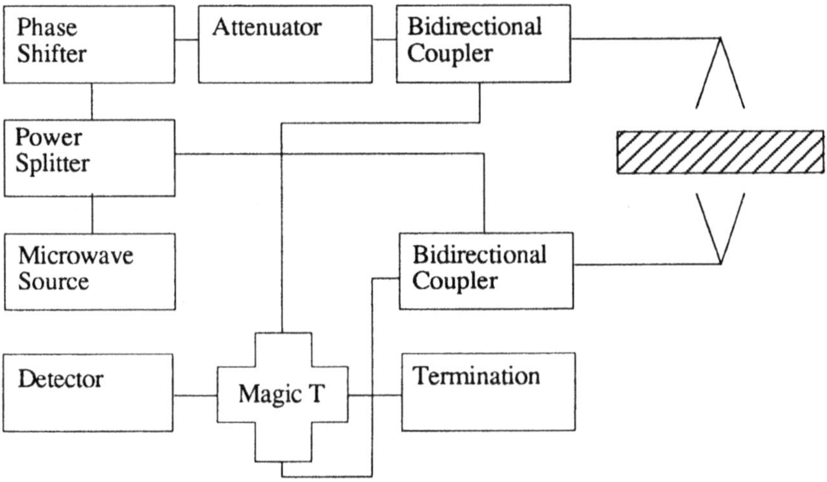

Figure 6.13 A bridge detector for thickness gaging. The attenuator and phase shifter are used to zero the bridge.

It is also possible to use a bridge without the use of attenuators and phase shifters as in Figure 6.14. In this case, balancing of the bridge is done by properly designing the distances of the arms of the bridge. If initially the arms are equal in length (or, for that matter differ by multiples of half-wavelengths) the bridge is nulled. Insertion of the sample is done so that it is at a fixed distance from one antenna (for example, the sample may be placed on the lower antenna or flush with a waveguide opening) while the length in the other path (path 2 in Figure 6.14) is adjusted for a null.

Both of these methods can be used for dielectrics and lossy dielectrics.

All three methods described above can also be used in conjunction with phase comparison rather than amplitude. As a rule, phase is more sensitive to small variation in thickness but it is difficult to separate phase and amplitude variations.

If the configuration in Figure 6.7 is adapted to phase comparison, the absolute phase can be measured as in Figure 6.15. This is not normally an acceptable method for small thickness variations because of the small changes in phase and the difficulty in measuring absolute phase. A slightly better approach is to shift the phase by introducing a phase shifter in the reflected wave path, such that when the sample is absent, the output is zero. Now, the phase introduced by the conductor, is directly proportional to thickness.

Figure 6.8 can also be adapted to phase measurements instead of amplitude. The required change in instrumentation is the addition of a phase shifter in each of the reflected wave paths. A simple configuration is shown in Figure 6.16. The phase meters can be absolute meters or, can be replaced by a phase discriminator.

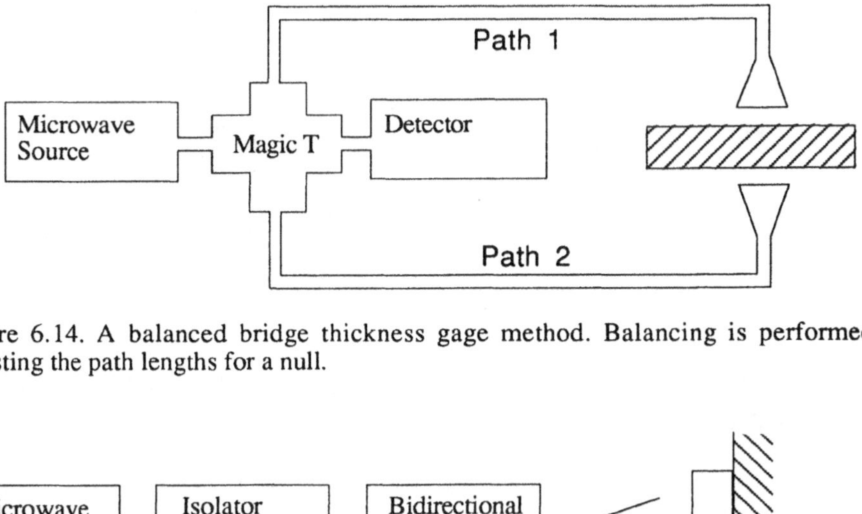

Figure 6.14. A balanced bridge thickness gage method. Balancing is performed by adjusting the path lengths for a null.

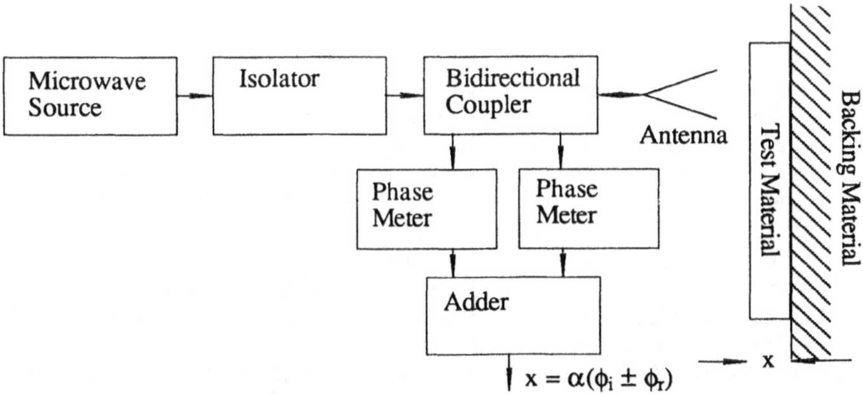

Figure 6.15. Basic phase measurement for thickness gaging.

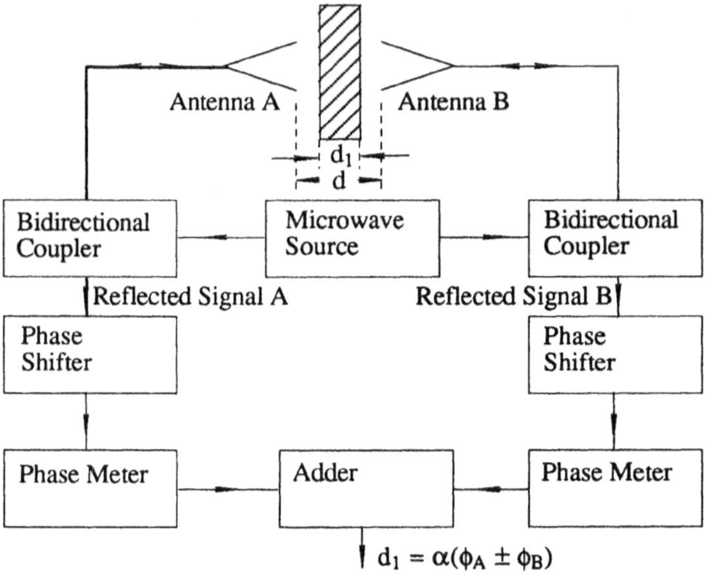

Figure 6.16. Thickness gaging using phase measurements.

6.3. Transmission and Attenuation Tests in Dielectrics and Lossy Dielectrics

The measurement of thickness in dielectrics and lossy dielectrics is similar to that of conductors but now the waves also penetrate into the dielectric. Thus, on the one hand, the methods discussed above can still be used but in addition, we can use the transmission and attenuation through the dielectric for testing. Also, there may be multiple reflections from both sides of the material which, under certain conditions may be used for measurement. Because of this, the use of interferometry is common in measurements in dielectrics. Even if all we measure is, for example, the reflected wave, the information on the material is much richer than for conductors. The reflection will give us information of the dielectric constant of the material, while attenuation measurements provide an indication of losses in the material. However, in addition to more opportunity for measurements, there are also more problems. Multiple reflections create both interference patterns as well as uncertainty as to what is being measured. In conductors, there was no reason to use any angle other than perpendicular incidence since we got total reflection in any case. With dielectrics a variety of angles may be used to provide the required reflection from the material or transmission into the material. In addition we must also define the polarization of the electric field as discused in chapter 3 since the transmission and reflection coefficients depend on the wave polarization. We may also be able to take advantage of the Brewster angle for the material (at which the reflection is zero). Total internal reflection is also possible for the waves inside the material (see section 3.4.1.). For this angle there is no transmission across the boundary and the far side of the dielectric may not be visible to the incoming wave. Although this does not occur with a single dielectric in free space (since all materials have a dielectric constant larger than that of free space) it may occur in layered materials and in immersed samples. Thus, it is obvious that we have many more parameters that can be used for testing.

In addition, flaws in the bulk of the dielectric can be detected based on scattering from the flaws but this aspect will be discussed separately in chapter 7.

In the following we discuss some of the methods applicable to testing and measuring dimensional aspects in lossless and lossy dielectrics. Before starting we follow the analysis of reflection and transmission in a dielectric as discussed in chapter 3. We assume first a lossless dielectric slab as shown in Figure 6.17 in free space with a wave incident from the left, reflected at both interfaces and propagating to the right. This produces a reflected wave to the left which accounts for all reflections and a transmitted wave. The reflected and transmitted waves were calculated in chapter 3 but,because of their importance, we will discuss here briefly the properties of the reflection and transmission coefficients.

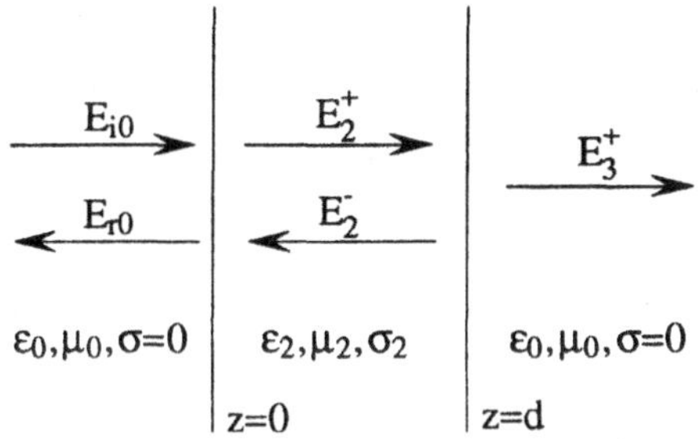

Figure 6.17. Reflection and transmission for a lossy dielectric slab in free space, at normal incidence.

For a lossless dielectric slab, the reflection and transmission coefficients are given by (see Eqs. (3.98) and (3.99))

$$\Gamma = -\frac{j\left[\eta_0^2 - \eta_2^2\right] \tan\beta_2 d}{2\eta_0\eta_2 + j\left[\eta_0^2 + \eta_2^2\right] \tan\beta_2 d} \qquad (6.7)$$

$$T = \frac{2\eta_0\eta_2\, e^{j\beta_2 d}}{2\eta_0\eta_2 \cos\beta_2 d + j\left[\eta_0^2 + \eta_2^2\right] \sin\beta_2 d} \qquad (6.8)$$

where d is the thickness of the slab and

$$\beta_2 = \omega\sqrt{\mu_2\varepsilon_2}$$

$$\eta_0 = \sqrt{\frac{\mu_0}{\varepsilon_0}} \qquad \eta_2 = \sqrt{\frac{\mu_2}{\varepsilon_2}}$$

From the reflection and transmission coefficient, it is obvious that either one can be calibrated to measure the thickness of the dielectric directly, since for given material properties and frequency this is the only parameter that varies in either the transmission or reflection coefficient. In fact, the transmission and reflection coefficient represent a "theoretical calibration" means for the test. In practice, there are be other parameters that need addressing. One is the attenuation in air. Others are noise, detection errors, and drift of sources. The behavior of the reflection and transmission coefficients for a lossless dielectric can be seen in Figure 6.18 where the reflection coefficient of a dielectric slab of thickness d and dielectric constant ε is shown at *10 GHz*. As the thickness is varied from zero to *1 cm*, the reflection coefficient varies from zero to a maximum and back to zero for a thickness of one quarter wavelength. Figure 6.18 shows the reflection coefficients of three materials: one of dielectric constant of *81* (water), the second with dielectric constant of *9* (e.g. alumina) and the third, with dielectric constant of *4* (e.g. quartz). Two important points should be noted. The first, is that any thickness slab can be measured but, if the slab thickness varies by more than a quarter wavelength, the measurement is not unique. Second, if the variations in thickness are small, and the thickness is much smaller than about $\lambda/8$, the change in reflection coefficient is more or less linear. This can be seen from Figure 6.19 where the plot for alumina is reproduced with a thickness varying between zero and *0.5 mm*. The maximum thickness in this case is about $\lambda/10$. The transmission coefficient behaves similarly, as shown in Figures 6.20 and 6.21, for the same dielectric constants and dimensions as in Figures 6.18 and 6.19.

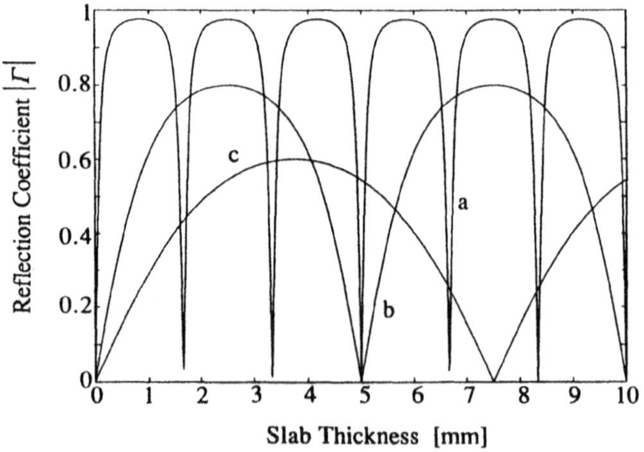

Figure 6.18. Absolute values of reflection coefficient at *10 GHz* for three lossless dielectrics. a. $\varepsilon_r=81$, b. $\varepsilon_r=9$, c. $\varepsilon_r=4$.

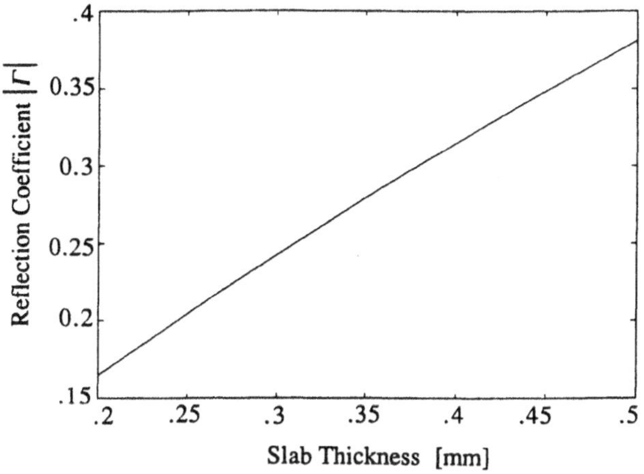

Figure 6.19. Absolute value of the reflection coefficient for an alumina slab *($\varepsilon_r=9$)*, of thickness between *0.2* and *0.5 mm*. The curve is almost linear and variations due to changes in thickness are also linear.

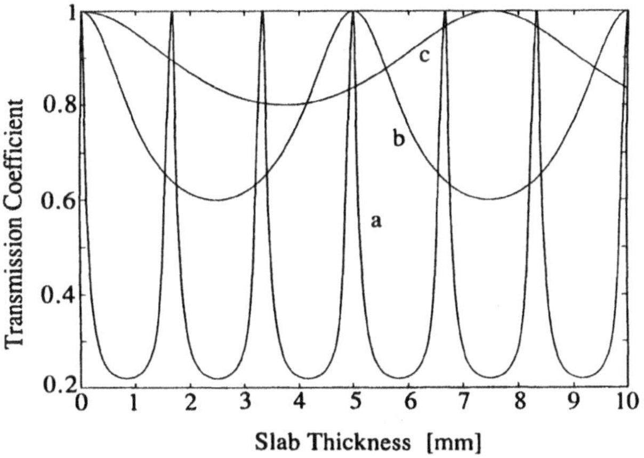

Figure 6.20. Transmission coefficient at *10 GHz* for three lossless dielectrics. a. $\varepsilon_r=81$, b. $\varepsilon_r=9$, c. $\varepsilon_r=4$.

If the dielectric is lossy, the attenuation in the slab must also be taken into account. Assuming a low loss dielectric, we can use the same relations as above but now the attenuation constant is part of the formula. If we use the same steps, we obtain the transmission and reflection coefficients as (see Eqs. (3.100) and (3.101))

$$\Gamma = -\frac{j\left[\eta_0^2 - \eta_2^2\right] \tan(\beta_2 - j\alpha_2)d}{2\eta_0\eta_2 + j\left[\eta_0^2 + \eta_2^2\right] \tan(\beta_2 - j\alpha_2)d} \tag{6.9}$$

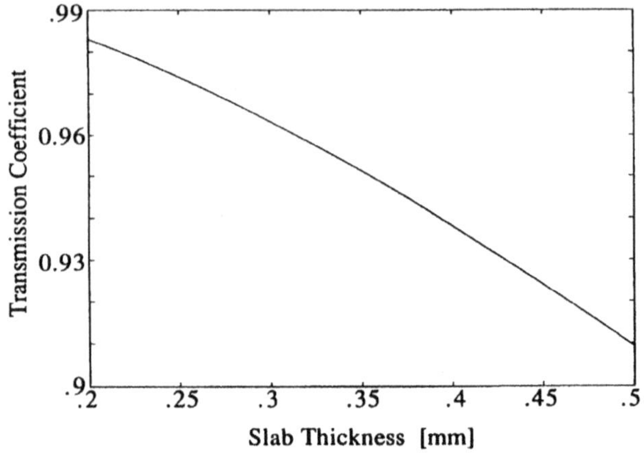

Figure 6.21. Transmission coefficient for an alumina slab ($\varepsilon_r=9$), of thickness between *0.2* and *0.5 mm*. The curve is almost linear as are changes in thickness.

and the transmission coefficient through the slab is

$$T = \frac{2\eta_0\eta_2 \, e^{\alpha_2 d} e^{j\beta_0 d}}{2\eta_0\eta_2 \cos(\beta_2 - j\alpha_2)d + j\left[\eta_0^2 + \eta_2^2\right] \sin(\beta_2 - j\alpha_2)d} \tag{6.10}$$

where, as previously, d is the thickness of the slab and:

$$\beta_2 \approx \omega\sqrt{\mu_2\varepsilon_2}$$

$$\alpha_2 \approx \frac{\sigma_2}{2}\sqrt{\frac{\mu_2}{\varepsilon_2}}$$

$$\eta_2 \approx \sqrt{\frac{\mu_2}{\varepsilon_2}}\left(1 + j\frac{\sigma}{2\omega\varepsilon}\right)$$

In a high loss material, the reflection coefficient is large while the transmission coefficient is rather small. The material may behave very close to a conductor and transmission measurements are not always possible. The reflection and transmission coefficients are given by (see Eqs. (3.102) and (3.103))

$$\Gamma = -1 - j\frac{\eta_2}{\eta_0}\frac{1 + j\tan\beta_2 d \tanh\alpha_2 d}{\tan\beta_2 d - j\tanh\alpha_2 d} \tag{6.11}$$

$$T = -j\frac{\eta_2}{\eta_0}\frac{2e^{j\beta_0 d}}{sin(\beta_2 - j\alpha_2)d} \qquad (6.12)$$

where

$$\alpha_2 = \beta_2 = \sqrt{\pi f \mu_2 \sigma_2} \qquad \beta_0 = \omega\sqrt{\mu_0\varepsilon_0}$$

$$\eta_2 = (1+j)\frac{\alpha_2}{\sigma_2}$$

and η_0 is as given above. The results here are for a high loss dielectric. If the dielectric has low losses, the results in Eqs. (3.100) and (3.101) must be used.

As is obvious from the formulas above, the reflection and transmission coefficients are frequency dependent. Figure 6.22 shows the reflection coefficient for two lossy materials. One has permittivity of $81\varepsilon_0$ *[C/m]* and conductivity of *4 [S/m]* (sea water) while the other has permittivity of $9\varepsilon_0$ *[C/m]* and conductivity of *0.1 [S/m]*. The attenuation effect is immediately obvious, especially in comparison with Figures 6.18 and 6.19. The frequency dependence of the reflection coefficient is shown in Figures 6.23 through 6.25. In Figure 6.23, the reflection coefficient for *10 cm* slabs of the two materials above are shown in the frequency range *1 MHz* to *100 GHz*. Figure 6.25 shows frequency dependence for lossless materials (same permittivities as above) in the range *1 MHz* and *10 GHz*. Although it may seem as if the reflection coefficient increases with frequency, this is misleading. The slab thickness in this plot is *1 cm* and, since this is less than a quarter wavelength, the reflection coefficient is monotonic. Figure 6.25 shows the same result for a *10 cm* slab. The reflection coefficient now changes and looks similar to that of Figure 6.18. Similar results may be obtained for low loss dielectrics or for the transmission coefficients of various materials.

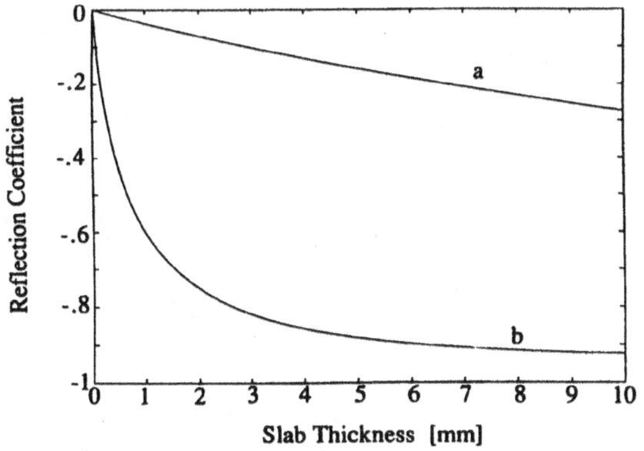

Figure 6.22. Reflection coefficient versus thickness for lossy dielectric slabs. a. ε_r=*81*, σ=*4 [S/m]*. b. ε_r=*9*, σ=*0.1 [S/m]*.

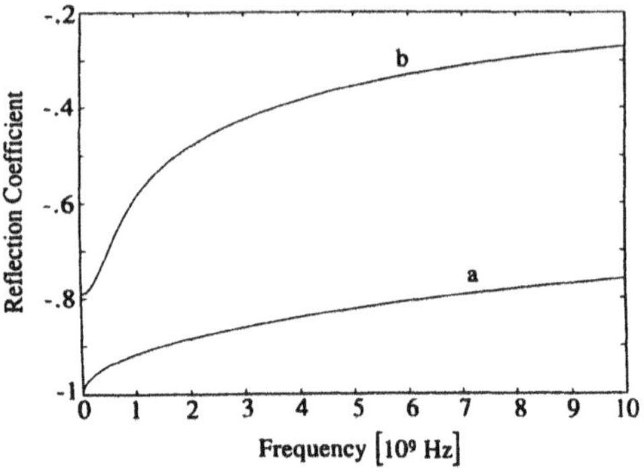

Figure 6.23. Reflection coefficient versus frequency for lossy dielectric slabs. a. $\varepsilon_r=81$, $\sigma=4$ *[S/m]*. b. $\varepsilon_r=9$, $\sigma=0.1$ *[S/m]*.

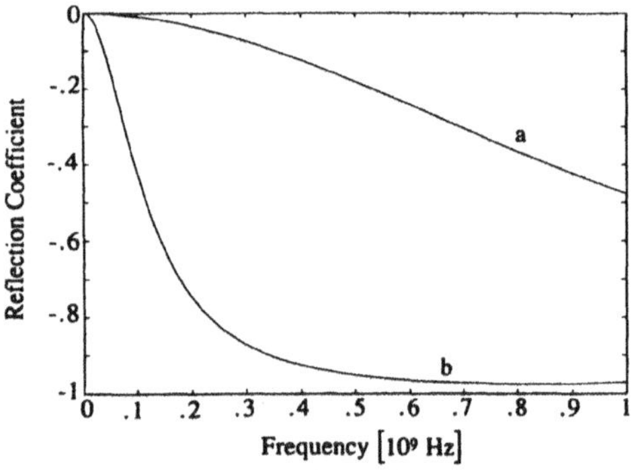

Figure 6.24. Reflection coefficient versus frequency for lossless dielectric slabs. Slab thickness is *1 cm*. a. $\varepsilon_r=81$, $\sigma=0$ *[S/m]*. b. $\varepsilon_r=9$, $\sigma=0$ *[S/m]*.

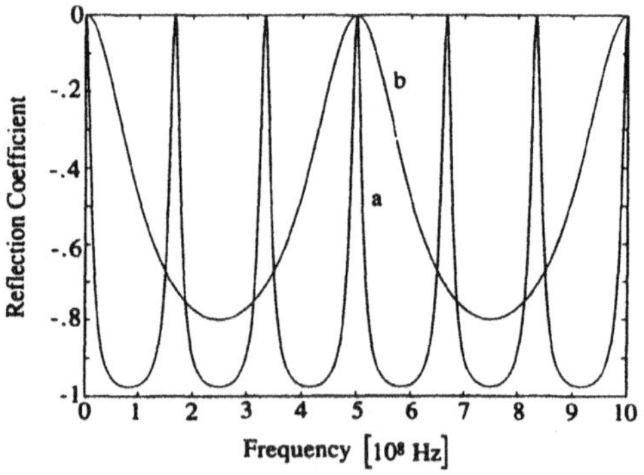

Figure 6.25. Reflection coefficient versus frequency for lossless dielectric slabs. Slab thickness is *10 cm*. a. $\varepsilon_r=81$, $\sigma=0$ *[S/m]*. b. $\varepsilon_r=9$, $\sigma=0$ *[S/m]*.

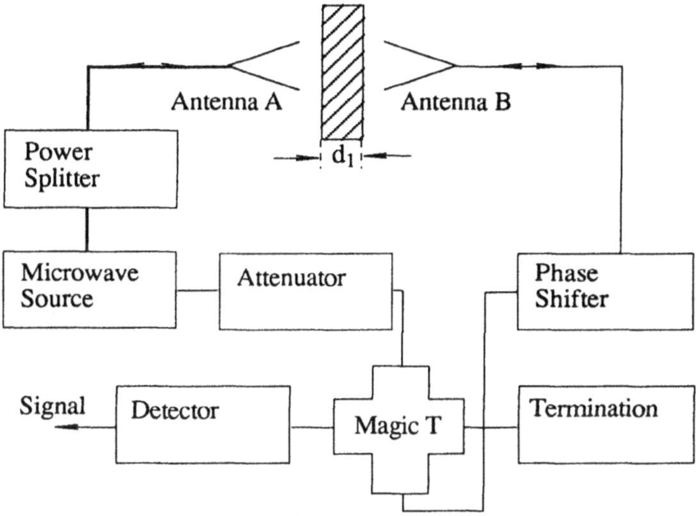

Figure 6.26. A modified transmission test using a balanced bridge.

Thickness of a dielectric piece may be measured from the transmission of waves through the dielectric. A very basic arrangement is shown in Figure 6.2. The dielectric is placed between the antenna and detector and the amplitude at the detector is a measure of the thickness of the material. This arrangement is not the best method of testing, and in fact, unless the changes in thickness are relatively large, it may not be practical. However we will use it here to show the principles involved, before proceeding further. For a perfect (lossless) dielectric, the detected signal depends only on the reflection and transmission coefficients at each of the two material boundaries. There may also be standing waves inside the material but we may assume that for an amplitude E_0 of the source (say of the electric field), there will be an amplitude at the detector, say E_1. This amplitude is a rather complex function of the incident field, the thickness of the dielectric, its permittivity and permeability as well as frequency and distance between antenna and detector but, for a given setup, changes in the detector output are proportional to material thickness, as shown in the equations and figures above.

The main problem with this arrangement is the need for absolute measurement of distance. Also, if the attenuation is relatively high, the measured signal may be very small, leading to large errors in measured thickness.

A more accurate method is shown in Figure 6.26, and is based on a bridge measurement of signals. The signal from the source is passed to one arm of the bridge while the signal after passing through the dielectric is passed to the other arm. This is a very basic measurement but, to make it practical, the signals must be equal in magnitude and phase. This is done by an attenuator in one arm and a phase shifter in the other. The attenuator and the phase shifter are adjusted to zero output of the bridge (in Figure 6.26, a magic *T* hybrid junction is used). Now, any variation in the amplitude (or phase) of the received signal due to passage through the material is a measure of material properties. If the material properties do not change, then the output is a measure of material thickness.

As with any other method of measurement, the output must be properly calibrated. In some cases, the output is linear in a small range and this can be employed in the test.

Under most conditions, there will be a reflected wave at the interface which alters the incident field both in the vicinity of the material and at the antenna itself. In fact, the reflected wave will create standing waves inside the waveguide leading to the antenna unless a directional coupler is installed near the antenna. In some cases these standing waves can be monitored to give an indication of the reflection coefficient, and therefore, of the dielectric that caused them. The interdependencies of the waves and material for lossless and lossy dielectrics was discussed at the beginning of this section. From these, it is obvious that the material thickness can be measured from the reflection coefficient. In practice, all that is required is to isolate the incident and reflected waves by use of directional couplers. A test setup is shown in Figure 6.27. In this, the incident wave is reflected back into the antenna (which, in fact may be just an open waveguide). This reflected wave is detected and measured as is the incident wave. The ratio between the two gives the reflection coefficient. Alternatively, if the two signals are mixed and they are equal in magnitude and phase for a given thickness, the output will be proportional to variation in thickness. The two signals can be nulled by introducing an attenuator in the incident wave section and a phase shifter in the reflected wave section. This is shown in Figure 6.28.

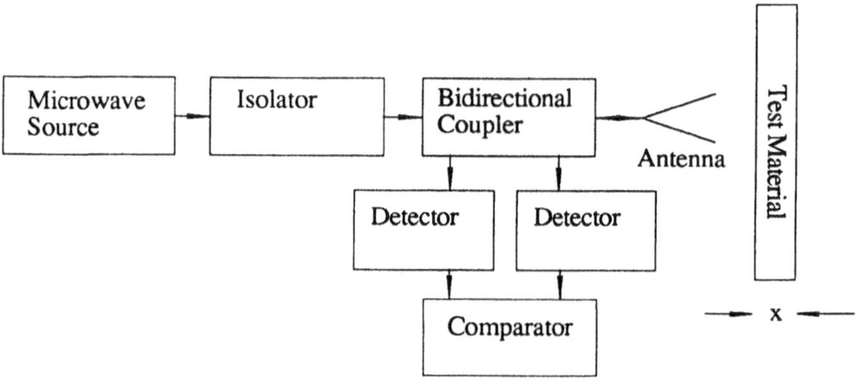

Figure 6.27. Thickness gaging of dielectric slabs using a reflection method.

The magic *T* hybrid can be used in a bridge configuration as shown in Figure 6.29. The attenuator and phase shifter are still required to balance the bridge, however, in the magic *T* configuration this can be easily accomplished by the adjustable short shown in Figure 6.29. The variable short is, a shorted section of waveguide in which the shorting plate can be moved and is, in effect, a variable impedance that allows balancing of the bridge. The output now is directly proportional to the thickness variation.

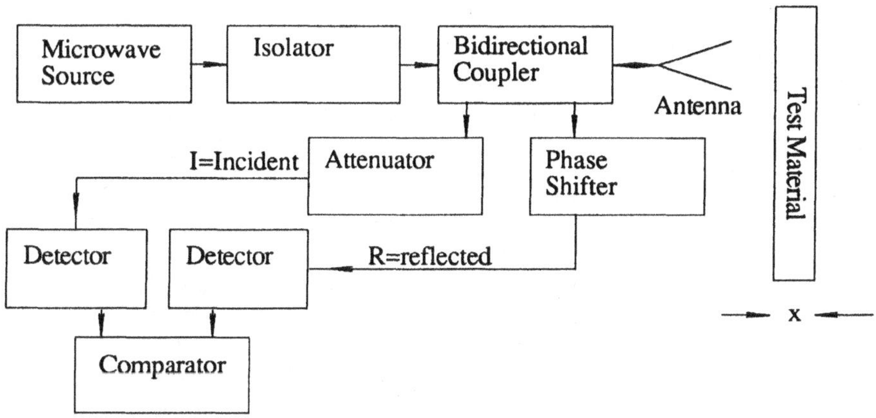

Figure 6.28. Thickness gaging with a null detector.

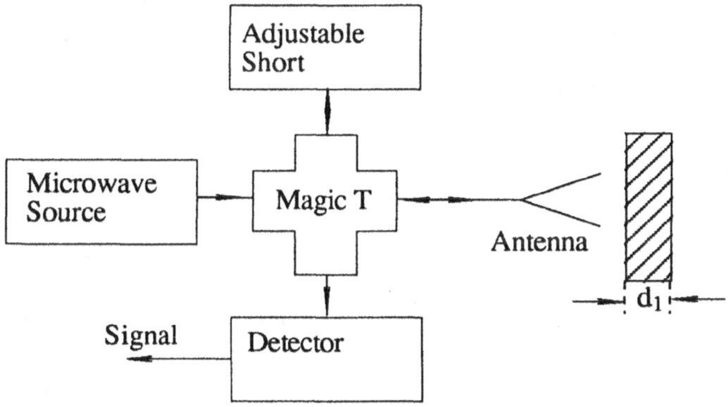

Figure 6.29. A balanced bridge reflection test for dielectric slabs.

6.4. Standing Wave Measurements

The setup of standing waves inside and outside the dielectric is another means of testing that can be utilized. Since the reflected wave interferes with the incident wave, there will be a standing wave pattern in the space between material and antenna and inside the waveguide. The principle of measurement is shown in Figure 6.30. In this simple arrangement, a slotted line is shown because it is one of the simplest methods of detecting maxima and minima in the standing wave pattern as well as measuring the distance between them. The standing wave pattern has been described in chapter 2. In dielectrics, the minima are never actual nulls (an exact null requires total reflection and zero losses) but the minima are usually easy to detect. For a given thickness, one of the minima (or maxima) is detected. Any change in thickness will shift the minimum to a new location. The distance required to move the probe in the slotted line to re-establish the minimum is a direct measure of the material thickness. Since minima (of the electric or magnetic field) are spaced a distance $\lambda/2$ apart, (see Figure 6.31) a change Δl in thickness must move the minimum the same distance (either positive or negative). Unless the reflections are very small (and, therefore, the maxima or minima are difficult to identify), this measurement can be very accurate. The same method can be used with conductors as well. However, it is extremely sensitive to location of the interface. Lateral movement of the material will be interpreted as variation in thickness. Another difficulty with standing wave measurements is the fact that there are multiple reflections. In particular, the two interfaces of the material will each produce a standing wave pattern. Thus, maximum or minimum is due to the total wave and separation between individual reflections may not be feasible. Also, and perhaps most important, the use of the standing wave measurement is difficult to automate because of the need to measure fields in a spatial sense. Thickness variations of up to $\lambda/4$ can be normally measured but, since the standing wave minima repeat at $\lambda/2$, it is possible to measure variations of up to $\lambda/2$. If this is done, it should be noted that the amplitude is ambiguous over $\lambda/4$ and unless it can be determined that the range is above $\lambda/4$, measurements should be restricted to $\lambda/4$. This is not normally a problem in the microwave region but can be a restriction in the *mm* wave region.

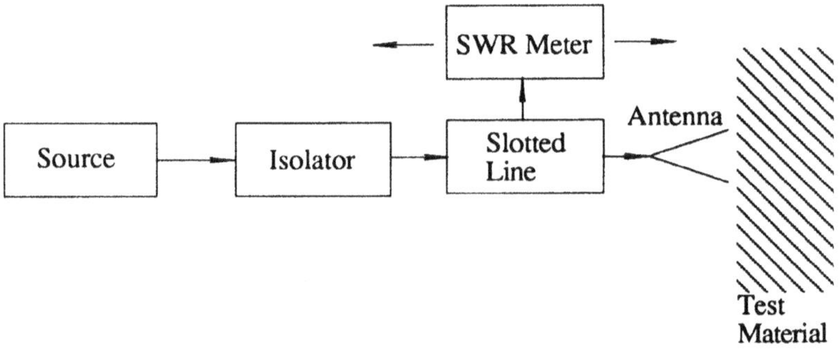

Figure 6.30. Standing wave measurement.

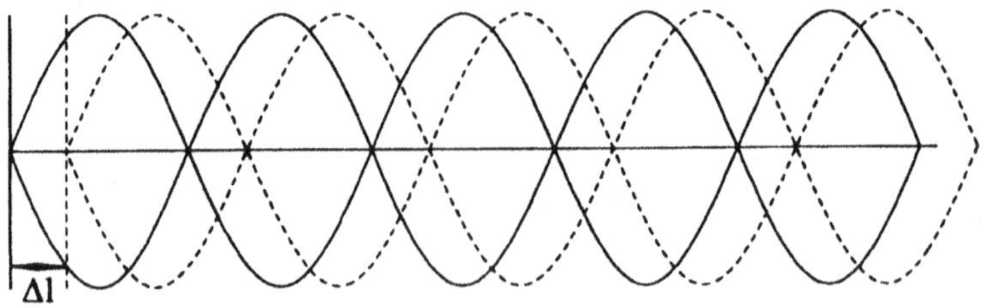

Figure 6.31. Standing wave patterns and their variation due to thickness.

6.5. Phase Measurements on the Transmission and Reflection Coefficients

While the reflection and transmission can either be used in measuring the thickness of a material, we can choose to monitor either the amplitude or the phase of the coefficients. When measuring the reflection (as, for example in Figure 6.13), the bridge can be set to measure the phase difference between the incident and either the transmitted or reflected wave. A measurement of the phase of the transmitted wave is shown in Figure 6.32 and that for the reflected wave in Figure 6.33. The phase of the transmitted or reflected waves can be calculated from the reflection and transmission coefficients. These are calculated from Eqs. (6.7) and (6.8) for the lossless material, and are given as

$$\phi_R = tan^{-1}\left[\frac{2\eta_0\eta_2}{\left[\eta_0^2+\eta_2^2\right]tan\beta_2 d}\right] \tag{6.13}$$

and

$$\phi_T = tan^{-1}\left[\frac{2\eta_0\eta_2 cot\beta_2 d}{\left[\eta_0^2+\eta_2^2\right]}\right] \tag{6.14}$$

Thus, the phase is a good measure of thickness. The plots in Figures 6.32 and 6.33 were obtained from the theoretical expressions of the reflection and transmission coefficients as given in Eqs (6.7) and (6.8) for a lossless material. This type of plot can be obtained for any given material and then used as a calibration curve. Note in particular that if the frequency is such that the measurement is done in the middle range, the phase of the reflection coefficient varies almost linearly with material thickness. Also, if the material is over one quarter wavelength, the measurement is not unique as discused above.

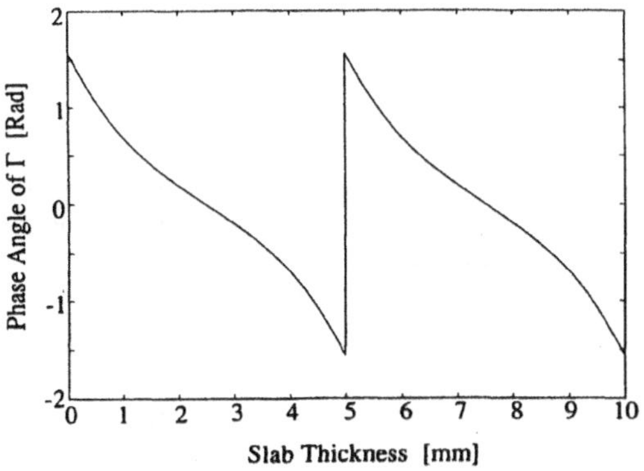

Figure 6.32. Phase of the reflection coefficient as a function of material thickness at *10 GHz* for a lossless material with $\varepsilon_r=9$.

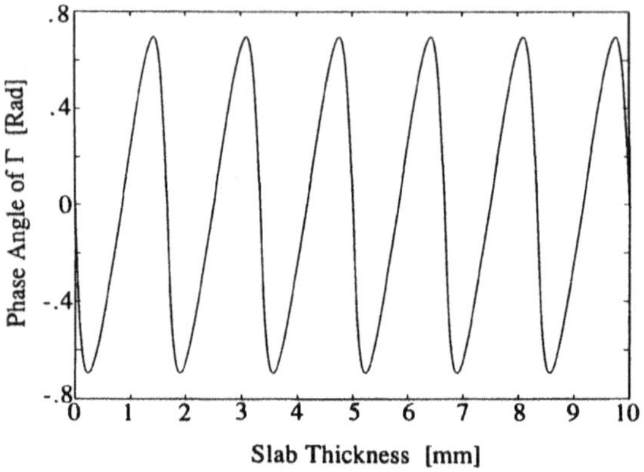

Figure 6.33. Phase of the transmission coefficient as a function of material thickness at *10 GHz* for a lossless material with $\varepsilon_r=9$.

6.6. Frequency Measurements

Since the reflection coefficient as well as the transmission coefficient are frequency dependent, (see Figures 6.23 through 6.25) the thickness of lossless and lossy materials may be measured by varying the frequency of the source. Material thickness as well as variations in thickness can be accomplished using frequency domain reflectometry. The principle is shown in Figure 6.34. A swept frequency source is used, allowing a continuous change in frequency at a constant rate. The incident and reflected waves will be of different frequencies because during the delay in time needed for the wave to travel to the target and back the source has shifted in frequency. Assuming the generator sweeps between a high frequency f_h and a low frequency f_l at a rate of df/dt, the thickness of the material is calculated as follows:

The time required for the wave to travel to the material interface and back to the antenna is ΔT and is given by

$$\Delta T = 2\Delta L\sqrt{\mu_0\varepsilon_0}$$

where the speed of light was used for the speed of propagation (the wave propagates in free space). The time required for the wave to travel from the surface of the material to the second interface and back to the surface is

$$\Delta t = 2x\sqrt{\mu\varepsilon}$$

The total time of travel is $\Delta T+\Delta t$. The shift in frequency is

$$\Delta f = f_i - f_r = \frac{df}{dt}(\Delta T + \Delta t) \qquad (6.15)$$

where f_i and f_r are the frequencies of the incident and reflected waves respectively.

Since the distance to the sample can be measured directly, the shift in frequency due to propagation in free space can be subtracted and the total shift in frequency due to travel of the wave from the surface of the material to the back surface is

$$\Delta f' = \frac{df\,\Delta t}{dt}$$

This frequency can be measured very accurately and from it the material thickness is calculated as

$$x = \frac{\Delta f'}{2\sqrt{\mu\varepsilon}\,\frac{df}{dt}} \tag{6.16}$$

In addition to the change in frequency, the phase of the wave also changes and these changes can also be used for testing. Thus, all that is required are the parameters of the swept source and the material properties of the sample. Alternatively, this method can also be used to monitor material properties. It is important to note that multiple reflection within the sample can occur. Each additional reflection increases the path of the wave by $2x$. This must be taken into account and only the lowest distance should be taken as the correct measurement. In terms of Figure 6.24, the frequencies obtained will be Δf, $\Delta f+\Delta f'$, $\Delta f+2\Delta f'$, etc.

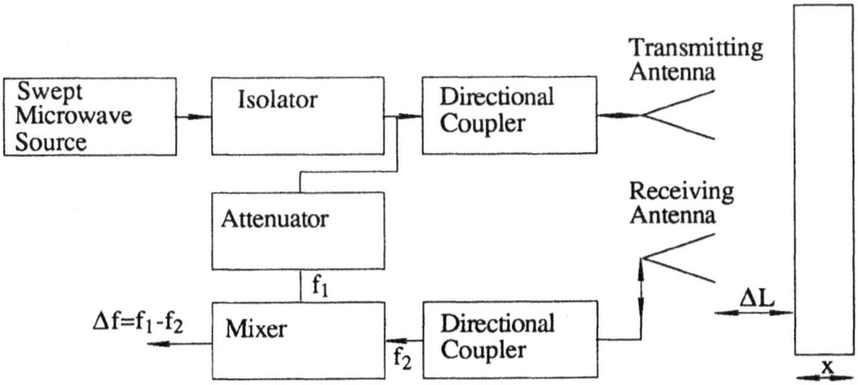

Figure 6.34. Swept-frequency testing of material thickness.

The frequency sweep rate should be relatively fast if high resolution is required. As an example, the shift in frequency due to water (ε_r=81), at a rate of sweep of *1 GHz/sec* (e.g.

the generator sweeps between *1 GHz* and *2 GHz* in one second) is *30 Hz/meter* depth. A sweep ten times as fast will produce a sensitivity of *300 Hz/meter* or *0.3 Hz/mm*. Since swept frequency generators with sweep rates well over *500 GHz/sec* are available, sensitivities of over *15 Hz/mm* or larger are possible, resulting in very accurate measurements.

This method is equivalent to the Doppler radar and is sometimes called a frequency domain Doppler method.

6.7. Measurement of Coated Conductors

On coated conductors, measurement of thickness is limited by the reflection of the conductor. However, some of the methods above can still be used. In particular, the reflection coefficient is a good measure of the coating thickness. Attenuation through the dielectric as well as standing waves can also be used. Standing waves must be used cautiously since the reflection coefficient at the metallic surface is equal to *–1* while at the dielectric interface the reflection coefficient is much smaller. The standing wave pattern is much more difficult to scan and interpret.

Measurement of the reflection coefficient is particularly useful since it is proportional to material thickness. This can be seen immediately from the reflection coefficient of a lossless dielectric backed by a perfect conductor, as shown in Figure 6.35.
The reflection coefficient for this configuration is (Eq. 3.104))

$$\Gamma = -\frac{\eta_0 - j\eta_2 \tan\beta_2 d}{\eta_0 + j\eta_2 \tan\beta_2 d} \qquad (6.17)$$

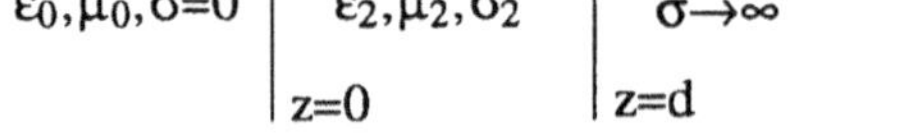

Figure 6.35. Geometry used in calculation of the reflection coefficient for a metal backed coating.

where η_0 is the intrinsic impedance of free space and η_2 is the intrinsic impedance of the dielectric. β_2 is the propagation constant in the dielectric. These three quantities are given by

$$\beta_2 = \omega\sqrt{\mu_2\varepsilon_2} \qquad \eta_0 = \sqrt{\frac{\mu_0}{\varepsilon_0}} \qquad \eta_2 = \sqrt{\frac{\mu_2}{\varepsilon_2}}$$

with ω the angular frequency, μ_0 and ε_0 the permeability and permittivity of free space and μ_2 and ε_2 the permeability and permittivity of the material.

As long as material properties and frequency remain constant, the ratio between the reflected and incident fields give the reflection coefficient and this only varies with the dielectric thickness.

Thus, an appropriate test using this method is the test setup in Figure 6.7a or 6.7b. The incident and reflected waves are measured and correlated to the thickness of the slab using a calibration curve for the material under test. The receiving and transmitting antennas are placed on the coating side. In all other respects, the measurement of coatings is the same as measurement of dielectric materials, other than the obviously different calibration curve and the fact that transmission measurements are not possible. As mentioned before, it is also possible to use standing wave measurements but this is more complicated than in testing of conducting materials because of the dielectric coating.

6.8 Bibliography

The use of microwaves for dimensional testing has been used extenssively for many applications, most of which are not related to *NDT*. The references given here are some of those that do relate to *NDT*. In particular, Bahr [1] summarizes many of the methods that have been used for dimensional testing. Other references describe specific aspects of testing and some related theoretical issues.

[1] A.J. Bahr, "Microwave Nondestructive Testing Methods, Gordon and Breach Science Publishers, 1982.
[2] R. Zoughi, "Review of NDT techniques at radio and microwave frequencies", QNDT Rev. of Progress, 1989, pp. 337-344.
[3] R. Zoughi and M. Lujan, "Nondestructive microwave thickness measurements of dielectric slabs", Materials Evaluation, Vol. 48, Sept. 1990, pp. 1100-1105.
[4] W.L. Rollwitz, "Microwave Inspection", Metals Handbook, Vol. 11, 8th edition, 1976, Nondestructive Inspection and Quality Control, ASM, Metals Park, OH., pp. 244-253.
[5] R. Hochschild, "Microwave nondestructive testing in one (not-so-easy) lesson", Materials Evaluation, Vol.26, No. 1, 1968, pp. 35a-42a.
[6] M. Rockowitz and L.J. McGuire, "A microwave technique for the detection of voids in honeycombed ablative materials", Materials Evaluation, Vo. 24, No. 2, 1966, pp. 105-108.
[7] T.M. Lavelle, "Microwaves in nondestructive testing", Materials Evaluation, Vol. 25, No. 11, 1967, pp. 254-256.
[8] A.J. Bahr, "Using electromagnetic scattering to estimate the depth of a rectangular slot", IEEE Transactions on Antenna and Propagation, Vol. AP-27, No. 6, Nov. 1979, pp. 738-746.

[9] A. Thansandote, S.S. Stuchly and J.S. Wight, "Microwave interferrometer for measurements of small displacements", IEEE Transactions on Microwave Theory and Techniques, Vol. No., 1982, pp.

[10] W.F.P. Chan and B. Chambers, "Measurement of nonplanar dielectric samples using an open resonator", IEEE Transactions on Microwave Theory and Techniques, Vol. MTT-35, No. 12, Dec. 1987, pp. 1429-1434.

[11] J.J. Holmes and C.A. Balanis, "Refraction of a uniform plane wave incident on a plane boundary between lossy media", IEEE Transactions on Antennas and Propagation, Vol. AP-26, No. 5, Sept. 1978, pp. 738-741. (MW 108)

Chapter 7. Testing for Discontinuities

7.1. Introduction

Discontinuities in materials affect the microwave energy in two basic ways, beyond the mere reflection and transmission of energy. One of these is scattering. Scattering means that the flaw, because of its shape and material composition scatters the beam in various directions. This scattered energy can then be detected and the scatterer properties (such as size, location, and composition, may be deduced from the scattered field. As a rule, scattering is more pronounced from abrupt, large changes in material properties. Small, continuous changes create little or no scattering. The general rule for strong scattering is that the scatterer should be larger than about one-half wavelength. Thus, higher frequencies are required for detection of smaller defects. As an example, at *10 GHz* (*X*-band), the wavelength is *3 cm*. Thus, delaminations in a composite material of about *1.5 cm* can be detected. This is reasonable for delaminations which can be extensive in size but detection of a flaw of this size is not considered a difficult problem in *NDT*. On the other hand, in the millimeter wave range, detection of defects can be useful since much smaller defects and higher resolutions are possible.

The other method of interaction of microwaves with discontinuities is through changes in attenuation or absorption in the material. This is a case of changes in material properties in which scattering plays no role. Thus, for example, an area of wash-off may not show strong scattering but will show a significant change in attenuation because of reduced fiber content. Similarly, small areas of delamination may affect the reflection and transmission coefficients as well as the attenuation but not the scattering pattern.

These basic interactions are used for defect detection and evaluation with microwaves. Many of the methods used are the same as those discussed in the previous chapter, but the emphasis, and the quantities of interest are different. In this chapter we discuss primarily scattering methods, because these methods are most often used in conjunction with flaw detection. This does not imply that other methods cannot be used or, that scattering methods are used more often than, say, resonant methods. All of these methods are used for flaw detection but, to avoid too much duplication, scattering methods are discussed here while resonant methods are discussed in the following chapter. Only short references are given to reflection and transmission methods since these were discussed in chapter 6.

7.2. Scattering Methods of Flaw Detection

One of the simplest methods of testing for defects of various types is to measure the scattered energy from flaws. The method is based on the fact that flaws are an irregularity inside, or on the surface of the material. Any flaw will change the reflection (as well as transmission) of energy through the material and will scatter some of the energy. Scattering from flaws can be used with perpendicular or oblique incidence and on almost any material. The principle is illustrated in Figure 7.1. In this arrangement, the beam is passed through the material (assuming a low loss dielectric). Under normal conditions, some of the wave is transmitted and some is reflected but there is no scattering. If a defect exists, there is also scattering in various directions. If the direction perpendicular to the direction of the beam, is available for measurement, this should be chosen as it gives the best rejection ratio with incident and reflected energy. If the perpendicular location is not available, or is not practical, any other angle can be used. A possible arrangement is shown in Figure 7.2 but in this case, the detector also detects energy that is not scattered. Scattered energy due to defects is indicated by change in the detected energy. As with other methods of *NDT*, the larger the defect, the larger the scattering. However, in microwave *NDT*, "larger" means larger electrically or larger in terms of a target or radar cross section. The radar cross section is a measure of the scattered field at a large distance (assuming Rayleigh scattering, see chapter 3)

$$\sigma = 4\pi r^2 \frac{P_s}{P_i}$$

where P_s is the scattered power density and P_i is the incident power density. The scattering cross section σ is a function of size, shape, and material properties of the object and, for a given frequency is constant. However, it varies with the orientation of the object.

Normally, we will be interested in the scattered field rather than the scattering cross section or

$$P_s = \frac{\sigma P_i}{4\pi r^2}$$

Thus, although a larger object will normally have a larger scattering cross section (and therefore larger scattered power density) a large, smooth defect, may scatter less than a small, sharp, faceted defect. Also, the orientation of the defect with the incident wave is very important. A defect, or discontinuity tends to have dominant dimensions and, if a polarized wave is used, with direction in polarization in the direction of the dominant dimension of the defect, the detection is enhanced considerably.

A similar method can be used for detection and evaluation of surface flaws and irregularities. In this case, the detector is placed above the point of incidence of the beam. Normally, little energy is coupled to the receiver, unless a flaw or irregularity is detected. The coupled energy is now increased due to scattering at the surface. This type of test is shown in Figure 7.3. The whole assembly can also be scanned sideways to obtain complete coverage say, of a strip of material.

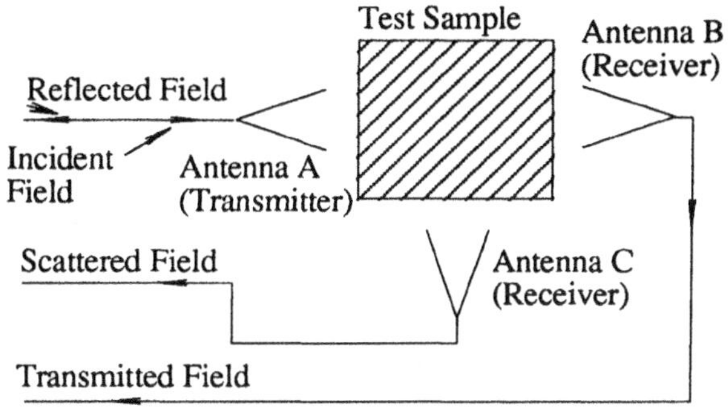

Figure 7.1. A basic scattering test. Antenna *B* receives transmitted waves, while antenna *C* receives scattered waves.

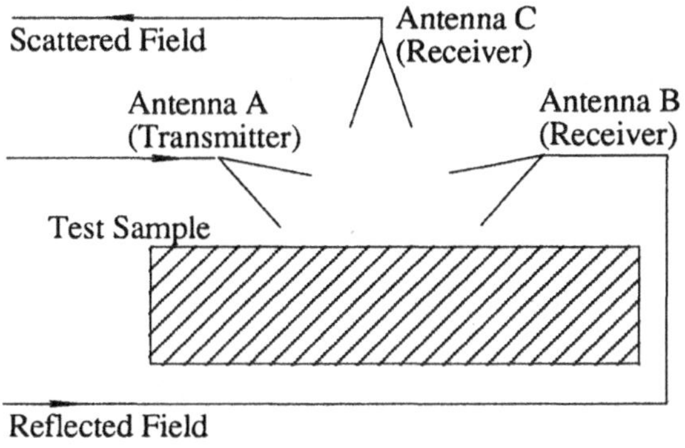

Figure 7.2. Scattering test with oblique incidence. This method is useful for surface defects.

The same method can be used to detect delamination and wash-off in composites. The situation is shown in Figure 7.4. In this case a simple absolute measurement is employed but bridge measurements are also possible as was done for reflection tests in chapter 6.

In principle, a pulsed method can be used for flaw detection as shown in Figure 7.5. This is discussed in more detail in the following section in terms of finding the location of defects. The method is similar with the pulsed method in ultrasonics and, in fact, is a standard "radar" method. The difficulty with this type of method is the very short ranges encountered in testing of materials and the high speed of propagation of the waves. Nevertheless, with proper instrumentation, short duration pulses can be used for testing for flaws in thick dielectric materials such as concrete.

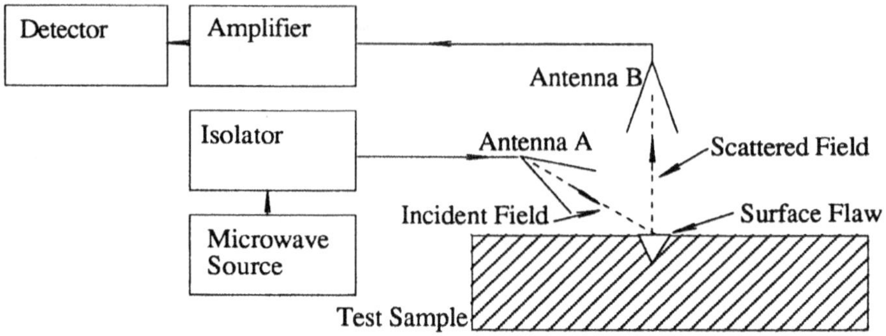

Figure 7.3. A scattering test for surface flaws.

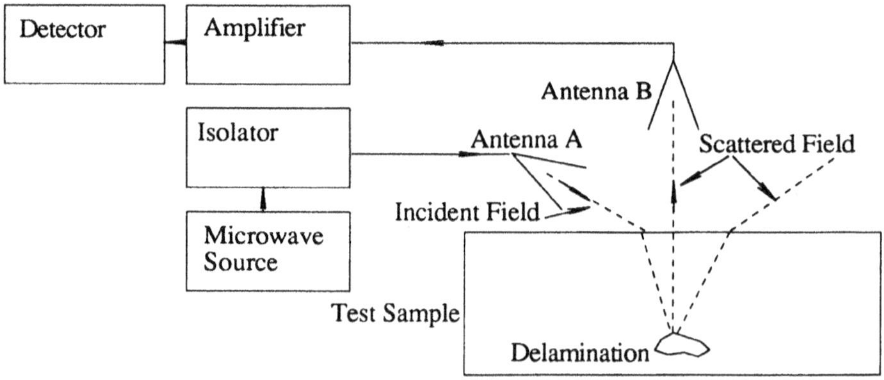

Figure 7.4. A scattering test for subsurface defects such as delaminations.

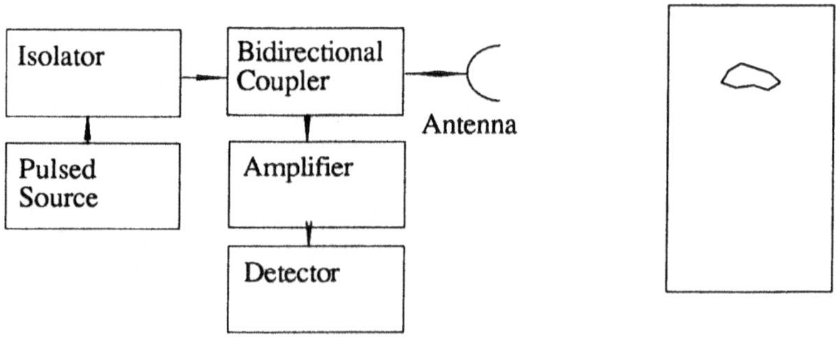

Figure 7.5. Scatterometry (radar) testing for defects.

A variation of the scattering methods described above is to use a continuous illumination of the test sample and a separate probe (or antenna) to scan the surface of the sample. This method is shown in Figure 7.6. If no flaws are present, there is only an incident and a reflected wave. These produce a field profile above the material which is known, and characteristic of the material. Any flaw inside the material or on the surface, changes the reflected wave because of the scattered field. The field profile now changes significantly.

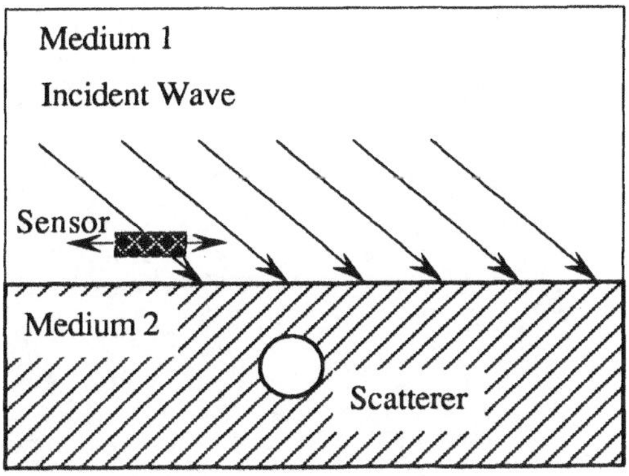

Figure 7.6. A method of scanning a test sample. The illumination is constant but the probe moves.

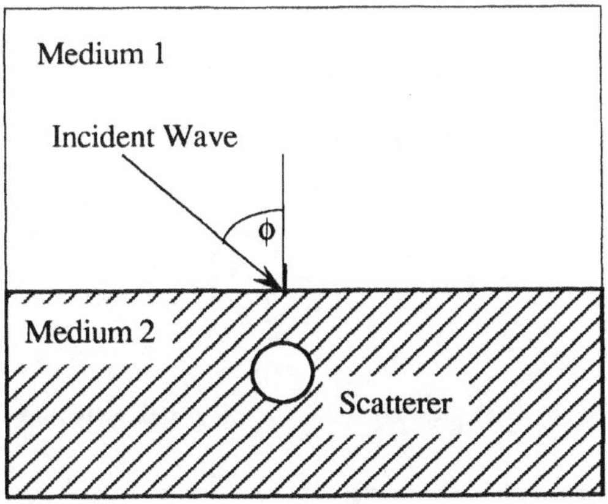

Figure 7.7. A scatterer in a lossy dielectric with oblique wave incidence.

As examples of this type of test consider the test geometry in Figure 7.7. It consists of a thick material with an embedded scatterer. The scatterer can represent, for example, rebars in thick concrete. Figures 7.8 and 7.9. show the field throughout the cross section of the geometry with and without the scatterer. Figure 7.10 shows a single scan above the surface for a geometry as in Figure 7.7 but with two scatterers. The difference between the two is significant. Figure 7.11 shows the scattered field only. This can be easily produced by subtracting the field without scatterers from the field with scatterers present. The location and shape of the scatterers is clearly visible. As with any other test, the wavelength must be of the same order of magnitude as the size of defects for best results. For small defects this is not always possible but for tests such as for voids or rebars in concrete this is easily accomplished with standard microwave equipment. For smaller scatterers, millimeter wave test may be necessary. This particular test shows not only what is possible in terms of detection of discontinuities but also the possibility of using microwave data for imaging of defects and other discontinuities in dielectrics and lossy dielectrics.

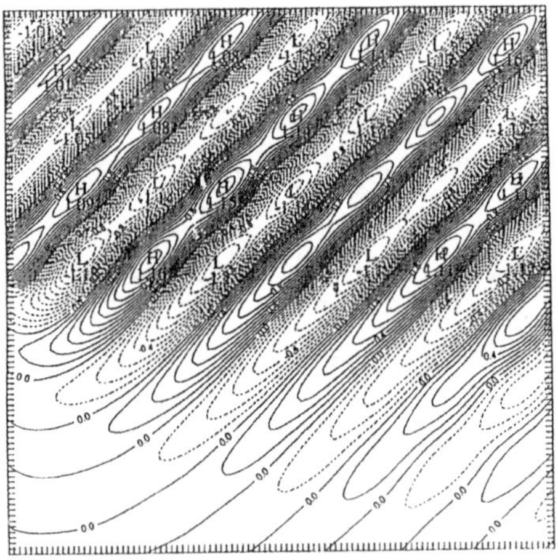

Figure 7.8. Field throughout the cross section of Figure 7.7, without the scatterer.

Figure 7.9. Field throughout the cross section of Figure 7.7, with the scatterer.

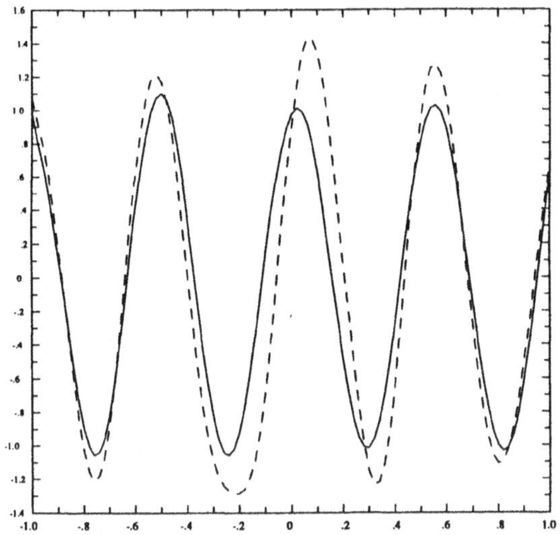

Figure 7.10. Single scan above the lossy dielectric in Figure 7.7, showing the signals with and without two scatterers.

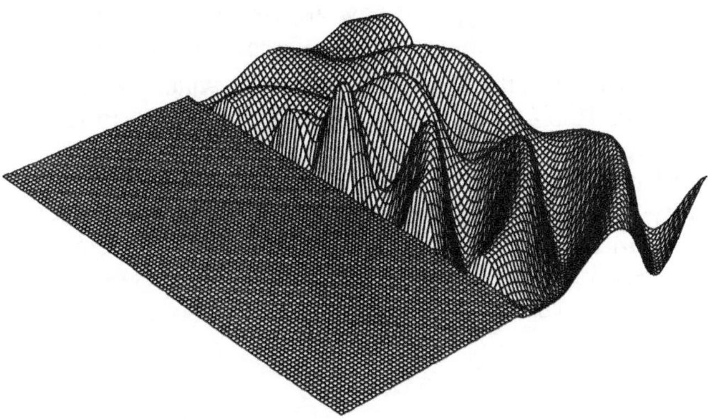

Figure 7.11. Scattered field due to two scatterers.

7.3. Location of Flaws

The accurate location of flaws such as delaminations is an important aspect of testing. In some cases, the location of the flaw is more important than its size. As a rule, location of the flaw should always be viewed as an integral part of the test.

There are three basic methods of locating a flaw in a material. Depending on the type of test, size of flaw and depth, type of material tested, and accuracy required, one of these methods or variations of the basic methods may be employed.

The most straightforward method is ranging. This method is often done with ultrasonic methods, and is the fundamental measurement in radar systems. The basic idea is shown in Figure 7.12. A pulse is generated at the source. This travels through free space, is reflected from the first material interface and reaches the source at a time t. Assuming the flaw also reflects the wave, a second pulse, normally smaller in magnitude reaches the source at a time $t+2\Delta t$. A third pulse is reflected from the back surface of the material and reaches the antenna at a time $t+2\Delta t_1$. The relations of these three times are shown in Figure 7.13 where the signal from the back surface is shown larger than the signal from the defect. This is often the case especially with small defects. The time Δt is the time needed for the wave to travel from the surface to the flaw. If this time can be accurately measured, the depth can be calculated as

$$x = v_p \Delta t = \frac{\Delta t}{\sqrt{\mu\varepsilon}}$$

where v_p is the phase velocity of the wave in the material, and x is the flaw depth from the surface. By measuring the time Δt, the thickness of the material can be evaluated.

This method is in fact employed in radar measurements such as into the ground radar. For most nondestructive testing applications, the distance x, and therefore the time Δt are very small, complicating measurements. As an example, if we were to choose the dielectric constant of sea water ($\mu_r = 81$), the phase velocity in sea water is equal to $c/9$ (c is the speed of light). At a depth of $10\ cm$, the wave reaches the flaw in 3 nanoseconds (starting from the surface). Although this is certainly a measurable time, the pulse width of the source must be of this order of magnitude or shorter to allow the reception of the scattered pulse. Thus the difficulty in this method. However, with proper instrumentation it is a viable method, especially in thick, low loss dielectrics and if the source can be located at a considerable distance away from the material. This condition is often satisfied in testing of concrete slabs and thick pavements.

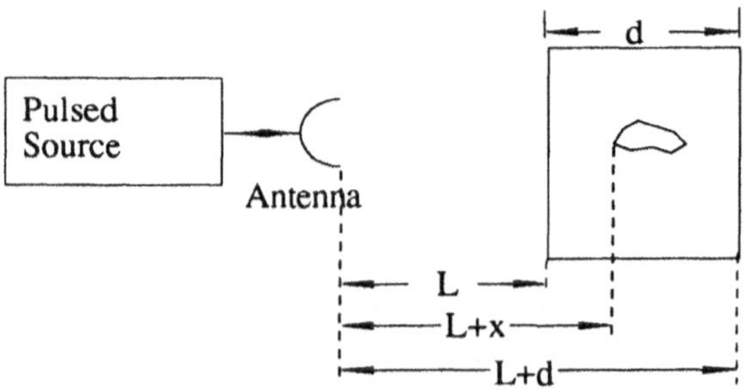

Figure 7.12. Flaw location using a scattering method.

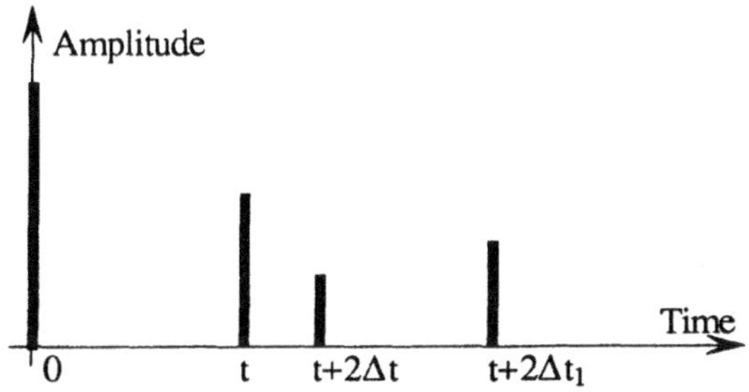

Figure 7.13 Relationships between reflected signals in Figure 7.12.

The method above can also be applied to low loss dielectrics because the phase velocity of low loss dielectrics is about the same as for lossless dielectrics and attenuation is low. Since we only measure time, the lossy aspect is not important other than the obvious attenuation and the need to measure lower amplitudes. With high loss dielectrics, this measurement is impossible although, because of the much lower phase velocities, time measurements would be simplified. But because of the high attenuation in the lossy dielectric, it is difficult to measure the scattered signal.

A second method which can be used, is to evaluate the depth from the angle of refraction in the material. The principle involved was discussed in chapter 1 in the context of refraction and is shown schematically in Figure 7.14.

Since the angle of incidence and the exit angle (that is, the transmission angle for waves propagating out of the material, after internal reflections) are the same, the depth of the flaw can be calculated from this angle and the distance between the incidence and exit points. The angle of refraction is given by

$$sin\theta_r = \frac{v_p}{v_0} sin\theta_i = \sqrt{\frac{\mu_1 \varepsilon_0}{\mu_0 \varepsilon_1}}\ sin\theta_i$$

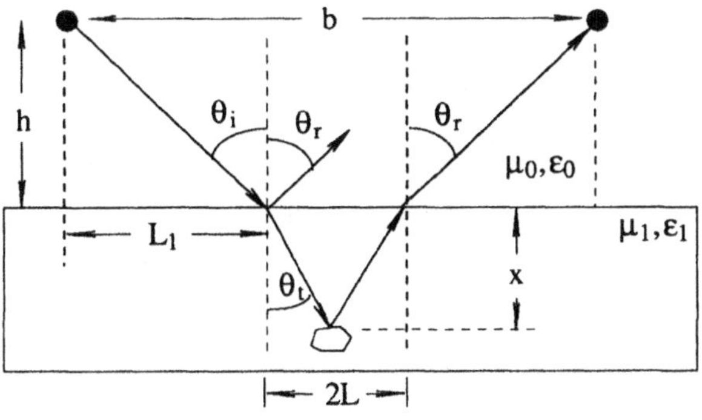

Figure 7.14. Location of defects using the refraction angle.

Therefore

$$\theta_r = arcsin\left[\sqrt{\frac{\mu_1 \varepsilon_0}{\mu_0 \varepsilon_1}} \; sin\theta_i\right]$$

Denoting the distance between incidence and exit points as $2L$, the depth is calculated as

$$x = \frac{L}{tan\theta_r}$$

If the measurement is to be of any use, the distance $2L$ must be measured accurately. This can be done by employing highly directive antennas and measuring the angle θ_i. If one antenna is kept stationary and the other scanned until maximum reading is obtained, the distance L is obtained from the distance between antenna and surface and the angle θ_i, both of which can be measured relatively accurately. The distance between antenna and point of incidence is given by

$$L_I = h \; tan\theta_i$$

The total distance between the two antennas is known and we can write

$$L = \frac{b - 2L_I}{2}$$

The third method uses frequency shift or heterodyning. This is rather similar to the frequency domain reflectometry discussed in chapter 6 but is discussed here in a different context. The principle is shown in Figure 6.34. A swept frequency source is used, allowing a continuous change in frequency at a constant rate. The incident and reflected waves are different in frequencies because during the delay in time needed for the wave to travel to the target and back the source has shifted in frequency. Assuming the generator sweeps between a high frequency f_h and low frequency f_l at a rate of df/dt, the depth of a flaw is calculated as follows:

The time required for the wave to travel to the material interface and back to the antenna is ΔT and is given by

$$\Delta T = 2\Delta L\sqrt{\mu_0 \varepsilon_0}$$

where the speed of light in free space was used for the speed of propagation.

The time required for the wave to travel from the surface to the defect and back is

$$\Delta t = 2x\sqrt{\mu \varepsilon}$$

The total time of travel is $\Delta T+\Delta t$. The shift in frequency is

$$\Delta f = f_i - f_r = \frac{df}{dt}(\Delta T + \Delta t)$$

Since the distance to the sample can be measured directly, the shift in frequency due to propagation in free space can be subtracted and the total shift in frequency due to travel of the wave from the surface of the material to the flaw is

$$\Delta f' = \frac{df\,\Delta t}{dt}$$

This frequency can be measured very accurately and from it the flaw depth is calculated as

$$x = \frac{\Delta f'}{2\sqrt{\mu\varepsilon}\,\dfrac{df}{dt}}$$

Thus, all that is required are the parameters of the swept source and the material properties of the sample.

The considerations given in section 6.6. regarding sweep rates and sensitivity apply here as well. The rule is that the higher the sweep rate, the larger the shift in frequency for a given depth and therefore, the more accurate the measurement.

In addition to the change in frequency, the phase of the wave also changes and these changes can also be used for measurement. However, for phase measurements, there is no need for a frequency sweep. To see this, consider a wave, propagating as shown in Figure 7.15. The electric field (as an example) can be written as

$$E = E_0(\cos\omega t - \beta z)$$

It is also possible to include attenuation but we are interested here in phase, therefore there is no need to do so since, for low loss materials, the phase constant changes very little. Assuming we start at $z=0$ (at the source in Figure 7.15, the wave at the surface is

$$E_i = E_0(\cos\omega t - \beta L_1)$$

where β is the wavenumber in free space. The reflected wave returns to the source and reaches the source as

$$E_r = \Gamma E_0(\cos\omega t - \beta 2L_1)$$

where Γ is the reflection coefficient at the interface. The transmitted wave reaches the flaw and is reflected through the interface, back to the source. The wave reaching the flaw is

$$E_{if} = TE_0\,(\cos\omega t - \beta L_1 - \beta_1 x)$$

where β_1 is the wavenumber in the material. This is reflected with a reflection coefficient Γ_f, and reaches the surface of the flaw as

$$E_{rf} = \Gamma_f E_{\iota\phi}\,(\cos\omega t - \beta L_1 - \beta_1 2x)$$

This is transmitted again through the interface and reaches the source as

$$E_r = \Gamma E_{rf}\,(\cos\omega t - \beta 2L_1 - \beta_1 2x)$$

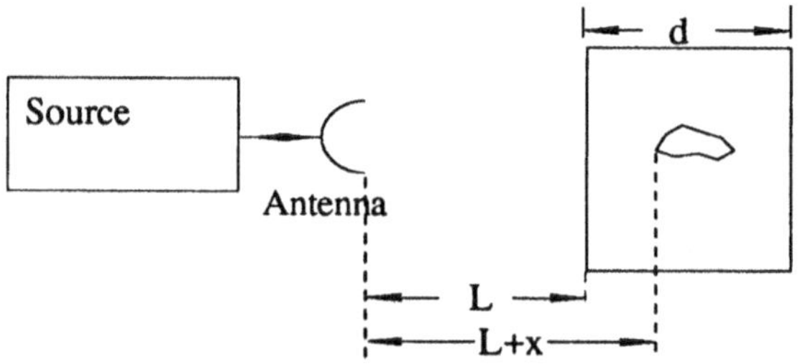

Figure 7.15. Distance relationships for measurement of defect location based on phase measurements.

Thus, the phase difference between the incident wave and the reflected wave from the surface is $\beta 2L_1$ and that between the incident and reflected wave from the flaw is $\beta_1 2x+\beta 2L_1$. These can be easily distinguished unless the distance from the surface to the flaw is extremely small. By measuring the difference between the two phases, the distance x is found as shown in Figure 7.15. Note that it was assumed above that the transmission coefficient at the interface is the same in either direction. This is almost never the case but, for low loss or lossless dielectrics, the assumption only affects the amplitude of the waves, not their phase.

7.4. Scanning Measurements

Any of the measurements above and, for that matter, almost any of the measurements in chapter 6, can be performed in a scanning mode. In particular, the data in Figures 7.8 through 7.11 were obtained by scanning. When tests are done on a continuous production line, the scanning aspect of the test is obvious. However, it is sometimes useful to perform a scan on a material for two basic purposes:

1. Location of a flaw is more easily identified from multiple measurements.
2. If the measurements are properly processed, the signals can be used to create an image of the flaw. This is relatively simple since any variation in a test parameter can be used for imaging.

When performing a scanning test with microwaves, the following should be considered:

1. It is normally not practical to move the test sources and sensors. These are often fed with waveguides which are normally rigid. It is possible to use flexible waveguides as well as a number of pivoting joints but this complicates the test setup. In many cases, it is more practical to move the test sample and keep the source and sensors fixed. A basic setup for scanning of a sample while measuring attenuation (e.g. moisture content or density) is shown in Figure 7.16.

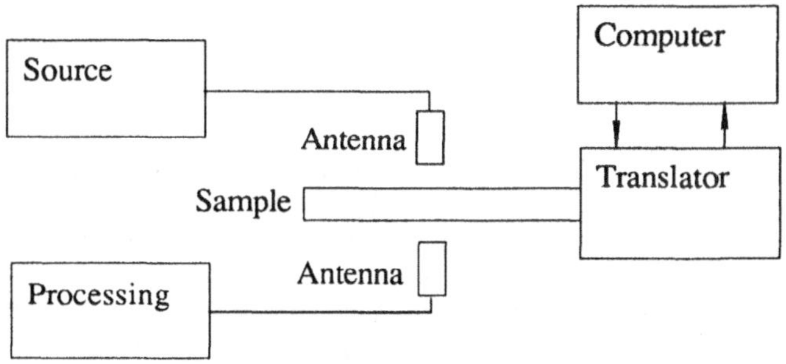

Figure 7.16. Scanning test. The sample is moved while antennas and all other equipment are stationary.

2. In cases when this is not possible, such as testing very large samples or testing fixed objects (i.e. testing concrete or pavements), the instrumentation should be mounted together on the translator. This may be as simple as a cart or a complete, accurate positioning system. When doing so, it is very important to keep vibrations of the sources and sensors to a minimum. Readings will be sensitive to vibrations as these alter the distance between antennas or between antennas and sample, introducing errors. A typical setup for this type of measurement is shown in Figure 7.17. This figure shows an into the ground radar used to test the thickness of a concrete pavement. The radar instruments are mounted together with the antenna on a translator and only the power supply cable (as well as any required signal cables) are connected to the translator.
3. The illumination area of the microwave beam on the scanned material depends on distance from the source. Even for a narrow beam, this is equal to or larger than the aperture of the antenna. Thus, unless special techniques of focusing the beam are used, the area illuminated is fairly large in comparison with the flaws we need to detect. The lateral resolution (perpendicular to the direction of the beam) is therefore lower than the

resolution along the beam. Also, in most cases, scans overlap each other and this should be taken into account when processing the data.

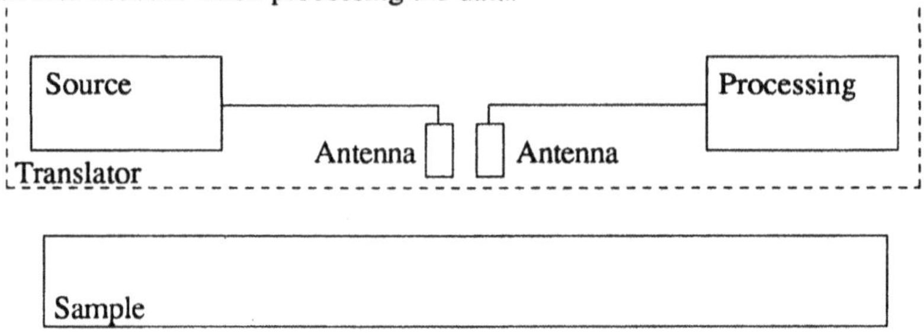

Figure 7.17. Scanning test. All equipment is installed on the translator. This is particularly useful for tests on stationary materials such as pavements.

7.5. Automatic Testing

Automatic testing can be done by any material handling system. As an example, Figure 7.16 can be used to automatically scan a sample. The variations possible on this basic setup are many and depend to a large degree on the type of test performed. It is even possible to set the system to scan at some preset widely spaced pattern and, when a defect is detected (such as an area of delamination) to change the scanning pattern to cover a smaller area around the delamination in a tighter pattern. If the initial pattern is spaced such that the smallest delamination required is detected, there is no possibility of missing a delamination. This type of test allows accurate identification both of defects and their location, and may also be used for imaging.

Scanning can also be achieved by scanning with the antenna (i.e. moving it on a pivot as is often done in radar antennas) but this obviously complicates both instrumentation and processing of data.

7.6. Bibliography

Testing for defects in the microwave domain can be found in a number of references although, most describe specific applications rather than general approaches to testing. Most texts on nondestructive testing contain a chapter dedicated to microwave testing. Bahr, in [1] summarizes a number of methods and specific applications of microwave testing. Other applications can be found in [2] through [9]. Theoretical aspects of scattering and testing can be found in [10] through [14]. References [15] through [19] discuss the use of ferromagnetic resonant probes for testing of surface flaws. This aspect of testing was not discussed in this chapter because it is actually an eddy current test in which eddy currents are generated in conducting materials by the fields of a resonant YIG (yitrium iron garnet) sphere. Additional references, [20-26], present a number of topics including general reviews of microwave *NDT*, microwave microscopy and microwave spectroscopy.

[1] A.J. Bahr, "Microwave Nondestructive Testing Methods, Gordon and Breach Science Publishers, 1982.
[2] T.M. Lavelle, "Microwaves in nondestructive testing", Materials Evaluation, Vol. 25, No. 11, 1967, pp. 254-258.
[3] R.J. Botsco, "Nondestructive testing of plastics with microwaves", Materials Evaluation, Vol. 27, No. 6, 1969, pp. 25a-32a.
[4] W.L. Rollwitz, "Microwave Inspection", Metals Handbook, Vol. 11, 8th edition, 1976, Nondestructive Inspection and Quality Control, ASM, Metals Park, OH., pp. 244-253.
[5] R. Hochschild, "Microwave nondestructive testing in one (not-so-easy) lesson", Materials Evaluation, Vol.26, No. 1, 1968, pp. 35a-42a.
[6] M. Rockowitz and L.J. McGuire, "A microwave technique for the detection of voids in honeycombed ablative materials", Materials Evaluation, Vol. 24, No. 2, 1966, pp. 105-108.
[7] R.C. Stinebring and R.H. Harrison, "Nondestructive testing of rocket components using microwaves and low frequency ultrasonics", Materials Evaluation, Vol. 23, No. 1, 1965, pp. 17-23.
[8] L. Feinstein and R.J. Hruby, "Surface crack detection by microwave methods", in Proc. of the 6th Symp. on NDE of Aerospace and Weapons Systems Components and Materials, San Antonio, TX. 1987, pp. 93-106.
[9] J.A. Bahr, "Nondestructive microwave evaluation of ceramics", IEEE Transactions on Microwave Theory and Techniques, Vol. MTT-26, No. 9, Sept. 1978, pp. 676-683.
[10] T.H. Krohn, "Scattering from composite materials: A first order model", IEEE Transactions on Antenna and Propagation, Vol. AP-37, No. 2, Feb. 1989, pp. 219-228.
[11] A.K. Dominek, H.T. Shamansky and N. Wong, "Scattering from three-dimensional cracks", IEEE Transactions on Antenna and Propagation, Vol. AP-37, No. 5, May 1989, pp. 586-591.
[12] S.M. Rao, D.R. Wilton and A.W. Glisson, "Electromagnetic scattering by surfaces of arbitrary shape", IEEE Transactions on Antennas and Propagation, Vol. AP-30, No. 3, May 1982, pp. 409-418.
[13] T.P. Sarkar, E. Arvas and S. Ponnapalli, "Electromagnetic scattering from dielectric bodies", IEEE Transactions on Antennas and Propagation, Vol. AP-37, No. 5, May 1989, pp. 673-676.
[14] L.N. Medgyeshi-Mitschang and J.H. Putnam, "Electromagnetic scattering from extended wires and two- and three-dimensional surfaces", EEE Transactions on Antenna and Propagation, Vol. AP-33, No. 10, Oct. 1985, pp. 1090-1100. [15] H. Zijlstra, "Experimental methods in magnetism", North Holland Publishing Company, Amsterdam, 1967.
[16] B.A. Auld, "Theoretical characterization and comparison of resonant probe microwave eddy current testing with conventional low frequency eddy current methods", in Eddy-Current Characterization of Materials and Structures, Birnbaum, G. and Free, G., Eds., ASTM, Philadelphia, 1979, pp. 332-347.

[17] A.J. Bahr, "Microwave eddy-current techniques for quantitative nondestructive evaluation", in Eddy-Current Characterization of Materials and Structures, Birnbaum, G. and Free, G., Eds., ASTM, Philadelphia, 1979, pp. 311-331.
[18] B.A. Auld and D.K. Winslow, "Microwave eddy current experiments with ferromagnetic resonance probes", in Eddy-Current Characterization of Materials and Structures, Birnbaum, G. and Free, G., Eds., ASTM, Philadelphia, 1979, pp. 348-366.
[19] A.J. Bahr, "Using electromagnetic scattering to estimate the depth of a rectangular slot", IEEE Transactions on Antenna and Propagation, Vol. AP-27, No. 6, Nov. 1979, pp. 738-746.
[20] D.S. Dean and L.A. Kerridge, "Microwave techniques", in Research techniques in nondestructive testing, R.S. Sharpe, ed., Academic Press, London, 1970, pp. 417-441.
[21] R. Zoughi, "Review of NDT techniques at radio and microwave frequencies", QNDT Rev. of Progress, 1990, pp. 337-344.
[22] M.S. Wang and J.M. Borrego, "High resolution scanning microwave electric field probe for dielectric constant uniformity measurement", Materials Evaluation, Vol. 48, Sept. 1990, pp. 1106-1109.
[23] A. Husain and E.A. Ash, "Microwave scanning microscopy for non-destructive testing", Proceedings of the European Microwave Conference, 1976, pp. 213-217.
[24] A.D. Lucian and R.W. Cribbs, "The development of microwave NDT technology for the inspection of nonmetallic materials and components", in Proc. of the 6th Symp. on NDE of Aerospace and Weapons Systems Components and Materials, San Antonio, TX. 1987, pp. 199-232.
[25] P.J. Moser and H. Uberall, "Radar spectroscopy", in RADAR 82, pp. 271-273.
[26] P.C. Ely, "Swept-frequency techniques", Proceedings of the IEEE, Vol. 55, No. 6, June 1967, pp. 991-1002.

Chapter 8.
Testing by Monitoring Material Properties

8.1. Introduction

Various methods for testing for material composition, material properties and chemical properties are used as indicators in nondestructive testing. The major effect is that of changes of the dielectric constant of the material. If losses in materials do not change (as, for example, with curing degree in some materials), then only the real part of the dielectric constant causes measurable changes. More often, both real and imaginary parts vary (i.e. the loss tangent varies) and can be used for testing. Although the permeability of materials may also change this is less common than changes in dielectric constant. Thus, measurements based on monitoring material properties are most often based on variations dielectric constant. This, in turn, causes changes in reflection, transmission and attenuation constants. These can be tested using the methods described in the previous two chapters or, as will be shown shortly, using resonant techniques.

Sometimes, the state of a material may be deduced from its emissions or its chemical composition rather than from its electrical properties. Drying and curing materials emit various gases as well as water. These can all be detected and monitored using, for example, the shift in resonant frequency of a tuned cavity. A microwave mass spectrometer is nothing more than a calibrated, tuned cavity with the appropriate instrumentation. One classic example for this type of test is the detection of explosives based on measurement of emitted gases. In this type of measurement, the material tested is either sampled and measured in a closed cavity, or monitored continuously in an open cavity.

8.2. Transmission Tests

A transmission measurement technique is shown in Figure 6.1. The transmitting antenna (A) is placed on one side of the material while the receiving antenna (B) is placed on the other side. The transmitted energy (as detected) is compared with the incident energy giving a direct reading of the transmission coefficient. This coefficient is proportional to the material properties. If we assume a material of thickness d, the transmission coefficient can be calculated from the transmitted and incident waves as was shown in chapters 3 and 6. The transmission coefficient through the slab is

$$T = \frac{2\eta_0\eta_2\, e^{j\beta_0 d}}{2\eta_0\eta_2 \cos\beta_2 d + j\left[\eta_0^2 + \eta_2^2\right] \sin\beta_2 d} \qquad (8.1)$$

where

$$\beta_2 = \omega\sqrt{\mu_2\varepsilon_2} \qquad \eta_0 = \sqrt{\frac{\mu_0}{\varepsilon_0}} \qquad \eta_2 = \sqrt{\frac{\mu_2}{\varepsilon_2}}$$

Similarly, for a low loss dielectric, the transmission coefficient is

$$T = \frac{\dfrac{4\eta_0\alpha}{\sigma}\left[1 + j\dfrac{\sigma}{2\omega\varepsilon}\right] e^{j\beta_0 d}}{\dfrac{4\eta_0\alpha}{\sigma}\left[1 + j\dfrac{\sigma}{2\omega\varepsilon}\right]\cos\beta_2 d + j\left[\eta_0^2 - \dfrac{4\alpha^2}{\sigma^2}\left[1 + j\dfrac{\sigma}{2\omega\varepsilon}\right]^2\right]\sin\beta_2 d} \qquad (8.2)$$

where

$$\alpha_2 = \frac{\sigma}{2}\sqrt{\frac{\mu_2}{\varepsilon_2}} \qquad \eta_2 = \frac{2\alpha}{\sigma}\left[1 + j\frac{\sigma}{2\omega\varepsilon}\right]$$

For a high loss material (essentially a conductor), the following is obtained

$$T = -j\frac{\eta_2}{\eta_0}\frac{2e^{j\beta_2 d}}{\sin(\beta_2 - j\alpha_2)d} \qquad (8.3)$$

where

$$\eta_2 = (1+j)\frac{\alpha_2}{\sigma_2}$$

where σ_2 is the conductivity of the material under test. The expression for the transmission coefficient in Eq. (8.3) is only marginally useful since in high loss materials the transmission coefficient is very small (i.e. η_2 is very small compared to η_0) and transmission tests on lossy materials yield poor results if they are possible at all.

In all these expressions, the important effect is that for a given material thickness, the transmission through the material is dependent on the intrinsic impedance of the material. As was mentioned above, for most materials, this means variations in the dielectric constant.

As an example of how the transmission through the material changes with drying (or curing), consider the test in Figure 8.1 where a foam material is tested. Initially, the foam is saturated with water, with an effective permittivity of about *80*. As the water dries, the effective permittivity is reduced and the transmission increases. In this case, the final (dry) permittivity is equal to *2*. Permittivity for this material is assumed to vary as $\varepsilon'=(2+80c)\varepsilon_0$, where c is the fraction of water content. The transmission coefficient for this material is shown in Figure 8.2 where Eq. (8.1) was used. In a given test, the "degree of cure" is measured over time. A curve of this type is shown in Figure 8.3. As the material dries up or as the curing process progresses in time, the transmission coefficient increases. This is then a measure of the rate of curing or rate of loss of moisture, depending on calibration.

This method of testing, and many of its variations (including direct measurement without the bridge), primarily for monitoring of degree of curing for composite materials and polymers has been used extensively because it is simple and accurate. In some cases, the test is part of the process, as in cases in which microwaves are used for the drying process itself. The main difficulty one should encounter in applying this method is in the stability of the microwave source if the curing process is long. If long curing periods are expected and a stable source cannot be used, only relative measurements can be taken. As an example, complete curing may be indicated by lack of change in the transmission of microwaves through the material. Also, the material thickness must be small compared with the wavelength if absolute measurements are required. If only relative measurements are needed this is not so important. However, the electrical thickness (electrical length is equal to βd) of the dielectric changes with curing and this must be taken into account while testing.

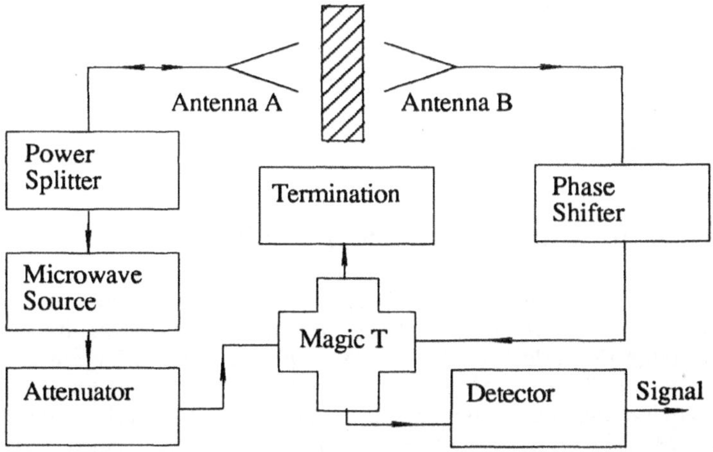

Figure 8.1. Transmission test for monitoring of water content in a dielectric or lossy dielectric.

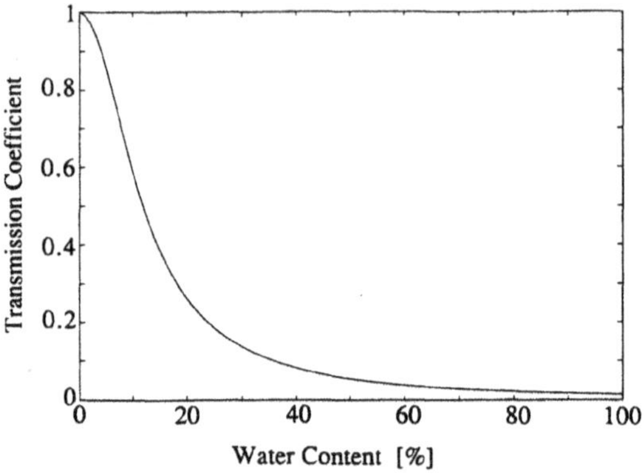

Figure 8.2. Dependence of transmission coefficient on water content for a *0.1 mm* thick foam at *1 GHz*.

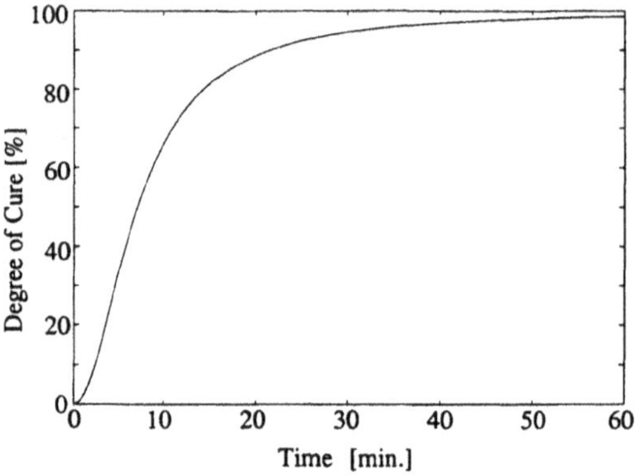

Figure 8.3. Degree of curing with time.

8.3. Reflection and Attenuation Tests

Although we chose to start with transmission measurements, reflection measurements can be used just as well. Any effect measurable by the transmission through the material is also measurable through the reflection since the transmission and reflection coefficient are related. Moreover, if the transmission coefficient is small, as for lossy materials, the reflection coefficient may be a better choice for measurements. Reflection measurements were discussed extensively in chapter 6 and therefore will not be repeated here.

Water (or any other dielectric or lossy dielectric fluid such as alcohols and solvents) in a material can be monitored by reflection, transmission, attenuation, and by resonant cavity methods. The measurement of moisture is important in curing and drying processes (i.e. in monitoring) and in testing for curing and drying in composites, wood products, paper products and the like. In some cases, only the dielectric constant changes while losses remain negligible. This is the case with alcohols. In others, the loss tangent is the measurable quantity as is the case in many drying products such as paper and some polymers.

Attenuation measurements are based on comparison of the wave transmitted through the dielectric and the wave transmitted without the dielectric. The complex dielectric constant of a lossy dielectric is given by

$$\varepsilon_c = \varepsilon' - j\varepsilon'' \tag{8.4}$$

where ε' is the real part of permittivity and ε'' represents losses as:

$$\varepsilon_c = \varepsilon' - j\frac{\sigma}{\omega} \tag{8.5}$$

From these, the loss tangent is defined as

$$tan\delta = \frac{\varepsilon''}{\varepsilon'} \approx \frac{\sigma}{\omega\varepsilon} \tag{8.6}$$

where ε is the dielectric constant of the material.
The attenuation constant is

$$\alpha = \frac{\omega\varepsilon''}{2}\sqrt{\frac{\mu}{\varepsilon'}} \tag{8.7}$$

For a given thickness d of the material, the electric field after attenuation through the material is

$$E = E_0 e^{-\alpha d} \tag{8.8}$$

where E_0 is the amplitude before attenuation. The ratio for the electric fields (or the magnetic fields) is

$$\frac{E}{E_0} = e^{-\alpha d} \tag{8.9}$$

Similarly, the power ratio between the power entering and leaving the material is

$$\frac{P}{P_0} = e^{-2\alpha d} \tag{8.10}$$

These relations assume low reflections at the interfaces. Either the relation in Eq. (8.9) or (8.10) can be measured. Figure 8.4 shows a basic measurement technique. In this test the material normally fills the gap between the antennas or, more often, two open waveguides. Here the transmitted power through the material is detected and, if a calibration curve exists, the attenuation of the material can be correlated with the loss tangent of the material. This in turn is proportional to a variety of effects such as moisture content. An example of a calibration curve is shown in Figures 8.5 and 8.6. In these, the real and imaginary parts of permittivity are shown as a function of water content for a lossy dielectric spheroid (a soybean in this case) The relative permittivity for this material is given as $\varepsilon=9+(70-j25)c$ where c is the water content. For the measurement in Figure 8.4, the power at antenna A is known and the power at B is measured.

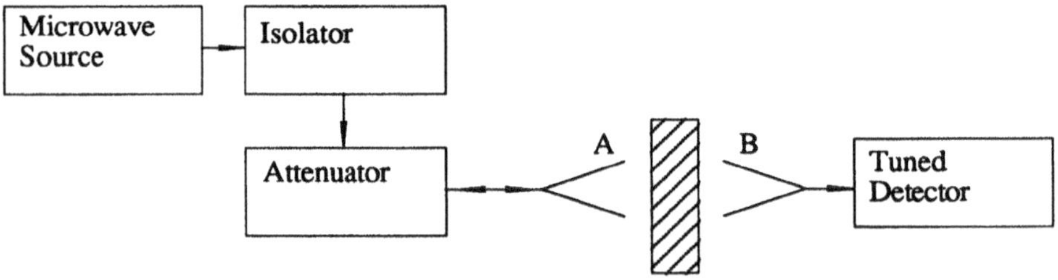

Figure 8.4. Attenuation test arrangement for low loss dielectrics.

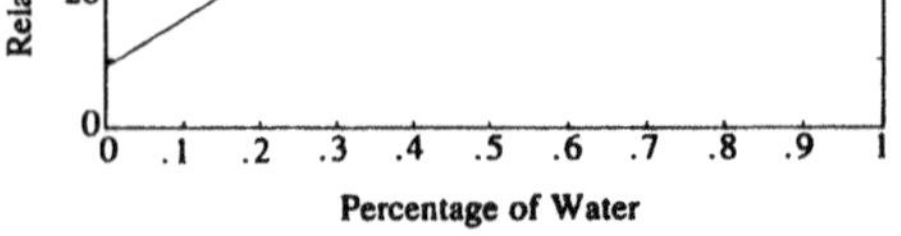

Figure 8.5. Real part of permittivity for a low loss dielectric as it depends on water content.

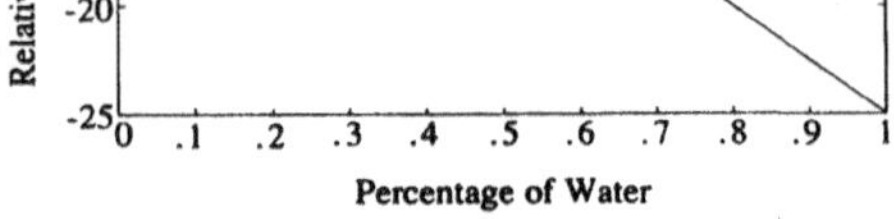

Figure 8.6. Imaginary part of permittivity for a low loss dielectric as it depends on water content.

A simpler arrangement is to measure the power at the detector as in Figure 8.4. This gives P in Eq. (8.10). Then the sample is removed and the power measured again to give P_0. If properly calibrated, the measurements can be processed to read moisture content directly (say as a percentage). Alternatively, the second measurement is done by varying the attenuator until the power at the detector is equal to the power with the sample, to provide a direct attenuation measurement. A method better suited for automatic and continuous measurements is shown in Figure 8.7. Here P_0 is measured separately using an identical distance. If there is significant attenuation through air (i.e. if measurements are made in excessively moist environments), this can be compensated by calculating the attenuation without the sample.

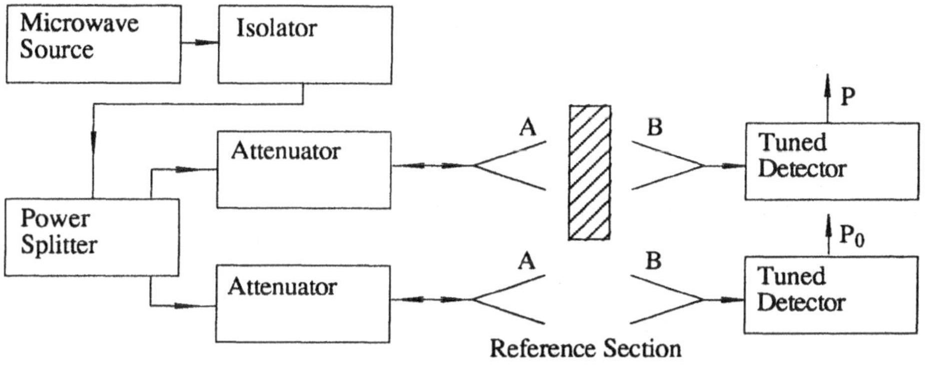

Figure 8.7. Attenuation test with reference measurement for low loss dielectrics.

8.4. Resonant Methods

Water content as well as other effects on the dielectric constant are very accurately tested in microwave cavity resonators. This includes lossy effects as well as defects in the dielectric. There are a number of basic methods to consider.

The simplest in terms of instrumentation is to introduce a sample inside a cavity of some predetermined shape, operating at a given mode. The introduction of a material changes the resonant frequency of the cavity as well as its Q-factor. The changes depend on the mode and location of the sample in the cavity. The basic testing procedure is shown in Figure 6.4. The microwave generator is swept (manually or automatically) in a narrow frequency range around the resonant frequency of the empty cavity and the peak in the field is detected by a wavemeter. The method as shown uses a transmission type cavity wavemeter. The cavity behaves as a short off-resonance and therefore, a dip in amplitude at the exit port indicates resonance. If the sample is small, the change in frequency is proportional to the volume of the sample and its dielectric constant or the magnetic permeability, depending on the location in the cavity (see section 2.4). Thus, the resonant frequency is a measure of the dielectric constant and the quality factor is a measure of the loss tangent. The resonant frequency may be measured by other types of wavemeters or by any other means of measuring frequency.

The change in resonant frequency is given by

$$\frac{\omega-\omega_0}{\omega_0}=-\frac{\Delta W_m+\Delta W_e}{W_m+W_e} \tag{8.11}$$

where W_m is the stored magnetic energy, W_e the stored electric energy and ω_0 is the resonant frequency of the empty cavity (see section 2.4 for details of derivation). Thus, if the sample affects a region of the cavity where the electric field is large, the change will be mostly due to the dielectric constant. If the sample is in a domain where the magnetic field is dominant, the change in the resonant frequency is mainly due to changes in material permeability.

The quality factor is defined as

$$Q=\frac{\omega_0 W_0}{P_d} \tag{8.12}$$

where W_0 is the energy stored in the cavity without the sample (empty cavity) and P_d is the power loss due to the loss tangent of the material. Thus, the higher the loss, the lower the quality factor.

Measurement of the resonant frequency is accomplished using any of the frequency measurement methods described in chapter 4, most often by a wavemeter or a heterodyne frequency meter. The Q factor is measured as described in section 4.4. A basic measurement method for loss tangent is shown in Figure 8.8. In this test, the frequency of the source is varied and the half bandwidth frequencies are found as described in section 4.4. The *VSWR* on the waveguide leading to the cavity is measured and the *3 dB* points (f_1 and f_2) in *VSWR* are found. The Q-factor is then given by

$$Q=\frac{f_0}{f_2-f_1} \tag{8.13}$$

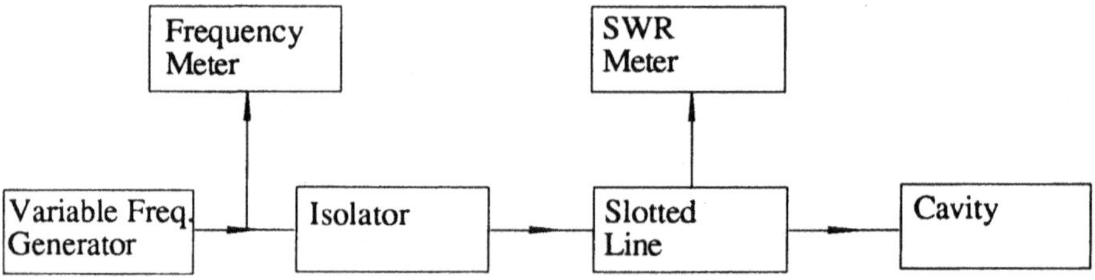

Figure 8.8. Measurement of loss tangent by measuring the Q-factor of a cavity. Reflection measurement based on *VSWR* is used.

The method described above is extremely inconvenient since it requires samples and therefore, in most cases, cannot be a true nondestructive method. However, a cavitiy can be easily adapted for on line measurements by "opening" the cavity. This results in open cavities which allow insertion of materials in the cavity. In some cases, simple apertures are created to allow the passage of material through the cavity. Two examples are shown in Figures 8.9 and 8.10. In Figure 8.9, the material passes through a coaxial tube in a cylindrical cavity. If the tube is of small diameter, the aperture is small compared to the cavity diameter and we may view this as a regular cavity. If the hole is large, then this is an open cavity and must be treated as such. This method is used extensively, particularly for measurements on fluids and gases, but also for production of pastes and other materials that can be made to flow or pass through the cavity. Figure 8.10 shows a rectangular cavity, which has been split in two to allow passage of a flat material (for example, rubber or plastic sheets. The slot may be only a small aperture or it may actually split the cavity. If the cut does not modify the basic properties of the cavity this becomes a useful method. By this is meant that the introduction of the material in the cavity, either as a sample or as a continuous material, only perturbs the existing modes in the cavity but does not alter them.

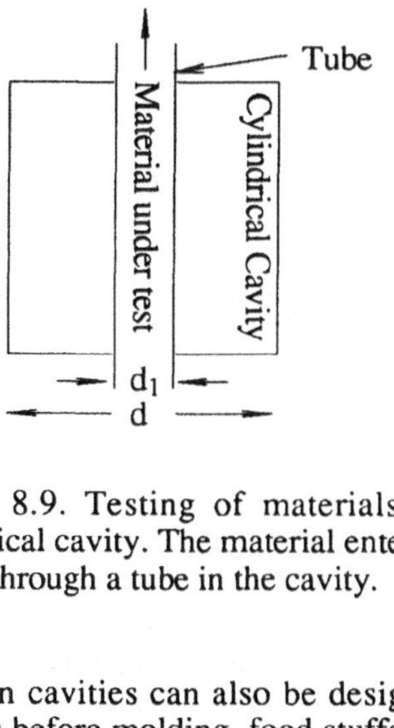

Figure 8.9. Testing of materials in a cylindrical cavity. The material enters the cavity through a tube in the cavity.

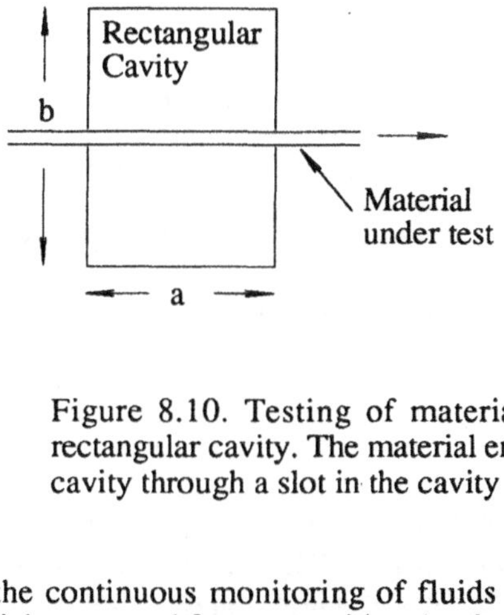

Figure 8.10. Testing of materials in a rectangular cavity. The material enters the cavity through a slot in the cavity walls.

Open cavities can also be designed for the continuous monitoring of fluids such as plastics before molding, food stuffs, etc., which are tested for composition (such as water content, percentage of fat) or, for example, existence of foreign materials. These may be inclusions of air bubbles that may not be desirable or lumping of the material because of problems in the process. An example of a cylindrical cavity resonator used for this purpose is shown in Figure 8.11. The cavity is actually a tube with two co-axial tubes on either side of the central cylinder as in Figure 8.11b. It is based on the properties of waveguides as shown in Figure 8.11a. If a wave is generated in the larger waveguide, it will not propagate into the smaller waveguide unless the frequency is above the cutoff frequency of the smaller waveguide. The cavity in Figure 8.11b (indicated by the dashed lines) acts as a cavity resonator because, for the designed tube diameters, the co-axial sections are below cut-off. The cavity is fed through the wall by a loop to excite, say, *TE* modes below the cutoff frequency of the smaller waveguides but above the cutoff frequency of the larger waveguide. Material is fed through the open ends of the cavity and fills the whole cavity.

The resonant frequency as well as the Q-factor are then used to monitor or detect variations in material composition. There are many variations on this basic concept, all of which are based on allowing the cavity to resonate while one or more of the cavity walls are removed to allow material handling. An open cavity of this type has a relatively narrow range of frequencies ($f_{c2}<f<f_{c1}$) but, because only resonance shift is normally required, and this shift is small, the narrow range is not normally a problem.

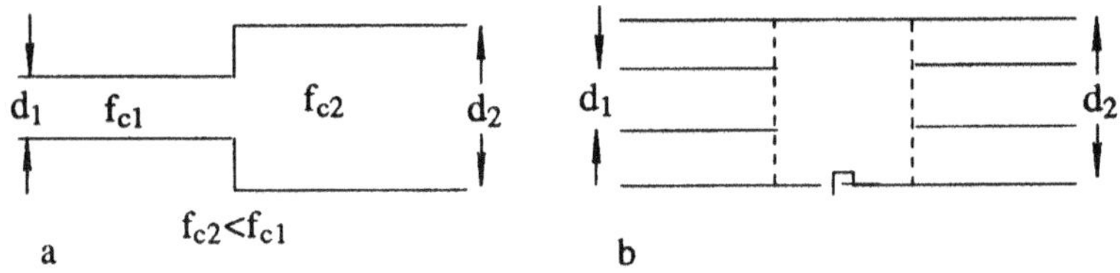

Figure 8.11. Construction of an open resonant cavity out of two waveguide sections.

Perhaps the easiest way do design an open cavity is to construct a resonator made of two parallel surfaces as shown in Figure 8.12a. Since a resonator can be built in any geometry that allows total reflections at the boundaries, (and therefore, a standing wave pattern can be produced) the normal, enclosed geometry is only required for simple, parallel surface boundaries. Although, the surfaces are parallel, and finite in size (say, rectangular), the system resonates at given modes, depending on the size and distance between the plates. For infinite, parallel plates, the resonant frequencies are given by

$$f = \frac{n}{2l\sqrt{\mu\varepsilon}} \tag{8.14}$$

where n is a mode index, l the distance between the plates and μ, ε, the material properties between the plates. Thus the distance between the plates is equal to half wavelength at resonance. This structure supports transverse electromagnetic *(TEM)* modes.

The use of infinite plates is not practical. If a resonant structure of this type is required, rectangular plates are used. Although the analysis and calculation of open cavities with finite size plates is much more complex, rectangular plate, cylindrical and spherical mirrors of finite size can be used for this purpose. Rectangular, parallel plate, cylindrical and two types of spherical mirror open resonators are shown in Figure 8.12. The resonant frequencies of this type of resonator depend on the size of the mirrors, the distance between them and type of mirror. We will use, however, the resonant frequency for infinite plate resonator as a guide. The resonant frequency of finite size open resonators varies form this because of the shape factor and the inevitable diffraction from the edges of the mirrors.

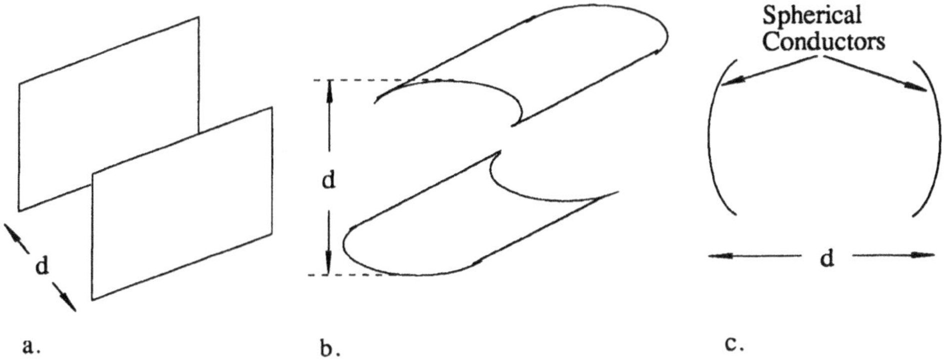

Figure 8.12. Open cavities built of: a. parallel plates, b. cylindrical surfaces and c. spherical surfaces.

The circular, spherical mirrors resonator is particularly easy to understand. The advantage of such an arrangement is that a complete set of modes can be excited. In the geometry shown in Figure 8.12c, the cavity resonates because the fields are focused towards the center of the resonator and are not directed towards the open ends. The reflecting surfaces are sections of spherical mirrors. The cavity can be fed in a number of ways but most often through a small aperture in one of the surfaces, to which a waveguide is connected. The beam produced in the cavity is narrowest at the center. The resonant frequency can now be altered by the introduction of materials between the mirrors. A test arrangement is shown in Figure 8.13. The material to be tested is at the center of the cavity where the wavefronts of the beam are essentially parallel. If the material is much closer to one of the mirrors, the wavefronts of the beam and the surface of the material are not parallel and there will be scattering from the material. This does not happen with parallel plate resonators and parallel surface samples.

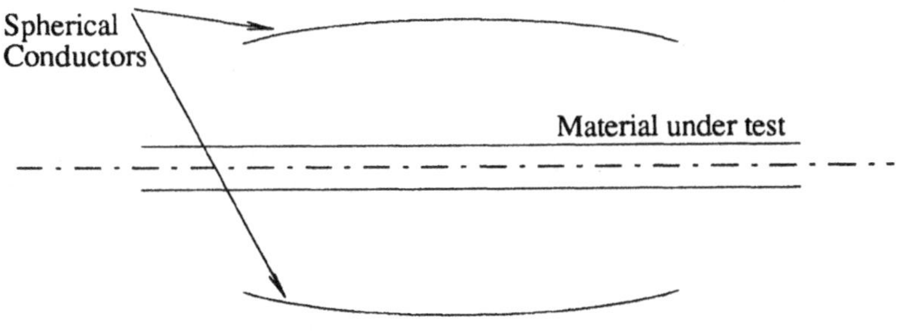

Figure 8.13. Testing of a material in an open cavity built of spherical conductors.

The main advantage of this type of open resonator, in contrast with an opening in a closed resonators is that access to the inside of the cavity is easily facilitated. In addition, unless there are reflections or scattered waves from the test material there is no coupling of energy from the cavity to outside. On the other hand, an aperture in a closed cavity may radiate, depending on its location and orientation. As with any cavity method described above, the shift in resonant frequency is a measure of the dielectric constant (or permeability) and the change in Q is a measure of the loss tangent.

Open cavity methods are also useful for monitoring gases wether in processes or in passive systems such as "sniffers" for detection.

8.4.1. Transmission Line Methods and Transmission Line Resonators

Another useful form of resonating structures is based on transmission line theory and is normally associated with transmission line methods of testing. We describe these here because of the common principles involved.

There are a number of measurements that can be made using transmission lines. These are usually done at lower frequencies than normal microwaves but are often excellent methods for a variety of parameters. The basic principle involved is the change of line parameters with loading of the transmission line. By loading is meant any change that will vary the line parameters as measured. As a simple example, insertion of a dielectric between the two conductors of a parallel conductors transmission line changes its capacitance per unit length and therefore its impedance.

Also, transmission lines of various types can be easily made into resonant structures and are used for measurements similar to those performed by cavity resonators.

The type of transmission line that can be used varies by application but it is possible to use parallel strip lines, coaxial lines, parallel conductors and a variety of other combinations. For a given transmission line and termination the parameters of the line are known and any variation must be due to the test sample. As an example, for a parallel strip transmission line as shown in Figure 8.14, the capacitance per unit length, and inductance per unit length are given as:

$$C = \frac{\varepsilon w}{d} \qquad L = \frac{\mu d}{w} \tag{8.15}$$

The intrinsic impedance of the line is

$$Z_0 = \sqrt{\frac{L}{C}} = \sqrt{\frac{\mu_0}{\varepsilon_0}}\frac{d}{w} = \eta_0 \frac{d}{w} \tag{8.16}$$

where η_0 is the intrinsic impedance of free space.

This line can now be connected to a matched load and a matched source as shown in Figure 8.15. Under these conditions the impedance of the line at the generator is equal to the intrinsic impedance. Also, the time it takes the wave to propagate from generator to load is

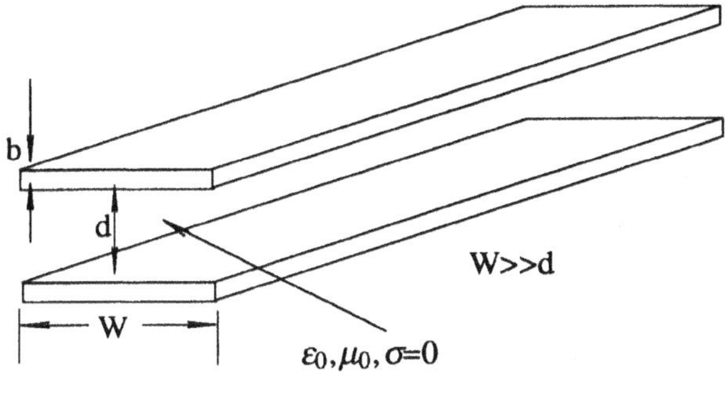

Figure 8.14. A parallel strip transmission line.

$$t = l\sqrt{LC} = l\sqrt{\mu_0\varepsilon_0} = \frac{l}{c} \tag{8.17}$$

where l is the length of the line and c the speed of light.

Now, as an illustration to a measurement process, a material is introduced between the parallel plates as shown in Figure 8.16. The impedance of the line section in which the material is introduced has changed to

$$Z_l = \eta_0\sqrt{\frac{\mu_r}{\varepsilon_r}}\frac{d}{w} \tag{8.18}$$

where ε_r and μ_r are the relative permeability and permittivity of the material. The time of propagation in the section has changed to

$$t = l'\sqrt{\mu_0\mu_r\varepsilon_0\varepsilon_r} = \frac{l'\sqrt{\mu_r\varepsilon_r}}{c}$$

where l' is the length of the section in which the dielectric is introduced. The time of propagation is therefore proportional to the relative permittivity and relative permeability of the material between the plates. Similarly, the characteristic impedance and any reflections from the interface between the two sections of the line are dependent on these material properties. The transmission line under these conditions is shown in Figure 8.16.

In addition, we can expect reflections from the interface between free space and dielectric and dielectric and load. Thus, the presence and properties of the dielectric are measurable either from the time of propagation (time domain reflectometry) or from impedance (or, more commonly, the reflection coefficient). This type of measurement is particularly easy to perform using network analyzers in which all these parameters can be measured and displayed. In the case above, the reflection coefficient to the left of the test material (at a distance a from the interface) is given by

Figure 8.15. A matched, parallel strip transmission line.

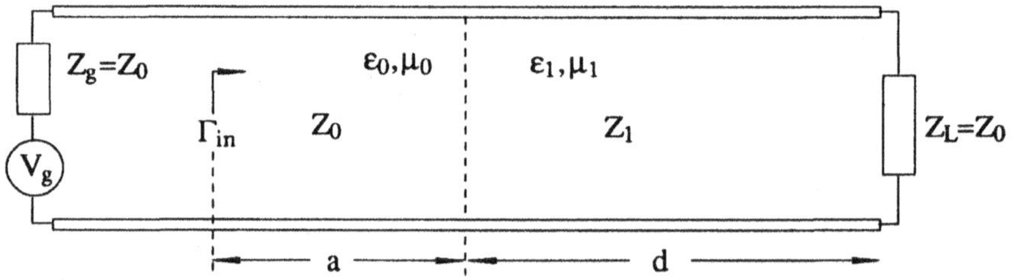

Figure 8.16. A parallel strip transmission line with a test material inserted between the strips.

$$\Gamma_{in} = \frac{Z_0(1+\Gamma_L A) - Z_1(1-\Gamma_L A)}{Z_0(1+\Gamma_L A) + Z_1(1-\Gamma_L A)} B \tag{8.19}$$

where

$$\Gamma_L = \frac{Z_L - Z_1}{Z_L + Z_1}$$

is the reflection coefficient at the load and

$$A = e^{-j2\beta_1 d} \qquad B = e^{-j2\beta_0 a}$$

$$\beta_0 = \omega\sqrt{\mu_0\varepsilon_0} \qquad \beta_1 = \omega\sqrt{\mu_1\varepsilon_1}$$

Note that if the load reflection coefficient Γ_L is zero (matched load), the only reflection is at the interface between the two materials. If further, Z_1=Z_0, there is no reflection and we are back to the circuit in Figure 8.15.

The reflection coefficient in Eq. (8.19) is measured and material properties are correlated with the coefficient. Similarly, the line impedance, which was initially equal to Z_0, has changed to

$$Z_{in} = Z_0 \frac{Z_l(1 + \Gamma_L A)(1 + B) + Z_0(1 - \Gamma_L A)(1 - B)}{Z_l(1 + \Gamma_L A)(1 - B) + Z_0(1 - \Gamma_L A)(1 + B)} \qquad (8.20)$$

Although this relation may seem complicated it reduces to $Z_{in}=Z_0$ if the reflection coefficient at the load is zero (matched load), and $Z_l=Z_0$.

Thus, the input impedance, or the reflection coefficient can be used directly for characterization of the material inside the transmission line. Either of these can be calibrated to read the material thickness for constant properties or any of the varying properties for constant thickness. The advantages of this type of measurements are that they are simple, easily instrumented using standard equipment and accurate. On the other the line itself must be made to fit the measurements needed.

In other transmission lines, this calculation may not be as simple but the reflection coefficient is still a measure of changes in permittivity or permeability. Obviously, thickness of samples and density of materials can also be measured as are loss related parameters such as water content.

A particularly useful method is to short the transmission line instead of using a load. The considerations above remain the same but the reflection coefficient at the load is equal to $\Gamma_L=-1$. Similarly, an open transmission line can be used with $\Gamma_L=1$. In particular, open and shorted transmission lines can be used as transmission line resonators.

Transmission line resonators can be best understood from the condition on a shorted transmission line. As we have seen in chapter 3, the impedance anywhere on a shorted transmission line is given by:

$$Z_{in} = jZ_0 \tan\omega\sqrt{\mu\varepsilon}d = jZ_0 \tan 2\pi f\sqrt{\mu\varepsilon}d \qquad (8.21)$$

where Z_0 is the intrinsic impedance of the line, and d the length to the short. This quantity is purely imaginary, and can be either negative or positive. A negative reactance indicates a capacitor, a positive reactance indicates an inductor. Thus, a resonator can be constructed by connecting two sections of a transmission line as shown schematically in Figure 8.17. One section is chosen of such a length that the input impedance is capacitive, the other such that it is inductive. An equivalent circuit is also shown in Figure 8.17.

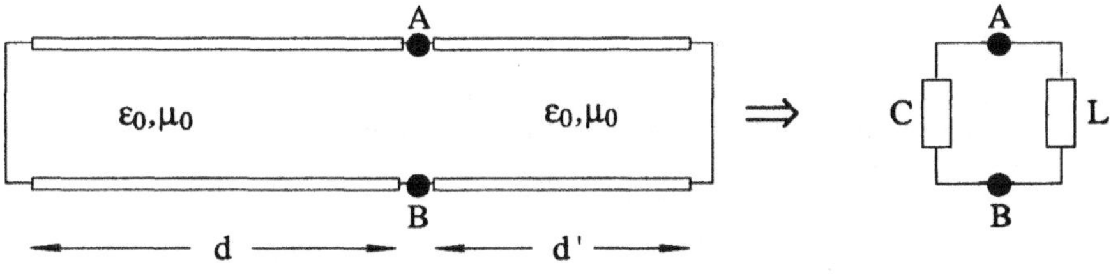

Figure 8.17. Two shorted transmission line sections used as a resonator and their equivalent circuit.

For this to happen, the function $tan(2\pi f(\mu\varepsilon)1/2d)$ must be negative for a capacitive line and the impedance must tend to $-\infty$. This gives

$$2\pi f\sqrt{\mu\varepsilon}d = \frac{3\pi}{2}$$

or

$$d = \frac{3}{f\sqrt{\mu\varepsilon}4}$$

In terms of the wavelength $(\lambda = v_p/f)$, where v_p is the speed of propagation of the wave) and we get

$$d = \frac{3\lambda}{4}$$

For the second line, $tan(2\pi f(\mu\varepsilon)1/2d')$ must be positive and the impedance must tend to $+\infty$. The condition is satisfied if

$$2\pi f\sqrt{\mu\varepsilon}d' = \frac{\pi}{2}$$

or

$$d' = \frac{1}{f\sqrt{\mu\varepsilon}4}$$

and, in terms of the wavelength:

$$d' = \frac{\lambda}{4}$$

Thus, the minimum lengths necessary for the structure to resonate are

$$d = \frac{3\lambda}{4} \qquad d' = \frac{\lambda}{4} \tag{8.22}$$

Now consider the resonator in Figure 8.18. This particular resonator is made of parallel strips. Any variation in the material between the strips will alter the electrical lengths of the two sections and change the resonant frequency, as for a cavity resonator.

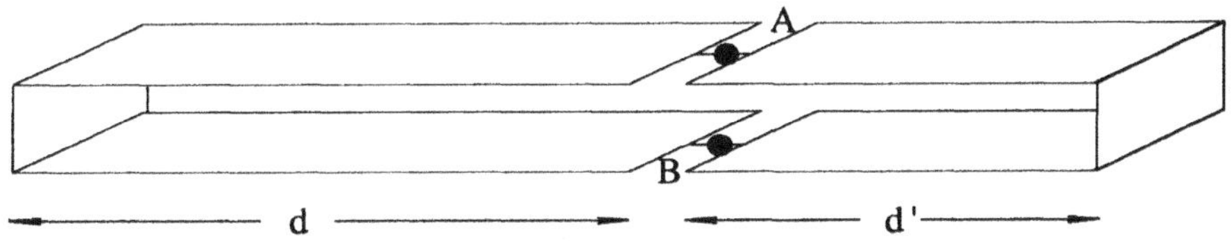

Figure 8.18. A parallel-strip transmission line resonator with short-circuited sections.

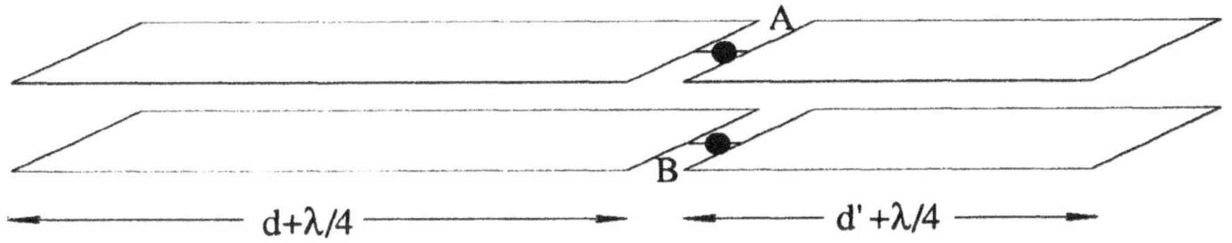

Figure 8.19. A parallel strip transmission line resonator with open-circuited sections.

We note from Eq. (8.21) that the impedance of a shorted transmission line tends to ∞ at $\lambda/4$ from the short (that is, at a phase angle that approaches $\pi/2$). If the line is cut or opened at this point, the impedance of the line does not change. This means that a resonator can be built with open transmission lines as shown in Figure 8.19. While there is no particular advantage of doing so in this case, it does simplify material handling.

A variety of structures of this type can be built for many applications and any frequency in the typical range for transmission lines (normally below *1 GHz*).

These resonators can be built in a variety of other combinations and lengths. Since the impedance along a line repeats at intervals of $\lambda/2$, addition of any multiples of $\lambda/2$ to any line does not change the properties of the line.

Another type of open transmission line resonator was discused in chapter 5 where it was introduced as a sensor. This is built from a coaxial transmission line, cut flat at one end. A discontinuity, or coupling gap is made approximately $\lambda/4$ from the open end. This creates a cavity with an open end for measurements. The cavity is actually a transmission type of cavity. The reflected waves from the test sample also pass through the cavity and are then detected. The geometry is shown in Figure 8.20. Testing with this type of cavity is performed in two ways. One, shown in Figure 8.21 requires the open end of the cavity to be in contact, or very close to the measured material. Measurements can be made on dielectrics, lossy dielectrics and the like. The second type of measurement is performed by introducing the materials in the cavity. This may include measurements of liquids and, especially, emission gases. This type of test is shown in Figure 8.22. Cavities of this type are relatively small (a few *mm* in diameter) and are very good for testing local variations of material (hence their introduction as sensors). However, the measurement is only sensitive in the resonator or near the open end. At short distances away, the sensitivity decreases considerably. In terms of tests, the test setups for other cavity resonators may be used although, since this particular resonator is a transmission line, the measuring equipment must be compatible with coaxial transmission lines.

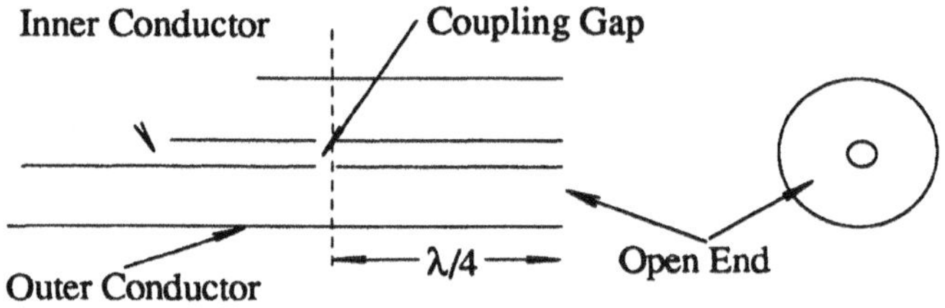

Figure 8.20. Open coaxial transmission line resonator.

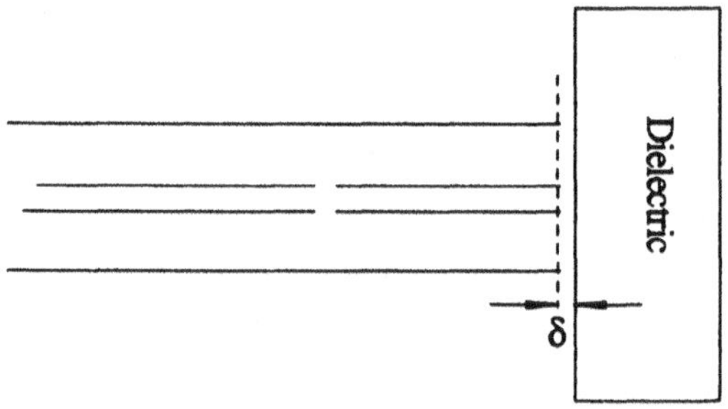

Figure 8.21. Method of testing with the coaxial transmission line resonator. The gap δ must be kept to a minimum.

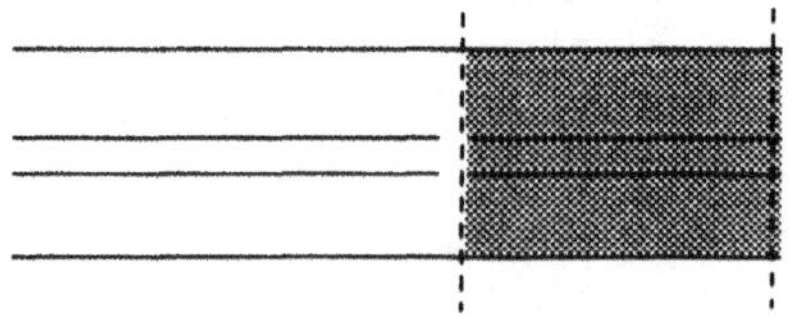

Figure 8.22. Method of testing gases or fluids in a coaxial transmission line resonator.

8.4.2. Density Tests

Testing density of materials is often associated with material conditions but can be a test in itself or an indirect measure of material characterization. In some cases, as in foams, density is a production parameter that must be monitored. Density of gases can be used to test for presence of materials or for state (for example monitoring of emission gases from drying or curing materials). In still other cases, density can be correlated to strength of materials or deterioration. Measurement of density is relatively simple and, in its basic form, relies on measurement of the dielectric constant which changes with density. For incompressible materials, this is normally easy to correlate with many of the material properties associated with density. In other materials, the conditions of the test must also be taken into account. These include such conditions as pressure, temperature and moisture content.

Density tests can be performed by any of the methods discussed above including the various reflection, transmission or attenuation methods because of the use of the dielectric constant. However, often the difference between the dielectric constant of free space and low density materials such as foams or gases is too low to allow sensitive measurements using these method. On the other hand, resonant methods are extremely sensitive to variations in ε and allow measurements of very low density materials. One of the uses of this method is detection of gases (such as ammonia or smoke) using cavity resonators as "sniffers".

There are two basic methods that can be employed. One is a sampling method, the other an open cavity or continuous method.

In a sampling method, a sample of the material, usually small, is inserted into a cavity and the shift in resonant frequency is recorded. Since this shift is entirely due to the change in the dielectric constant, and the dielectric constant is dependent on density, the reading can be easily calibrated to measure density. The basic process is shown in Figure 6.4. A cavity resonator is coupled to a waveguide and a wavemeter connected to the cavity. The sample is either small (i.e. a small piece of the material) or fills the whole cavity (for example, for a gas sample). In the first case, the dielectric constant of the material is calculated from the local perturbation conditions while in the second, the change in resonant frequency is due to the change in the dielectric constant throughout the cavity. If this change is small, perturbation methods can be used to calculate the shift in resonant frequency of the cavity. Under such conditions the shift in resonant frequency is entirely due to changes in material properties, primarily changes in dielectric constant.

If the sample is small, the shift in resonant frequency is calculated from the fact that under perturbation condition the shift in resonant frequency is directly proportional to the change in energy in the cavity. For a small sample we may assume the fields remain unchanged and the change in frequency is therefore

$$\frac{\omega - \omega_0}{\omega_0} \cong \frac{-\int_{\Delta v}\left(\Delta\varepsilon|E_0|^2 + \Delta\mu|H_0|^2\right)dv}{\int_{v0}\left(\varepsilon|E_0|^2 + \mu|H_0|^2\right)dv} \qquad (8.23)$$

In this relation, which was discussed in chapter 2, the nominator contains the change in electric and magnetic energy while the denominator is the total energy in the cavity. In our case, only the dielectric constant changes. Further, since the sample is small, we can write

$$\frac{\omega-\omega_0}{\omega_0} \cong \frac{-\Delta\varepsilon |E_0|^2 \Delta v}{\int_{v0} \left(\varepsilon |E_0|^2 + \mu |H_0|^2\right) dv} \tag{8.24}$$

where Δv is the volume of the material. Thus, the change in frequency is proportional to $\Delta\varepsilon$ and Δv of the sample. The minus sign simply indicates that an increase in ε decreases the resonant frequency. This then can be used to create a calibration curve.

If the sample is not small but does not fill the whole cavity, Eq. (8.23) should be used for calibration.

If the material fills the whole cavity, the resonant frequency is given as

$$f = \frac{f_0}{\sqrt{\varepsilon_r}} \tag{8.25}$$

where f_0 is the resonant frequency of the empty cavity.

Note for example, that for a change of 10^{-06} in the relative permittivity (that is: the relative permittivity has increased from *1.0* to *1.000001* due to presence of the test material), in a cavity that resonates empty at *10 GHz*, the shift in resonant frequency of the cavity is *5 kHz*; (the frequency is reduced by *5 kHz*). This is easily detectable with even the simplest measurement equipment. Thus it is obvious why resonant cavities can detect traces of gases and why they are preferred for density measurements, deterioration of materials and for monitoring of curing processes.

A continuous or open cavity method differs from the method described above only in that access to the cavity and material handling are much easier and the methods are more applicable to continuous testing. A simple example of density testing for a gas flowing through an open cavity is shown in Figure 8.23. A cylindrical foam is tested as it passes through the cavity. The cavity is cylindrical, with a cylindrical opening similar to that shown in Figure 8.9. The test uses an absorption cavity test but a transmission test as in Figure 6.4. can be used as well. Resonance is detected by a dip in the wavemeter. For continuous testing the fluctuations in resonant frequency are used as indications of density. The same test can be used for monitoring of thickness variations, moisture content, defects, and any other effect that changes the dielectric constant (or, with proper instrumentation), the loss tangent of the material.

The relations above allow the evaluation of the dielectric constant or, more commonly, change in the dielectric constant. From this the density is determined from empirical relations that exist for some materials. Some of these relations can be found in references at the end of this chapter. As an example, the Clasius-Mossotti equation relates the density to the dielectric constant as

$$d = \frac{3\varepsilon_0 m(\varepsilon_r - 1)}{a(\varepsilon_r + 2)} \tag{8.26}$$

where d is the mass density (in Kg/m^3), m the mass of the molecule, ε_0 the permittivity of free space, ε_r the relative permittivity of the material, and a a polarization factor for the molecules. This relation is empirical and is often used for gases and fluids. It is also applicable to some solid dielectrics. Thus, if a relation of this sort can be found, the density can be evaluated directly. In general relations of these sort are only useful if their parameters can be easily evaluated. In the case in Eq. (8.26) the polarization factor must be known for the material under test. This is not an easy task and in practice it is perhaps easier to create a calibration curve based on measurements.

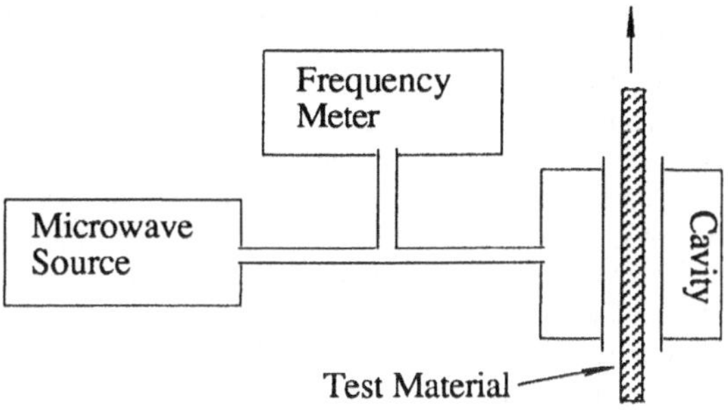

Figure 8.23. Density testing in an open microwave cavity. Density testing of a cylindrical foam is shown.

8.5. Scattering Methods

The fact that scattering can reveal location as well as size of flaws was discussed briefly in the previous chapter. Measurements depended on detection of scattered energy from the discontinuity. However, the detection mechanism was independent of the transmission antenna.

Probably the best developed of all microwave methods for general purposes is the radar (*RA*dio *D*etection *A*nd *R*anging). It is used in a variety of applications from the original application in aircraft detection and identification to weather prediction and remote sensing. It is in this capacity or remote sensing that we discuss radar methods here. The measurements with radar were discussed in the previous chapter, including the definition of radar cross section and the difficulties in measuring thin materials. We mention this method here again because of its usefulness in detecting dielectric variations in low loss materials. As an example, a local variation in the dielectric constant (in other words, a discontinuity or defect) in a thick material is be detected as shown in Figure 7.12. The source is a pulsed source and the detection is done by the same antenna (monostatic radar), allowing also for position detection from the antenna orientation. Figure 7.13 shows the detection of all discontinuities in the test but with proper processing the location and size of these discontinuities can be defined. Alternatively, the frequency domain method described in section (6.6) can be used for the same purpose. However, our interest here is not in detection of defects per se but in detection of material properties variations.

These types of scattering methods are very useful although, are normally more involved than standard tests. Part of the reason is that "radars" for nondestructive testing do not exist and must be built for the purpose. There are also special units, normally at lower frequencies than regular radars, specifically designed for testing into the ground, particularly for exploration and for mapping of underground objects. These are not adaptable for regular testing and, as with other radars, are not off the shelf units. Radar equipment and instrumentation is not adaptable for testing of the type described here because of their size, complicated instrumentation and, in particular, because the available units are designed for relatively long range operation. Not to be overlooked is their price.

Thus, in nondestructive testing radar (or, perhaps, scatterometry might be a better term) should be viewed as a pulsed source with an antenna and a directional coupler for detection of the return signal. Processing can also be included as required but it would normally be simpler than in radar equipment. Also missing is the sophisticated scanning mechanisms involved in other radar units. If scanning is used, this normally means that the sample is moved underneath the antenna. As the radar is pulsed, the signal is received and displayed. The size is normally inferred from the amplitude of the signal while location is found from timing considerations. To see this consider Figure 8.24. Here the radar is scanned over a flaw buried in a material. The antenna is moved, the source is pulsed and the signals received are plotted with time. Figure 8.25 shows the type of display obtained. The first pulse is due to the front surface of the material and is not shown in this figure. Normally this will be too close to the transmitter to be displayed (that is, the unit is still in transmit mode). The second pulse is due to the discontinuity or flaw. If the material is not very thick, or the attenuation in the material is low, there will also be a reflection due to the back surface of the slab (not shown in Figure 8.25, but the reflection from the front and back surfaces of the material were discused in section 7.3. in the context of location of flaws). Each scan represents a different location of the source while the time represents the propagation time to and from the scatterer. As the antenna approaches the defect, the signal becomes larger and the time it takes to propagate to the defect and back becomes shorter. If there are more than one scatterer, they produce superimposed signals and, unless they are very close to each other it is possible to detect all scatterers.

The scan presented above is a *B*-scan of the material. *C*-scans can also be obtained by properly moving the antenna to cover a test area. A simulated scan of this type is shown in Figure 7.9. If this is done, discontinuities can be imaged as was shown in Figure 7.9.

One of the main difficulties with any radar system and more so with the type shown above is that because the scattered signal is detected, the amplitude available is governed by the radar equation in which the amplitude is proportional to the distance to flaw to the negative *4th* power:

$$\frac{P_r}{P_t} = \frac{\sigma_b \lambda^2 G}{(4\pi)^3 r^4} \tag{8.27}$$

where P_r and P_t are the received and transmitted powers, σ_b is the backscatter cross section of the target, λ is the wavelength, G the antenna gain and r the distance between the antenna and target. The backscatter cross section is given by

$$\sigma_b = 4\pi r^2 \frac{P_s}{P_i}$$

where P_s is the power density at the receiver, P_i is the power density at the scatterer and r is the distance between scatterer and receiver.

Thus, the signals tend to be weak, especially if low level signals must be used for safety reasons. The scatterer must be in the far field (Fraunhofer) zone, further aggravating the problem. For this reason, most scatterometers of this type only work well in lossless and low loss dielectrics. This is also the case for into the ground radar where penetration is limited by the lossy nature of soil and water.

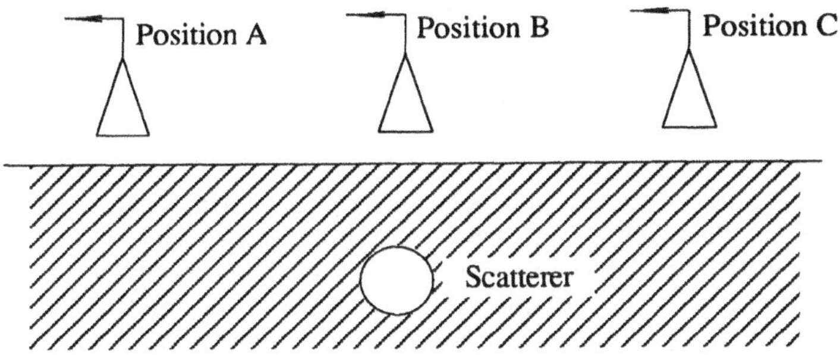

Figure 8.24. Testing with a pulsed scatterometer (radar). Three positions of the scanning antenna are shown.

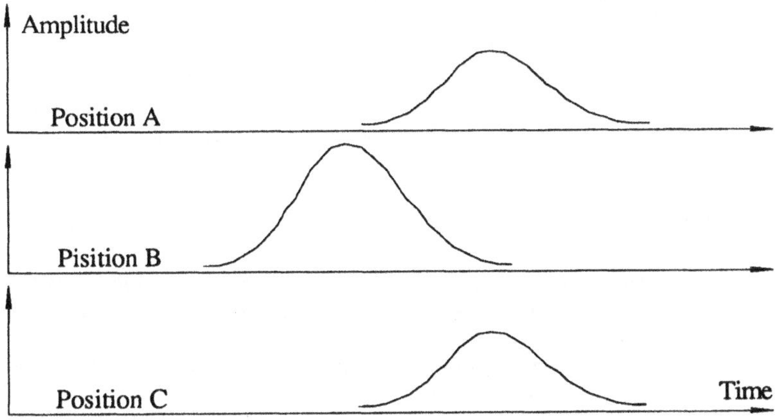

Figure 8.25. Signals obtained for the three positions of the antenna in Figure 8.24. The signal from the surface is not shown.

8.6. Bibliography

It was mentioned before that the number of measurements possible in the microwave domain is large since almost any condition in materials affects their electric and magnetic properties. General measurement techniques for permittivity variations are given in [1] through [14]. Special tests and instrumentation for measurements in biological material are given in [15] through [18]. Cavity resonator methods and perturbation techniques in closed cavities are discussed in [19 through 23]. Open cavity techniques are given in [24-35]. References [36-43] deal with measurement of moisture, water content and loss tangent. Additional information on transmission lines and resonators can be found in [44-48]. Density measurements are discussed in [49-50].

[1] E.L. Ginzton, "Microwave Measurements", McGraw Hill Book Company, Inc., NY, 1957.

[2] M. Sucher and J. Fox, Eds., Handbook of Microwave Measurements", Polytechnic Press, New York, 1963. (Vol. I, II, II).

[3] A.E. Bailey, "Microwave Measurements", IEE Electrical Measurement Series, Peter Peregrinus Ltd., London, 1985.

[4] F.E. Gardiol, "Introduction to Microwaves", Artech House, 1984.

[5] A.F. Harvey, "Microwave Engineering", Academic Press, London, 1963.

[6] A. Kumar and D.G. Smith, "The measurement of the complex permittivity of sheet materials at microwave frequencies using a resonant window in a rectangular waveguide cavity", proceedings of the 1975 Conference on Dielectric Materials, Measurements and Applications, The Institution of Electrical Engineers, 1975, pp. 151-154.

[7] D.D. Seltzer and Pless, W.M., "Correlative studies of microwaves to the physical properties of high density ablators and structural plastics", in Proc. of the 6th Symp. on NDE of Aerospace and Weapons Systems Components and Materials, San Antonio, TX. 1987, pp. 413-439.

[8] H.E. Bussey, "Measurements of RF properties of materials a survey", Proceedings of the IEEE, Vol. 55, No. 6, June 1967, pp. 1046-1053.

[9] M.F. Iskander and J.B. DuBow, "Time- and frequency-domain techniques for measuring the dielectric properties of rocks: a review", Journal of Microwave Power, Vol. 18, No. 2, 1983, pp. 143-152.

[10] C. Akyel, R.G. Bosisio, R. Chahine and T.K. Bose, "Measurement of the complex permittivity of food products during microwave power heating cycles", Journal of Microwave Power, Vol. 18, No. 4, 1983, pp. 355-365.

[11] A. Wehr, C. George and G. Lightsey, "Solvent recovery by microwave and radio-frequency processing", Journal of Microwave Power and Electromagnetic Energy, Vol. 25, No. 2, 1990, pp. 100-103.

[12] M. Baginski, R. Broughton, D. Hall, and L. Christman, "Experimental and numerical characterization of the radio-frequency drying of textile materials", Journal of Microwave Power and Electromagnetic Energy, Vol. 25, No. 2, 1990, pp. 104-113.

[13] G.G. Allan, B. B. Krieger and D.W. Work, "Dielectric Loss Microwave degradation of polymers: Cellulose", Journal of Applied Polymer Science, Vol. 25, 1989, pp. 1839-1859.

[14] A.M. Kurin and A. Yu, "Determining parameters of laminated materials having dielectric losses by measurements in a range of frequencies", Soviet Journal of Nondestructive Testing (Defectoskopiya), Vol. 22, No. 3, pp. 205-215.
[15] T.W. Athey, M.A. Stuchly and S.S. Stuchly, "Measurement of radio frequency permitivity of biological tissues with and open ended coaxial line: PartI and II, IEEE Transactions on Microwave Theory and Techniques, Vol. MTT-30, No. 1, Jan. 1982, pp. 82-91.
[16] A. Kraszewski, S.S. Stuchly, M.A. Stuchly and S.A. Symons, "On the measurement accuracy of the tissue permitivity in vivo", IEEE Transactions on Instrumentation and Measurements, Vol. IM-32, No. 1, March 1983, pp. 37-42.
[17] M.A. Stuchly and S.S Stuchly, "Dielectric properties of biological substances - Tabulated", V.A. Viktorov, B,V, Lunkin and A.S. Sovlukov, "Transmission line sections as sensors for radiofrequency measurements of nonelectrical quantities", Journal of Microwave Power, Vol. 15, No. 1, 1980, pp. 19-26.
[18] O.P. Gandhi, "Some basic properties of biological tissues for potential biomedical applications of millimeter waves", Journal of Microwave Power, Vol. 18, No. 3, 1983, pp. 295-304.
[19] S. Li, C. Akyel and R.G. Bosisio, "Precise calculations and measurements on the complex dielectric constant of lossy materials using TM010 cavity perturbation techniques", IEEE Transactions on Microwave Theory and Techniques, Vol. MTT-29, , 1981, pp.
[20] W. Meyer, "Dielectric measurements on polymeric materials by using superconducting microwave resonators", IEEE Transactions on Microwave Theory and Techniques, Vol. MTT-25, No. 12, Dec. 1977, pp. 1092-1099.
[21] W. Meyer, "Helical resonators for measuring dielectric properties of materials", IEEE Transactions on Microwave Theory and Techniques, Vol. MTT-29, No.3, 1981, pp. 240-247.
[22] S. Li and R.G. Bosisio, "Composite hole conditions on complex permitivity measurements using microwave cavity perturbation techniques", IEEE Transactions on Microwave Theory and Techniques, Vol. MTT-30, No. 1, Jan. 1982, pp. 100-103.
[23] F. Henry and A.J. Berteaud, "New measurement technique for the dielectric study of solutions and suspensions", Journal of Microwave Power, Vol. 15, No. 4, 1980, pp. 233-242.
[24] D. Xu, L. Liu and Z. Jiang, "Measurement of the dielectric properties of biological substances using an improved open ended coaxial line resonator method", IEEE Transactions on Microwave Theory and Techniques, Vol. MTT-35, No. 12, Dec. 1987, pp. 1424-1428.
[25] A. Denoth, A. Foglar et al, "A comparative study of instruments for measuring the liquid water content of snow", J. of Applied physics, Vol. 56, No. 7, 1984, pp. 2154-2159.
[26] I.J. Bahl, A. Thansandote and S.S. Stuchly, "Open-ended rectangular waveguides as antennas for medical diagnostics", V.A. Viktorov, B,V, Lunkin and A.S. Sovlukov, "Transmission line sections as sensors for radiofrequency measurements of nonelectrical quantities", Journal of Microwave Power, Vol. 15, No. 2, 1980, pp. 81-86.

[27] J.R. Mosig, J.C.E. Besson, M. Gex-Fabry and F. E. Gardiol, "Reflection of an open ended coaxial line and application to nondestructive measurement of materials", IEEE Transactions on Instrumentation and Measurements, Vol. IM-30, No. 1, March 1981, pp. 46-51.
[28] E. Tanabe and W. T. Joines, "A nondestructive method for measuring the complex permitivity of dielectric materials at microwave frequencies using an open transmission line resonator", IEEE Transactions on Instrumentation and Measurements, Vol. IM-25, No. 3, Sept. 1976, pp. 222-226.
[29] V.P. Maslov and E.M. Vorob'ev, "Single mode open cavity resonators", Soviet Physics-Doklady, Vol. 13, No. 3, Sept. 1968, pp. 223-225.
[30] J. Smetana and N.C. Wenger, "Instrumentation for liquid hydrogen density measurements using an open ended microwave cavity, Nasa report TN D-6415, July 1971.
[31] N.C. Wenger and J. Smetana, "Hydrogen density measurements using an open ended microwave cavity, IEEE Transactions on Instrumentation and Measurements, Vol. IM-21, No. 2, May 1972, pp. 105-114.
[32] L.A. Weinstein, "Open Resonators and Open Waveguides", The Golem Press, Boulder, CO., 1969.
[33] J. Zuercher, L. Hoppie, R. Lade, S. Srinivasan and D. Misra", Measurement of the complex permittivity of bread-dough by an open-ended coaxial line method at ultrahigh frequencies", Journal of Microwave Power and Electromagnetic Energy, Vol. 25, No. 3, 1990, pp. 161-167.
[34] E.D.M. El-Sayed and A.M.K. Hashem, "Ridged waveguide applicators for uniform microwave heating of sheet materials", Journal of Microwave Power, Vol. 19, No. 2, 1984, pp. 111-117.
[35] D. Misra, M. Chabbra, B.R. Epstain, M. Microtznik, K.R. Foster, "Noninvasive electrical characterization of materials at microwave frequencies using an open-ended coaxial line: test of an improved calibration technique", IEEEE Transactions on
[36] P.G.S. Castle and J. Roberts, "A microwave instrument for the continuous monitoring of the water content of crude oil", Proceedings of the IEEE, Vol. 62, No. 1, 1974, pp. 103-108.
[37] A. Kraszewski, "Microwave aquametry - a review", V.A. Viktorov, B,V, Lunkin and A.S. Sovlukov, "Transmission line sections as sensors for radiofrequency measurements of nonelectrical quantities", Journal of Microwave Power, Vol. 15, No. 4, 1980, pp. 210-220.
[38] N.C. Staub, "The measurement of lung water content", Journal of Microwave Power, Vol. 18, No. 3, 1983, pp. 259-263.
[39] W. Meyer and W. Schilz, "High frequency dielectric data on moist materials", Journal of Microwave Power, Vol. 17, No. 1, 1982, pp. 67-77.
[40] S. Okamura, "High-moisture content measurement of grain by microwaves", Journal of Microwave Power, Vol. 16, No. 3,4, 1981, pp. 253-256.
[41] A.E. Abara and M.A. Hill, "determination of the moisture content of potatoes by microwave energy", Journal of Microwave Power, Vol. 16, No. 3,4, 1981, pp. 249-252.
[42] N.P. Vasilakos and F. Magalhaes, "Microwave drying of polymers", Journal of Microwave Power, Vol. 19, No. 1, 1984, pp. 135-144.

[43] W.J. Chudobiak, B.A. Syrett and H.M. Hafez, "Recent advances in broad-band VHF and UHF transmission line methods for moisture content and dielectric constant measurement", IEEE Transactions on Instrumentation and Measurements, Vol. IM-28, No. 4, Dec. 1979, pp. 284-289.
[44] K.C. Gupta, R. Garg and I.J. Bahl, Microstrip Lines and Slotlines, Artech House, Dedham, MA., 1979.
[45] M. Brodwin and J. Benway, "Experimental evaluation of a microwave transmission moisture sensor", Journal of Microwave Power, Vol. 15, No. 4, 1980, pp. 261-265.
[46] M. Kent and T.E. Price, "Compact microstrip sensor for high moisture content materials", Journal of Microwave Power, Vol. 14, No. 4, 1979, pp. 363-365.
[47] V.A. Viktorov, B,V, Lunkin and A.S. Sovlukov, "Transmission line sections as sensors for radiofrequency measurements of nonelectrical quantities", Journal of Microwave Power, Vol. 15, No. 1, 1980, pp. 1-6.
[48] M. Kent and J. Kohler, "Broadband measurements of stripline moisture sensors", Journal of Microwave Power, Vol. 19, No. 1, 1984, pp. 173-179.
[49] S.O. Nelson, "Density dependence of the dielectric properties of wheat and whole-wheat flour", V.A. Viktorov, B,V, Lunkin and A.S. Sovlukov, "Transmission line sections as sensors for radiofrequency measurements of nonelectrical quantities", Journal of Microwave Power, Vol. 19, No. 1, 1984, pp. 55-64.
[50] S.O. Nelson, "Observations on the density dependence of dielectric properties of particulate materials", Journal of Microwave Power, Vol. 19, No. 1, 1984, pp. 55-64.

Chapter 9.
Methods of Modeling

9.1. Introduction

This chapter deals with the basics of numerical modeling as these relate to the microwave domain. None of the methods described is specific to *NDT*, but rather, general methods for modeling. The unique aspects of *NDT* are manifest in the way the techniques are used and the material properties and geometries.

The methods described here are described in rather simple and general terms. Only the basic aspects are discussed while the specific application of the methods are deferred to the following two chapters. After becoming familiar with the basic method, this will be modified to take into account the specifics of the model. The main goal is to show the available methods and point to basic differences between the various methods. As an example, we discuss here the finite difference method as it applies to a simple time dependent equation in one dimension. Later this will be modified to account for infinite space in two and three dimensions. Other aspects, such as stability criteria and economics of solution are also deferred.

Also to be noted is that we do not mention analytical solutions of the wave equations. The analytic solution of any equation, including the wave equations is understood to be the ultimate desire of any design process. If we could solve the equations exactly, there would be no reason for existence of numerical models. Thus, the lack of discussion of analytic solutions should be viewed as a statement of the difficulty in doing so rather than of their importance. The fact is that, other than for the most trivial of problems, we cannot obtain analytic solutions and must resort to numerical methods.

9.2. Purpose and Scope of Modeling

The basic premise of modeling is the simulation of the testing environment for purposes of design and prediction of parameters. In the testing environment, computer simulation has the following purposes:

1. Simulation of the physical environment of testing. This includes all system parameters, materials and material properties and sources.
2. Characterization of the test. All aspects of materials, method of excitation, and any other parameter needed for correct modeling.

3. Accurate definition of geometries of samples, sources and applicable boundaries and interfaces.

Modeling should be viewed as a method of performing an experiment without the need for laboratory equipment. It is therefore a very effective method of experimentation and if properly done can be accurate, inexpensive, and almost always faster than a real experiment. While most aspects of experimentation can be modeled, it is not an actual experiment and very often the test is idealized. For example, a microwave source may have frequency stability problems or power output variations. These can be modeled but normally frequency and power output are assumed to be constant. In this sense, the numerical experiment is "better" than the actual experiment. On the other hand, in the actual design these parameters must be addressed.

Computer modeling can be used for design purposes or for simulation of specific aspects of a test. In many cases it allows the user to preview an actual experiment before it is built. For example, if a small loop is used to test a piece of composite material as in Figure 9.1, both the near and far field effects can be shown. The field distribution in space as well as within the material can be seen, both of which are difficult to measure or visualize in the actual experiment. The propagation effects along the material may or may not be of interest in the experiment but the simulation will show these as well. As an example, consider Figures 9.2 and 9.3. The first shows the field distribution in the near field of the loop. The second shows the far-field of the loop in which the propagation nature of the field is obvious. In this case (which will be discussed in detail later) the material acts as a dielectric waveguide, something that may or may not be obvious during the actual test. This aspect of the test may in itself be of great value since when the test is actually carried out, the safety aspects of coupling to the dielectric may be considered, or the guiding nature of the test sample may be used for testing.

Finally, a good model should also indicate if an experiment is useful and if it should be performed, without the need for sample preparation, material handling equipment, and safety considerations. Only after simulations have shown that the test is feasible should one engage in actual testing.

9.3. General Approach to Modeling: Numerical

Modeling of the continuum environment entails the following general steps:

1. Formulation of the problem in a form amenable for computer solution.
2. Discretization of the continuum into subspaces over which an approximation can be applied.
3. Definition of a consistent method of approximation.
4. Solution of the system unknowns as defined by the approximation process over the discretized domain.
5. Analysis of data.

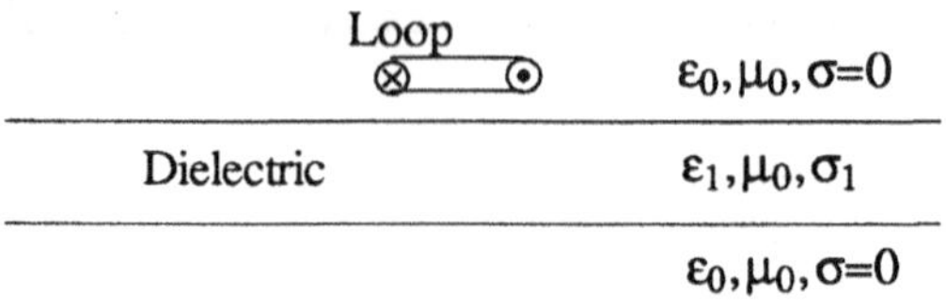

Figure 9.1. A small loop used for testing of composite materials.

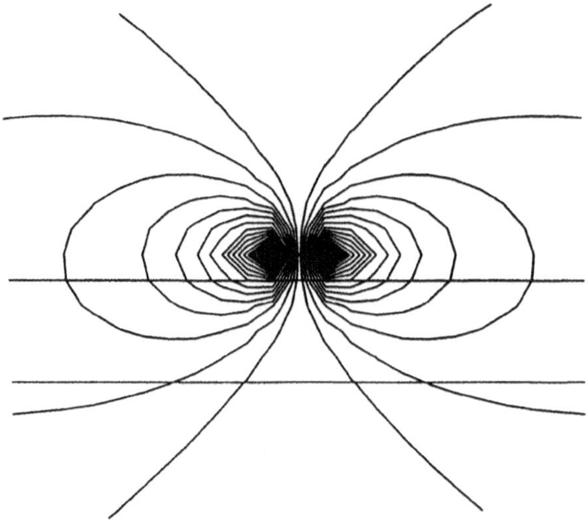

Figure 9.2. The near field due to a loop in the presence of a lossy dielectric.

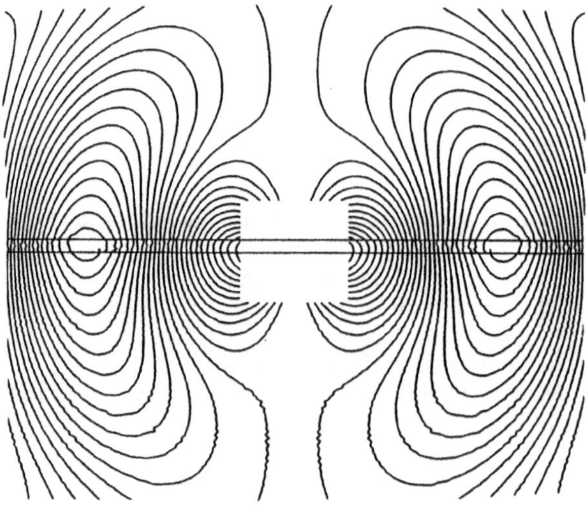

Figure 9.3. The far field solution of the field generated by the loop in Figure 9.1.

1. By formulation of the problem is meant a method of describing the physical relations within the domain under study. For example, Maxwell's equations are certainly adequate for modeling of microwave applications, provided that material constitutive relations are added. However, proper solution of the field variables requires solution of 12 variables: three components of each of the vectors **B**, **H**, **E** and **D**. Therefore, while the system of equations is correct it is not a convenient one. By formulation we will mean an equation, or system of equations in one or more variables that solves the same problem that Maxwell's equations solve but in a simpler form. While the process may introduce approximations and simplifications, this is not necessarily so. The formulation may in fact be exact in that it represents the same solution or, in some cases, may be approximate since it may neglect some aspect of the model. As an example, an approximate formulation may neglect free charges in the solution domain or, the boundary conditions may not be imposed exactly.
2. Discretization. By nature, a computer solution to a continuous problem is obtained by discretization. The solution domain is divided into surfaces, segments or volumes for the purpose of calculating the solution in each of these domain. To understand this suppose we wish to build a very large spherical dome say, as the roof of a house. One method would be to order a dome with exact spherical structure. This is the continuous problem solution. Alternatively, we may build the dome out of small tiles, each of a very simple shape and each planar in structure. This is an approximation to the continuous problem. We have discretized the solution domain and "solved" the problem by a very simple method. The most important point in the process is that the solution is correct in the sense that we may choose to use as many tiles as we wish of as small a size as possible. In the limit, the approximation approaches the exact solution.
3. The approximation: In the example above, we chose to use a planar approximation for a spherical surface. Even though this is not strictly correct, it becomes a good approximation if the size of the discretization element (tile) domain is not large. Obviously, if we chose to represent the dome by say, 4 tiles, our solution would not be as good as if we chose 40,000 tiles. Again, the important point is that the approximation itself is correct. What makes it useful is the number and size of domains we use. A more complex approximation might result in a better surface representation with fewer tiles but it would also complicate the solution process (in this case the manufacture and positioning of the tiles to create the dome).
4. Solution process: After the problem has been described, how do we intend to solve it?. This step is rather obvious but, in many cases is the most time consuming from the computation point of view. If our interest is in on-line testing of a material then computation cannot be long but, if all we want is to compute some testing parameters for the purpose of designing a test, a longer solution time is acceptable.
5. Results: After the problem has been solved and the results are obtained, (in the form of fields, geometrical data, etc.) we must analyze the results to help in the decision process. Perhaps these results will show that our measurements are wrong, or that the experiment is not physically possible (e.g. the data measured is too noisy or the magnitudes are so small as not to be measurable).

The following sections deal with the general aspects of modeling. The more common methods of discretization and approximation are discussed in general terms, as general methods for solution. The discussion of formulation of specific problems are deferred to the following two chapters but the basic ideaa are described here.

9.4. The Finite Difference Method

We discuss the finite difference method first, because it is rather simple and because it is based on direct approximation of the derivatives of the functions. Therefore, no special techniques are needed for formulation. The formulation of a partial differential equation, at the basic level, consists of replacing the partial derivatives by the finite difference approximation. There are modifications and improvements that can be made but these are not important for this discussion.

The finite difference method, (*FDM*) is not a new method -it has been traced back to Gauss- but has found significant utilization in recent years as a general method for the solution of partial differential equations, in particular for time-dependent applications. The main reasons for its widespread application are its relatively simple implementation and its generality. It is applicable to *dc* fields, quasi static or transient fields and to linear and nonlinear problems. In its simplest form, the formulation of the field equations consists of simply replacing the partial derivatives by appropriate difference formulas. A solution is then obtained for the dependent variable at discrete points within the solution region either by an iterative process or by the solution of a system of algebraic equations, depending on the finite difference formula chosen.

On the other hand, application of the finite difference method is complicated by problems of convergence and stability of the solution and by restrictions on the discretization process. For example, regular sets of discretization points (grids) are easy to handle, but irregular grids are either not practical or much more difficult. Discretization of complex geometries into regular grids is not practical and irregular grids may in some cases render the solution nonconvergent. In field problems the inability to properly discretize small areas such as air gaps or defects is critical and is detrimental to the use of the finite difference method. In addition, the finite difference method does not take into account distributed parameters such as current densities, conductivities and permeabilities. These have to be described as equivalent nodal quantities with all the associated errors. The solution obtained is only valid at the nodal points.

9.4.1. The Finite Difference Representation

As mentioned above, the main step in a finite difference approximation is replacement of the partial derivatives by difference formulas. These are in effect, approximations to derivatives based on discrete points. Rather than discuss this in a general sense, we show here two methods of defining the finite difference approximation. One is based on an "intuitive" approach (actually based on the definition of the derivative), the second on truncated Taylor series.

Considering Figure 9.4, where a general function is described, the true derivative dy/dx at a point x_i is the tangent to the curve at this point. An approximation to the derivative can be found by taking any two points, say on both sides of the point x_i and passing a straight line through them. If the two points are chosen to be equally spaced about the point x_i, as in Figure 9.4, the following expression for the slope of the line is obtained

$$\frac{dy(x_i)}{dx} = \frac{y(x_i+\Delta x) - y(x_i-\Delta x)}{2\Delta x} \tag{9.1}$$

Denoting in short form $y(x_i+\Delta x)$ as y_{i+1} and $y(x_i-\Delta x)$ by y_{i-1} a simple expression linking the approximation of the derivative to the function value at x_{i+1} and x_{i-1} is written as

$$\frac{dy_i}{dx} = \frac{y_{i+1} - y_{i-1}}{2\Delta x} \tag{9.2}$$

The same result can be obtained in a formal way by using Taylor series expansions. By expanding the function $y=f(x)$ about the point x for $x=x_i-\Delta x$ and $x=x_i+\Delta x$ the following are obtained

$$y(x_i+\Delta x) = y_i + \frac{dy_i}{dx}\Delta x + \frac{d^2y_i}{dx^2}\frac{(\Delta x)^2}{2!} + \frac{d^3y_i}{dx^3}\frac{(\Delta x)^3}{3!} + \dots \tag{9.3}$$

$$y(x_i-\Delta x) = y_i - \frac{dy_i}{dx}\Delta x + \frac{d^2y_i}{dx^2}\frac{(\Delta x)^2}{2!} - \frac{d^3y_i}{dx^3}\frac{(\Delta x)^3}{3!} + \dots \tag{9.4}$$

By subtracting Eq. (9.4) from Eq. (9.3) and rearranging the terms, the first derivative can be written as

$$\frac{dy_i}{dx} = \frac{y(x_i+\Delta x) - y(x_i-\Delta x)}{2\Delta x} - \left[\frac{1}{6}\frac{d^3y_i}{dx^3}(\Delta x)^2\right] + \dots \tag{9.5}$$

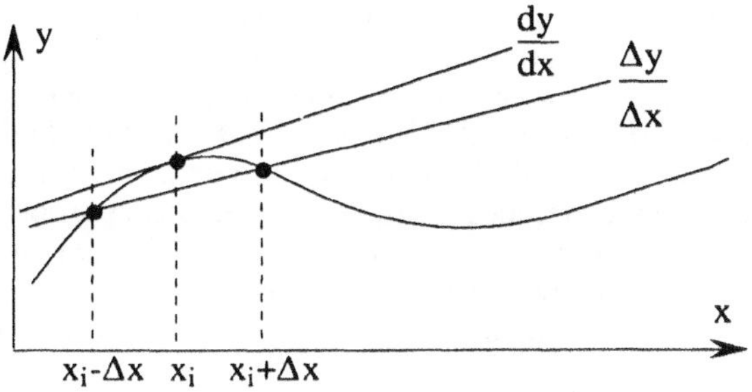

Figure 9.4. Definition of the finite difference approximation.

Then, neglecting terms with $(\Delta x)^2$ and higher powers of (Δx), the expression in Eq. (9.2) is obtained. This method, although less intuitive than the one used to derive Eq. (9.2) shows two important points.

1. The error produced by the approximation is of the order of $(\Delta x)^2$. Thus, a simple way of estimating the errors is available and, at the same time the solution can be improved by reducing the size of Δx.
2. The finite difference formula in Eq. (9.5) was obtained by choosing an appropriate expansion so as to cancel specific terms of the expansions. This indicates that higher order derivatives and different approximations to the same derivatives can be obtained. Indeed, many difference formulas have been derived using this method.

An approximation to the second derivative is obtained by adding the two expansions in Eq. (9.3) and Eq. (9.4) and rearranging the terms.

$$\frac{d^2 y_i}{dx^2} = \frac{y(x_i+\Delta x) - 2y(x_i) + y(x_i-\Delta x)}{(\Delta x)^2} - \left[\frac{1}{12}\frac{d^4 y_i}{dx^3}(\Delta x)^2 + \dots\right] \qquad (9.6)$$

This particular approximation also introduces an error of the order of $(\Delta x)^2$.

The finite difference expressions derived here both use values at points on both sides of the point at which the derivatives are calculated. They are therefore called central difference expressions. Backward and forward difference formulas may also be used as discussed below.

Other finite difference formulas may be obtained as required. Depending on the truncation of the Taylor series, the accuracy may be better or worse. In particular, finite difference formulas for the so called backward or forward approximations are important. Before defining any of these formulas, we define a forward difference scheme as any scheme that uses only values at the current level and future levels (for time derivatives) or values in "front" of the current value for spatial derivatives. Thus, in the context of the definitions above, only values of the function at point i, j, and $i+1, j+1$, (or $i+n, j+m$)are allowed, in any combination. The most obvious application is in the time domain with values at time t and $t+\Delta t$ (or $t+n\Delta t$). Similarly, backward difference schemes use only values at the present and past levels. These difference schemes are used to formulate derivatives in cases in which the central difference schemes above are either not applicable or inconvenient. As an example, in approximating the time derivative of a function we may choose to use a backward difference scheme since it only requires knowledge of "past" values as shown in Figure 9.5. Similarly, if we wish to calculate the derivatives on the boundary of a domain, central differences may not be the most appropriate approach. For example, in Figure 9.6 a backward difference approach may be used on the left boundary and a forward difference approach on the right boundary.

In its simplest form, a forward difference scheme may be defined for a point x and $x+\Delta x$ as

$$\frac{dy_i}{dx} = \frac{y_{i+1} - y_i}{\Delta x} \tag{9.7}$$

Similarly, a backward difference scheme can be written as

$$\frac{dy_i}{dx} = \frac{y_i - y_{i-1}}{\Delta x} \tag{9.8}$$

However, unlike the central difference scheme in Eq. (9.2), these schemes have an accuracy of the order of (Δx) instead of $(\Delta x)^2$.

Forward and backward difference schemes can be written in a variety of alternative forms both for first and second order derivatives (or, for that matter, for a derivative of any order) but we will not discuss these here. Rather, we will introduce these schemes as necessary. The references at the end of the chapter give ample information on this subject.

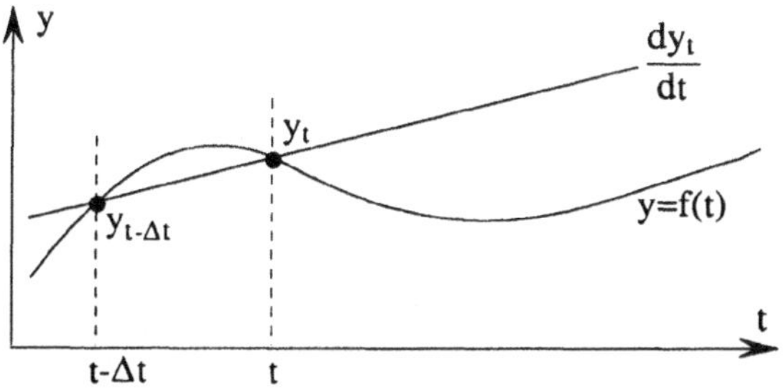

Figure 9.5. Definition of a backward finite difference scheme.

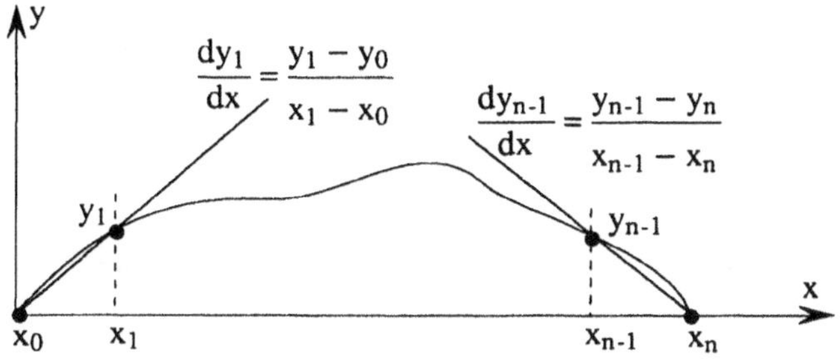

Figure 9.6. Forward and backward difference schemes in space.

9.4.2. Finite Difference Formulation For the 1-D Wave Equation

To demonstrate the use of finite differences for a simple problem we use the one dimensional wave equation. This is equivalent to assuming we have a plane wave. To simplify solution we will look at the wave solution between two very large conducting parallel plates. This allows simple specification of boundary conditions. Solutions are obtained as propagating waves or standing waves, depending on the wavelength and the distance between the plates. However, we will emphasize the solution rather than the physical meaning so that we may concentrate on the finite difference aspects rather than on the wave equation. A more detailed discussion of the wave equation and its discretization in finite differences is given in the following chapter.

The wave equation we are trying to solve is

$$\frac{\partial^2 E_x}{\partial z^2} = \mu\varepsilon \frac{\partial^2 E_x}{\partial t^2} \tag{9.9}$$

For convenience we will rewrite this as

$$\frac{\partial^2 E_x}{\partial t^2} = \frac{1}{\mu\varepsilon}\frac{\partial^2 E_x}{\partial z^2} \tag{9.10}$$

The first step in the field formulation consists of replacing the partial derivatives by a difference equation. To do so we use a finite difference grid (or mesh) with equally spaced points in space (Δz) and in time (Δz). The grid is bounded in space by $z=0$ and $z=z_0$ but not bounded in time. For practical solution we will also have to limit the solution in time but there is no need to do so now. Refering to the grid in Figure 9.7, and denoting time by the index j and space by the index i we can write for the partial derivatives in space, at a general point i,j

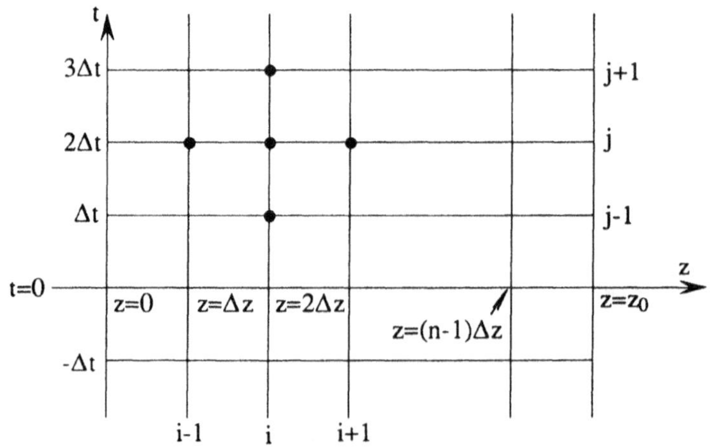

Figure 9.7. Finite difference mesh for a one-dimensional time dependent problem.

$$\frac{\partial^2 E_x}{\partial z^2} \approx \frac{E_{z-\Delta z,t} - 2E_{z,t} + E_{z+\Delta z,t}}{(\Delta z)^2}$$

Similarly, for the time derivative at this point:

$$\frac{\partial^2 E_x}{\partial t^2} \approx \frac{E_{z,t-\Delta t} - 2E_{z,t} + E_{z,t+\Delta t}}{(\Delta t)^2}$$

Substitution of these terms in Eq. (9.10) gives

$$\frac{E_{z,t-\Delta t} - 2E_{z,t} + E_{z,t+\Delta t}}{(\Delta t)^2} = \frac{1}{\mu\varepsilon}\frac{E_{z-\Delta z,t} - 2E_{z,t} + E_{z+\Delta z,t}}{(\Delta z)^2} \tag{9.11}$$

To simplify the expression we write this as

$$E_{z,t-\Delta t} - 2E_{z,t} + E_{z,t+\Delta t} = \frac{(\Delta t)^2}{\mu\varepsilon(\Delta z)^2}\left[E_{z-\Delta z,t} - 2E_{z,t} + E_{z+\Delta z,t}\right]$$

or, using the indices i, j, to simplify both the expression and its implementation in a computer algorithm:

$$E_{i,j-1} = 2E_{i,j} - E_{i,j+1} + \frac{(\Delta t)^2}{\mu\varepsilon(\Delta z)^2}\left[E_{i-1,j} - 2E_{i,j} + E_{i+1,j}\right] \tag{9.12}$$

This expression indicates that if we know the solution at time t and $t-\Delta t$ for all values of z, then a solution can be obtained for a time $t+\Delta t$. Thus, a solution in time can be obtained.

There are two methods of solving this problem: one simply assumes that none of the solutions are known but that the initial and final condition in time are known. This implies that the boundary conditions in space and time can be applied on a finite-size grid. These boundary conditions are Dirichlet type boundary conditions (value of the function is known). The result is that we can write Eq. (9.12) for all interior nodes of the mesh in Figure 9.7 and solve the resulting system of equation. This is called an implicit solution. Unless the boundaries in time can be easily evaluated, (that is: the steady state solution for a transient problem or the solution at a given time for the wave equation) it is not an appropriate method of solving the problem. As an example, in a transient problem, this would imply that the initial conditions in time and the steady state solutions are known explicitly, including the time required to reach steady state, and we want to evaluate some intermediate result.

A more feasible method for time dependent problems is to apply initial conditions to the problem and solve the problem explicitly in time, so that a solution is obtained at each time step. Not only there is no need to solve a large system of equations but the solution at each time step is evaluated and the propagation in time may be stopped based on the progression of solution rather than requiring any previous knowledge about the solution.

Before we do so, the initial conditions must be known at time t and $t-\Delta t$, that is: the solution for each value of z at time t and $t-\Delta t$ must be known. The simplest way to specify these values is to assume a solution at time t and to assume that the solution at $t+\Delta t$ is equal to the solution at time $t-\Delta t$. That is:

$$E_{z,t-\Delta t} = E_{z,t+\Delta t} \tag{9.13}$$

This is also equivalent to saying that

$$\frac{\partial E_{z,t}}{\partial \tau} = 0 \tag{9.14}$$

or, equivalently, that the solution also propagates in negative time. Since this latter solution is not actually used there is nothing wrong in this assumption. Now we can rewrite Eq. (9.12) as

$$2E_{z,t+\Delta t} = 2E_{z,t} + \frac{(\Delta t)^2}{\mu\varepsilon(\Delta z)^2}\left[E_{z-\Delta z,t} - 2E_{z,t} + E_{z+\Delta z,t}\right]$$

or

$$E_{z,t+\Delta t} = E_{z,t} + \frac{(\Delta t)^2}{2\mu\varepsilon(\Delta z)^2}\left[E_{z-\Delta z,t} - 2E_{z,t} + E_{z+\Delta z,t}\right] \tag{9.15}$$

With specified values of E at $t=0$, the values at $t=\Delta t$ are calculated using Eq. (9.15) as a first step. Now Eq. (9.12) can be used to calculate the solution at time $2\Delta t$, $3\Delta t$, etc. The solution can be stopped at any time step, depending on the solution we wish to obtain.

A final point that must be considered is the so called Courant stability criterion. For an equation of the type shown in Eq. (9.10) (i.e. a hyperbolic equation), the following must be satisfied:

$$0 < \frac{(\Delta t)^2}{\mu\varepsilon(\Delta z)^2} \leq 1 \tag{9.16}$$

for a stable solution. Also, to obtain an accurate solution we must have

$$\frac{(\Delta t)^2}{\mu\varepsilon(\Delta z)^2} \approx 1 \qquad (9.17)$$

Thus, we must choose Δt and Δz such that this condition is satisfied. We will discuss this again in Chapter 10 but it is obvious that this condition imposes severe restrictions on our choice for discretization. We must choose Δz such that the required resolution in space is obtained. Since μ and ε are give, the time step Δt must be chosen to satisfy this condition.

Example: We can now solve the problem given above by simply choosing some values for the problem. Assuming the two plates are separated by free space (μ_0,ε_0), we apply first the initial condition shown in Figure 9.8. Because the plates are conducting, the electric field must be zero at the plates. The distance between the plates (*1 m*) is divided into *10* equal segments, resulting in *11* points in the *z* direction. With the values of ε_0 and μ_0 and Δz, the time step must be equal or smaller than *0.33 nanoseconds*. As an example we may choose *0.1 nanoseconds*.

Using Eqs. (9.12) and (9.15), a solution is obtained. The solutions at times *t=1.0, 1.5*, and *2.0 nanoseconds* are shown in Figure 9.9. These are the solutions we would expect from a transient wave which bounces back and forth at the boundaries.

A simpler solution can be obtained by using the initial condition in Figure 9.10. This is equivalent to saying that the frequency of the wave is equal to *300 MHz*. Repeating the solution process as before, the solutions in Figure 9.11 are obtained. These are obviously standing wave solutions at times *t=0.2, 0.3, 0.5*, and *1.0 nanoseconds*. Again, these solutions are as expected based on our initial assumption. Particularly important in these solutions is that the solutions are rather good in spite of the large distance between the points and the very small number of points. A more accurate, and much smoother solution can be obtained by increasing the number of points (with appropriate readjustment of the time step).

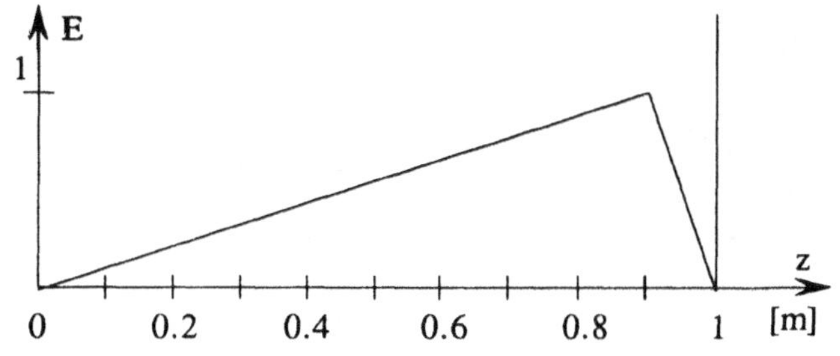

Figure 9.8. Initial conditions for transient wave solution.

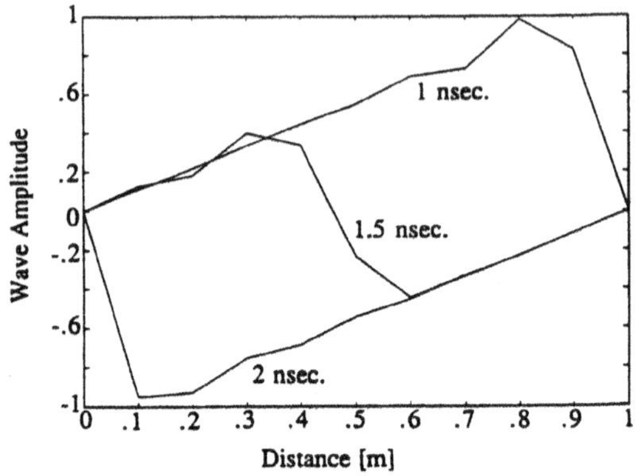

Figure 9.9. Wave solutions at *1.0*, *1.5*, and *2.0 nanoseconds*.

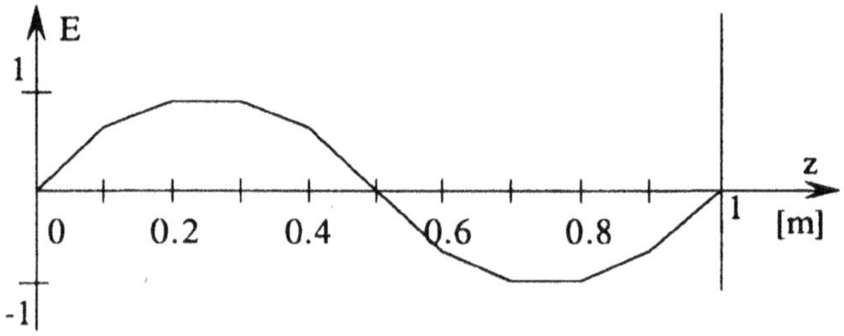

Figure 9.10. Initial conditions for standing wave solution.

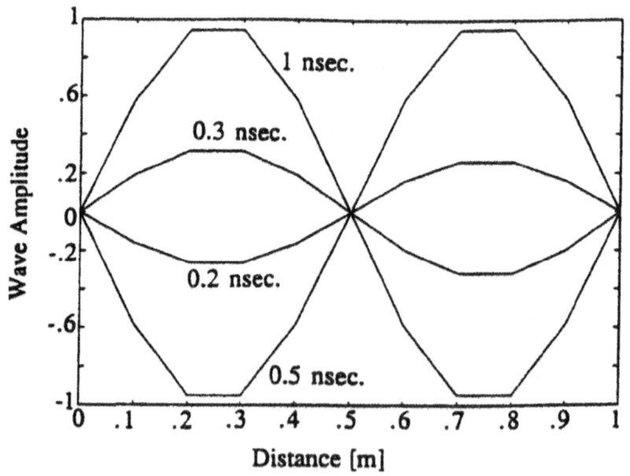

Figure 9.11. Standing waves at *0.2*, *0.3*, *0.5*, and *1.0 nanoseconds*.

The finite difference solution to this problem proved to be rather simple. Extension to two dimensions is relatively easy to implement as is the extension to three dimensions. However, we have deferred discussion of many aspects of solution such as interface conditions between different materials in the solution domain, problems with large number of time steps that may be required, as well as that of curved or nonuniform boundaries, to obtain the simple solution above. We have done so to concentrate on principles rather than problem solving. These aspects will be reintroduced in chapter 10. Also, we mention here that the finite difference method is equally well adapted for solution in the frequency domain or for that matter, for static problems. However, following the accumulated experience in field computation we will only discuss finite difference methods in the context of time dependent problems while using the finite elements and method of moments in the frequency domain. Static problems will not be discussed here.

9.5. Finite Element Methods (*FEM*)

The finite element method has a briefer history than the finite difference method. It evolved in the late 1950's as a numerical method in structural analysis but has since become a major analysis tool in engineering and the physical sciences.

Because of its success in modeling intricate geometries efficiently and accurately, the potential for its application to electrical and magnetic fields was recognized in the early 1970's and has been applied with great success to the study of *dc* and low frequency electromagnetic fields in electrical machines, magnet structures design and nondestructive testing of materials. The finite element method has considerable advantages over the finite difference method including the ease of handling boundary conditions and the ability to follow awkward shaped boundaries. The method is, by definition a "volumetric method" where various parameters are defined as being associated with the volume (or the surface in the case of *2-D* and axisymmetric formulations) and therefore it is very well adapted to the modeling of continuum problems. The finite element method is also quite flexible in terms of the discretization process. Being a discrete method it requires discretization of the solution region but no restrictions are imposed on the shape, size and number of finite elements and the solution process as well as the formulation are identical regardless of the size and shape of the elements used. Furthermore, while the finite difference method assumes linear relations between the unknowns, the finite element method can handle higher order relations as well.

On the other hand, finite elements are defined in terms of spatial function and, as a consequence, there are some difficulties in handling time dependent problems. It is not impossible to use finite elements in the time domain but finite differences are much more efficient for time discretization. For this reason we will view finite elements as applying to non time-dependent equations. In this section we will describe the finite element method as it applies to quasistatic equations, and to the wave equation in the frequency domain.

Finite element solutions require, in general, larger computer resources than finite differences, especially for nonlinear and time dependent problems.

9.5.1. The Finite Element Formulation

To introduce the method we present here the formulation of Poisson's equation in two dimensions. The purpose is to present the basis of the method without the complications associated with wave propagation or losses. For this purpose we use the magnetic vector potential as the field variable and a specific finite element. The description and solution are in the frequency domain. Other methods, and extensions to the method described here will be introduced in chapter 11.

Among the many methods available for finite element formulation of general problems, the weighted residual method (e.g. Galerkin's method) and the use of an energy functional stemming from a global energy balance concept, in conjunction with variational methods are the most commonly used. Both procedures allow a direct formulation based on the original equations and both are satisfactory in terms of the resulting solution. Galerkin's method is more convenient when only the differential equations and their boundary conditions are available or when an energy function cannot be defined. When physical interpretation of the problem is important, the energy balance formulation, offers an attractive alternative. While we will not be able to discuss here all aspects of the finite element formulation or implementation, the method is rather straightforward.

A general energy functional for electromagnetic field problems can be written as

$$F(A) = \int_v \left(E_s - E_i + E_d\right) dv \tag{9.18}$$

where the stored energy (E_s) is due to the magnetic and electric fields, the input energy (E_i) is derived from applied current densities and the dissipated energy (E_d) arises from induced current densities in the conducting parts of the geometry, excluding sources or perhaps from dielectric losses. In some cases, one term or the other may not exist. For example, in lossless media, E_d is zero. In other cases, the source does not exist (i.e. it is not in the solution domain and therefore need not be taken into account directly).

The derivation that follows is limited to *2-D* geometries. It uses a particular type of isoparametric element (four node quadrilateral element) to demonstrate the method but it is completely general and applies to other types of elements as well. Another isoparametric element will be discussed in chapter 11.

9.5.2. Energy Functional for a Quasi-Static Magnetic Problem

The general energy functional in Eq. (9.18) applies to any field situation regardless of dimensionality since it is merely a statement of energy balance in the system. Although there are equations for which an energy functional is not available, this is a very common method of analysis in field problems. For those equations for which an energy functional is not available, the direct, Galerkin method is normally used. In terms of the magnetic vector potential the functional for magnetic fields in *2-D* geometries can be written as

$$F(A) = \int_v \left[\frac{1}{2\mu} \left(\left| \frac{\partial A}{\partial x} \right|^2 + \left| \frac{\partial A}{\partial y} \right|^2 \right) + \frac{j\omega\sigma}{2} A^2 - J_s A \right] dv \qquad (9.19)$$

Note that we chose to use the time-harmonic representation here. This is consistent with the previous statement of the usefulness of finite elements in time-harmonic problems. The functional above is equivalent to the field equation

$$\nabla^2 \mathbf{A} + \mu \mathbf{J}_s - j\omega\mu\sigma \mathbf{A} = 0 \qquad (9.20)$$

Minimization of the functional in Eq. (9.19), results in a condition for energy balance and is equivalent to the solution of the original field equation in Eq. (9.20). This can be shown formally to be the case by deriving the so called Euler's equation but we will not do so here and, rather, assume that this is correct. A discussion of the Euler equation is given in Chapter 12 together with the method used to verify the correctness of a functional.

From the principles of variational calculus it can be shown that a correct solution to Eq. (9.18) or Eq. (9.19) can be obtained by minimizing the energy functional throughout the solution region. Although proof of this statement is not included here (see references 55 for example) it is a very important step since it also defines the natural boundary conditions of the problem (i.e. boundary conditions that are implicit in the formulation and need not be explicitly applied. The most important of these are on the interfaces between different materials) and external, Neumann boundary conditions.

9.5.3. Finite Element Discretization

The solution of the variational formulation of the eddy current equations presented in the previous section is performed by finding a set of functions which minimize the functional $F(A)$. As it is not possible to minimize the functional everywhere it is minimized at discrete points (nodes) in a bounded region (solution region). The discretization of the solution region is therefore a very important step in the finite element method since the number of nodes as well as their location in the solution region has an impact on the solution. A discretization (mesh) with few nodes in a region of high gradients in the solution is bound to introduce errors while too many nodes will unnecessarily complicate and lengthen the solution. It is at this stage in the finite element process where efficient discretization methods combined with correct user's judgment can have a significant impact on the efficiency and accuracy of the solution.

A large variety of volumetric finite elements can be used to discretize the region but, depending on the equations to be solved, some are more useful than others. The most common finite elements used for *2-D* field calculations are the triangular and quadrilateral elements. Linear or first order elements are defined as having a node at each vertex of the element as in Figure 9.12. Parabolic or second order elements are defined by including a node between each two vertices as in Figure 9.13. Higher order elements can similarly be defined. Three-dimensional elements are defined similarly and these will be discussed in chapter 11.

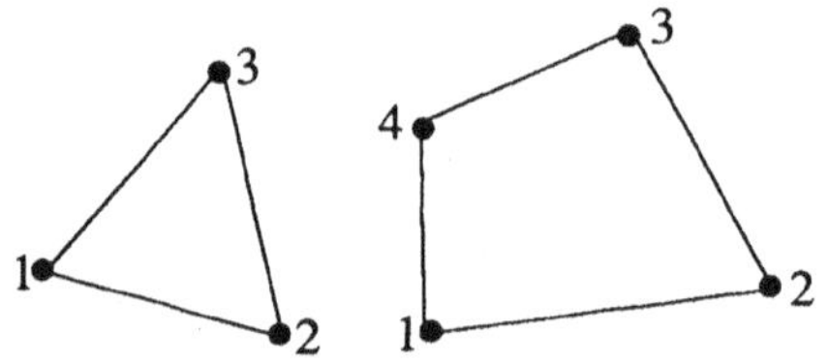

Figure 9.12. Two-dimensional, first order or linear finite elements.

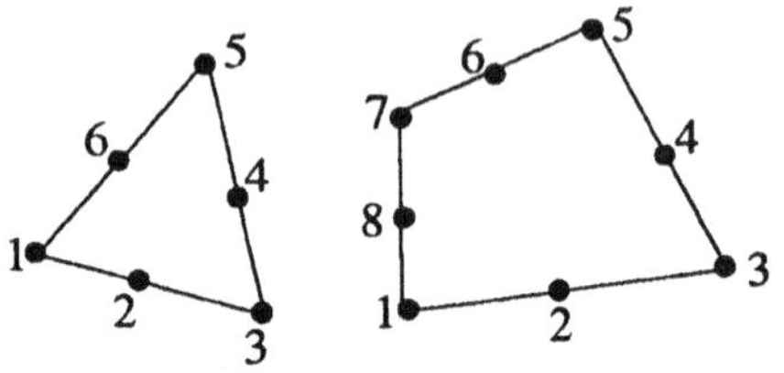

Figure 9.13. Two-dimensional, second order or parabolic finite elements.

Automatic mesh generators for a variety of element shapes are available and offer a simple and efficient way to define the necessary input data. The details of the discretization process are discussed elsewhere (see bibliography) but the following steps and assumptions are general and applicable to the discretization of a region into finite elements.

1. The solution region is subdivided into finite elements. The number and shape of the elements is not restricted in any way as long as the element can be defined in terms of a finite number of nodes. The element density however must be decided upon depending on the geometry of the region and expected gradients in the solution. Small, dense elements must be used in regions of high curvature or high gradients.
2. Material interfaces within the solution region and on the boundaries must coincide with element boundaries. An element cannot cover more than one material.
3. The current density and each conductivity and permeability component are assumed to be constant within the element. Calculated quantities are either nodal values as is the case for the magnetic vector potential or quantities associated with the element such as flux densities or energy. In this case the calculated value is associated with the volume (energy) or with the area of the element (flux density).
4. At the outer boundaries of the solution region the magnetic vector potential is either zero by ensuring that the discretized region extends far enough so as to have negligible flux density on the boundary or is otherwise prescribed from known or calculated conditions. In some cases, symmetry or Neumann boundary conditions can be used.
5. The discretization for *2-D* and axisymmetric geometries is identical. The difference between the two formulations manifests itself in the integration over the element volume and in the form of the functional itself.

9.5.4. Finite Element Formulation

The discretization of the solution region is a geometrical procedure and by itself is not sufficient to ensure that the elements chosen can be used for finite element solution. A set of special functions, called interpolating or shape functions N_i must be chosen for the element. An approximation to the unknown function inside the element is written as

$$A = \sum_{i=1}^{n} N_i A_i \tag{9.21}$$

where n is the number of nodes in the element, N_i the shape function at node i and A_i the unknown function at node i..

These must meet two conditions to ensure convergence of the solution as the size of the elements decreases:

1. At element interfaces C^r continuity must be maintained (compatibility requirement) and,
2. Within an element C^{r+1} continuity must be met (completeness requirement).

where r is the necessary continuity order of the function at the element boundaries and C^r continuity means that the function and its first r derivatives are continuous. For field problems formulated in terms of the magnetic vector potential only C^0 continuity is necessary meaning that only the function is continuous. (Its first derivatives define the flux density components which are not necessarily continuous).

The shape functions for a given element can be obtained by a variety of methods, each leading to a separate sets of functions. One common method is to define polynomials to fit the sides of the element based on a "natural" system of coordinates. The method used here is to define the shape functions in a convenient, local system of coordinates and then to map the functions into the Cartesian or cylindrical system in which the solution is required (parametric elements).

9.5.5. Quadrilateral Isoparametric Elements

The interpolation functions for quadrilateral elements are simple polynomials. The elements are generated in a local system of coordinates, where the functions are found by inspection. Considering Figure 9.14, the shape functions for the four nodes are written in the following form

$$N_i = K(1 + \xi\xi_i)(1 + \eta\eta_i) \tag{9.22}$$

where ξ_i, η_i are either $+1$ or -1 and K is a constant. The shape functions for quadrilateral elements are shown in Figure 9.14b for a particular choice of local coordinates.

These shape functions are next used to map the global coordinate system using the shape functions themselves for the mapping (isoparametric mapping). The elements then become curvilinear as shown schematically in Figure 14c.

The variation of the magnetic vector potential within the element can now be written in terms of the shape functions and the nodal unknowns as

$$A(\xi,\eta) = \sum_{i=1}^{4} N_i(\xi,\eta) A_i \tag{9.23}$$

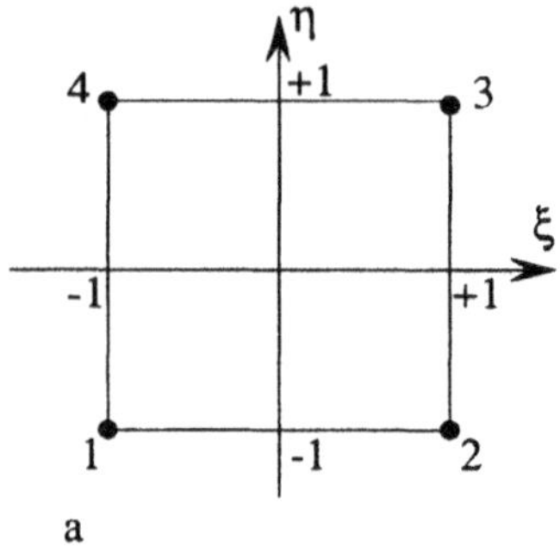

$$N_1=\frac{1}{4}(1-\xi)(1-\eta)$$
$$N_2=\frac{1}{4}(1+\xi)(1-\eta)$$
$$N_3=\frac{1}{4}(1+\xi)(1+\eta)$$
$$N_4=\frac{1}{4}(1-\xi)(1+\eta)$$

b

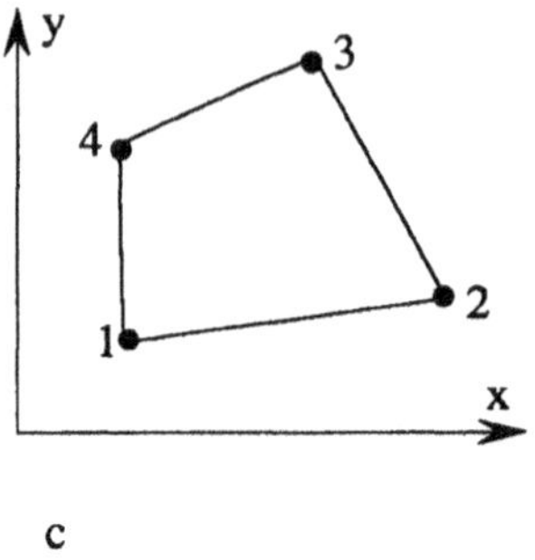

Figure 9.14. Definition of a finite element. a. Finite element in local system of coordinates. b. Shape functions for the element. c. The same element in the global system of coordinates.

Thus, a complete description of the magnetic vector potential within the finite element has been obtained in terms of the shape functions and the unknown values of the magnetic vector potential at the nodes of the element. The energy functional in Eq. (9.19) requires in addition the definition of the first derivatives of A with respect to x and y. These can be written as

$$\frac{\partial A}{\partial x}=\sum_{i=1}^{4}\frac{\partial N_i}{\partial x}A_i \tag{9.24}$$

$$\frac{\partial A}{\partial y}=\sum_{i=1}^{4}\frac{\partial N_i}{\partial y}A_i \tag{9.25}$$

The shape functions are derived in local coordinates. Their derivatives are easily obtained in the local system by using the chain rule of differentiation

$$\frac{\partial N_i}{\partial \xi}=\frac{\partial N_i}{\partial x}\frac{\partial x}{\partial \xi}+\frac{\partial N_i}{\partial y}\frac{\partial y}{\partial \xi} \tag{9.26}$$

and

$$\frac{\partial N_i}{\partial \eta}=\frac{\partial N_i}{\partial x}\frac{\partial x}{\partial \eta}+\frac{\partial N_i}{\partial y}\frac{\partial y}{\partial \eta} \tag{9.27}$$

Rewriting these equations in matrix form and inverting the system to obtain the derivatives of the shape functions in the global system the following is obtained

$$\begin{bmatrix} \frac{\partial N_i}{\partial x} \\ \frac{\partial N_i}{\partial y} \end{bmatrix} = [J]^{-1} \begin{bmatrix} \frac{\partial N_i}{\partial \xi} \\ \frac{\partial N_i}{\partial \eta} \end{bmatrix} \tag{9.28}$$

where $[J]$ is the Jacobian matrix calculated for Eqs. (9.26) and (9.27). Similarly we can write

$$dv = dxdy = [J]d\xi d\eta \tag{9.29}$$

Note that we have called the above a volume when in fact we show an area. The reason is that the element is assumed to be of unit depth and therefore the units of this relation are volume units although the evaluation is done on the surface of the element.

To evaluate the elemental contribution, equations such as (9.23) and (9.24) must be integrated over the element. In the local system of coordinates these are of the form

$$\int_{-1}^{+1}\int_{-1}^{+1} f'(\xi\eta)d\xi d\eta \tag{9.30}$$

for the *2-D* formulation.

9.5.6. Functional Minimization

Energy balance in the solution region is achieved by minimizing the energy functional in Eq. (9.19) with respect to every nodal unknown in the solution region. This is done by setting the first partial derivative of $F(A)$ with respect to each nodal value A to zero:

$$\frac{\partial F(A)}{\partial A_i} = 0 \qquad i=1,2,...,N \tag{9.31}$$

where N is the total number of nodes in the solution region. Instead of performing this operation over the entire region it is convenient to do it element by element and then sum the contributions of individual elements to obtain N simultaneous linear algebraic equations in N unknown values of the magnetic vector potential in the entire solution region. This way a repeatable process is performed on each element, one that is well adapted for automatic assembly on a computer. The size of the elemental matrix is 4×4 (for a four node element) and the individual contributions to the elemental matrix are summarized below

$$S_{i,j} = \frac{1}{\mu}\int_v \left(\frac{\partial N_i}{\partial x}\frac{\partial N_j}{\partial x} + \frac{\partial N_i}{\partial y}\frac{\partial N_j}{\partial y}\right)dv \tag{9.32}$$

$$r_{ij} = \omega\sigma \int_v N_i N_j dv \qquad (9.33)$$

and

$$q_i = -J_s \int_v N_i dv \qquad (9.34)$$

where the indices i, j, vary from 1 to 4.

Each coefficient above is numerically integrated over the volume of the element using Gaussian quadrature and then summed into a complex elemental matrix of the form

$$\left[[s]_e + j[r]_e\right][a]_e = [q]_e \qquad (9.35)$$

where $[s]$ is the real part of the elemental matrix, $[r]$ is the imaginary part and $[a]$ and $[q]$ are the 4×1 unknown and source vectors respectively.

The elemental matrices of all the elements in the solution region are summed into a global system of the form

$$[G][A] = [Q] \qquad (9.36)$$

This is a system of N equations in N unknowns, with N the total number of nodes in the solution region. The matrix is symmetric and banded, the bandwidth depending on the type of element, the number of nodes per element and especially on the way the nodes are numbered.

The system in Eq. (9.36) can be solved by any standard solution technique such as Gauss elimination or the Conjugate Gradient method to yield the nodal values of the magnetic vector potential.

9.5.7. Boundary Conditions

The field equations, formulated in terms of finite elements in Eq. (9.36) for eddy current problems can only be solved provided a correct set of boundary conditions is specified. Either Dirichlet boundary conditions (the function A is known on the boundary) or Neumann boundary conditions (the first derivative of A is known) can be specified. In the finite element analysis of magnetic field problems it is more convenient to specify Dirichlet boundary conditions since the global matrix in Eq. (9.36) can accommodate the function value A but not its derivative. Moreover, Neumann boundary conditions are natural (i.e. implicit) in the formulation in Eq. (9.36) and need not be specified.

To demonstrate the use of the finite element method, consider the example given in Figure 9.15. Two thick, current carrying lines are located above a thick, flat conductor with dimensions as shown. The first step in the solution is the discretization process. In this case, we will only use half the geometry, as shown in Figure 9.16 because of the symmetry in solution. Normally, a symmetry line implies that a Neumann boundary condition is imposed. In this particular case, the solution is antisymmetric (currents in the two bars are in oposite directions) and, therefore, the potential is zero midway between the two lines. Thus, a Dirichlet boundary condition is applied on the symmetry line.

A simple finite element mesh is shown in Figure 9.17 and the results are shown in Figure 9.18. The latter is a plot of A which represents the magnetic flux distribution in the solution domain. Other quantities are calculable from this result if these are required using the basic relations in chapter 1.

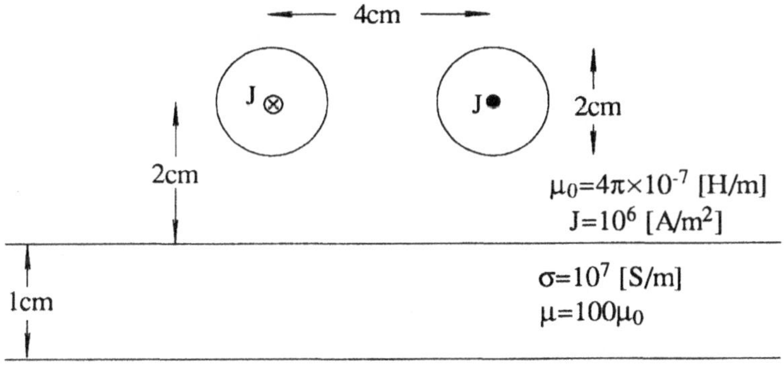

Figure 9.15. Two thick current carrying conductors over a conducting slab.

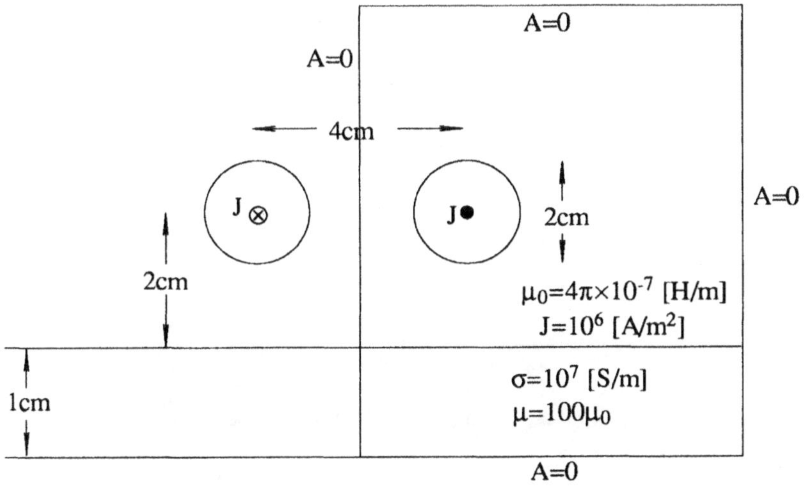

Figure 9.16. Solution domain for the problem in Figure 9.15. The symmetry line is set to zero magnetic vector potential because the currents flow in opposite directions.

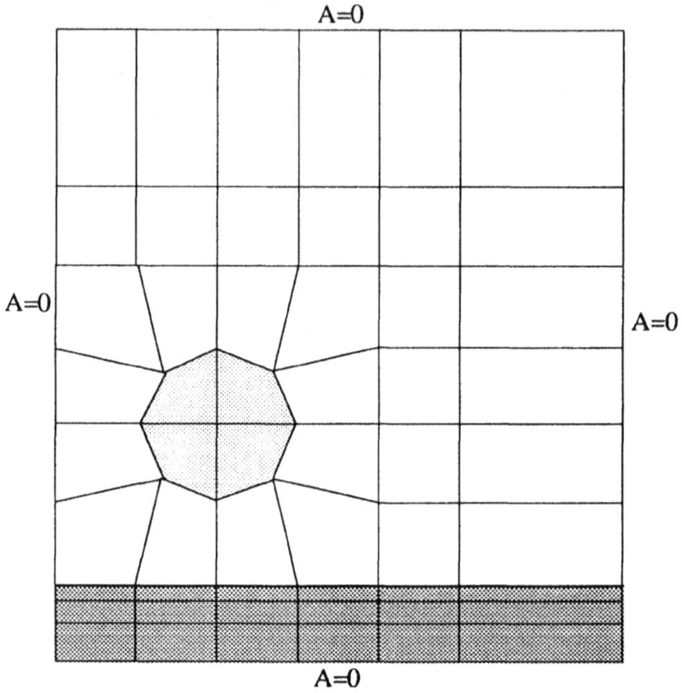

Figure 9.17. A finite element mesh for the solution domain in Figure 9.16. This is a rather coarse mesh.

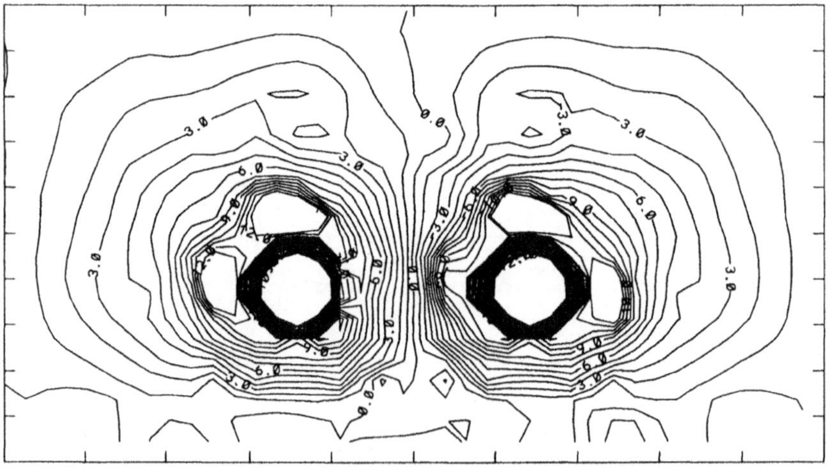

Figure 9.18a. Real part of magnetic vector potential for the geomentry in Figure 9.15.

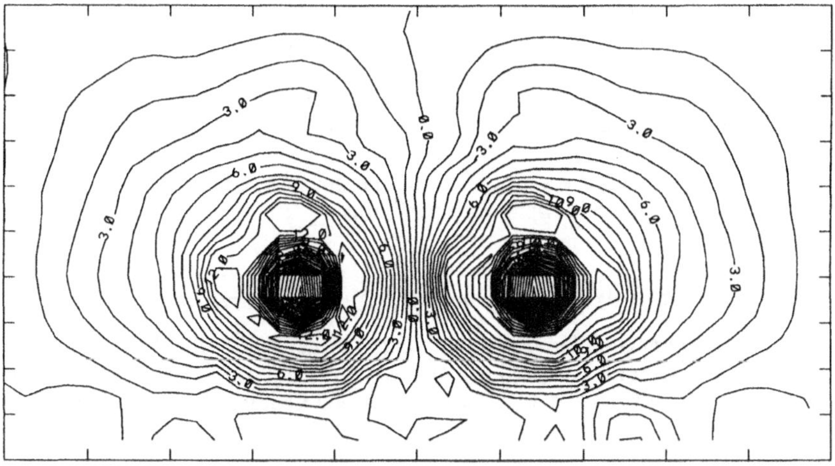

Figure 9.18b. Imaginary part of magnetic vector potential for the geomentry in Figure 9.15.

9.6. Boundary Integral Methods: The Method of Moments

As a simple example of a boundary integral method we will consider here the method of moments *(MoM)*. This method, which is a form of the weighted residual method has found many applications in calculation at high frequency in particular although, it is not specifically designed for, or limited to high frequency applications. The discussion here focuses again on the principles of the method as it relates to the more general boundary integral technique of solution for boundary value problems.

Perhaps the best way to view these methods is to contrast them with the so called differential methods like the finite element or the finite difference techniques. In differential methods we found a way to approximate the solution throughout the domain. Thus, for three dimensional problems this involved finding a solution at every point in the solution space. This solution then provided all the required information, including a detailed description of the fields in space.

Sometimes, this detailed solution is not required, or is not practical. For example, if we want to know the impedance of an antenna or the capacitance of a small plate, the detailed knowledge of the fields is not needed. Also, as an example, if we require the far field of an antenna, only the current distribution on the antenna is needed. From this we can calculate the field at any point in space.

The method of moments provides a method of evaluating the equivalent sources (current densities, charge densities, or magnetization) on the surface of a device. From this, the fields outside the sources can be calculated. One of the most attractive points of this method, as of any other integral method, is the fact that boundary integral methods reduce the dimensionality of the problem by *1*. Thus a *3-D* problem is reduced to the calculation of surface sources which is a *2-D* problem, considerably simplifying the solution.

We will first present the method of moments as a weighted residual method. This leads to a formal solution involving a matrix system that needs to be solved. This matrix is not unlike that of the finite element matrix.

After that, a very simple intuitive approach to solution by the method of moments is given which clearly indicates the basic properties of the method as a boundary integral method.

The method of moments has enjoyed particular success in high frequency applications, particularly for scattering and propagation problems.

9.6.1. The Method of Moments for Differential Operators

The method of moments can be viewed as a general method to transform a linear operator equations (e.g. a differential equation) into a matrix equation which can be solved by standard matrix methods with relative ease. The problem at hand is that of a solution of the operator equation

$$Ax=y \tag{9.37}$$

where A is a general linear operator, x is the unknown function and y is the excitation or known source function. An exact solution to this equation is in general not possible. We will seek to find an approximation to the solution in terms of a series of approximating functions with undetermined coefficients

$$x = \sum_{i=1}^{n} \alpha_i x_i \tag{9.38}$$

The summation in Eq. (9.38) can be either finite for an approximate solution or infinite for an exact solution. Substitution of this approximation into Eq. (9.37) introduces an error which should be minimized for the solution to be acceptable. The solution in Eq. (9.38) should be viewed as an approximation which, in addition to satisfying the operator (for example, it must be twice differentiable for second order partial differential equations) it must also satisfy the boundary conditions of the problem.

$$\sum_{i=1}^{n} \alpha_i A x_i \approx y \tag{9.39}$$

A residual can be defined as

$$R(x) = \sum_{i=1}^{n} \alpha_i A x_i - y \tag{9.40}$$

Next, a set of weighting functions are defined as W_n. These again can be defined in any consistent way and, depending on the choice result in different type of solutions. One example is to use the approximation itself as the weighting function

Minimization of this residual can be obtained by using a general weighted average with respect to some, yet undetermined, weighting functions w_m.

$$\int_{\Omega} R(x)W_1(x) = 0$$

$$\int_{\Omega} R(x)W_2(x) = 0$$

$$\int_{\Omega} R(x)W_n(x) = 0 \tag{9.41}$$

This will provide n linear equations in n unknowns.

In a more compact form, Eq. (9.41) is written as

$$\int_{\Omega} \sum_{\substack{i=1 \\ j=1}}^{n} W_j\left[\alpha_i A x_i - y\right] = 0 \tag{9.42}$$

To introduce the method in a less abstract way consider the solution to a simple, one dimensional problem. A parallel plate capacitor is held at a zero potential as in Figure 9.19. The distance between the plates is d and we assume a static solution. The potential distribution in the solution domain is required. The potentials on the two plates are zero but there is a charge density within the capacitor that generates a potential distribution between the plates.

The differential equation and the associated boundary conditions are:

$$\frac{d^2V}{dx^2} = -\frac{\rho_0 x}{\varepsilon d} \tag{9.43}$$

$$V(0) = 0., \quad V(d) = 0. \tag{9.44}$$

The analytical solution to this problem is found by directly integrating the equation twice:

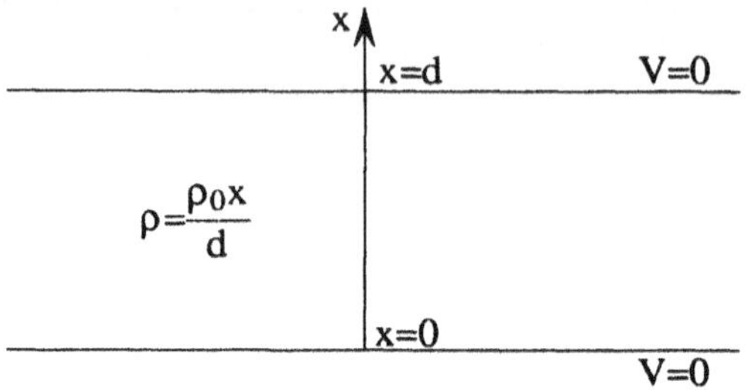

Figure 9.19. A parallel plate capacitor with a charge density distribution between the plates.

$$V(x) = -\frac{\rho_0}{6\varepsilon d}\left(-x^3 + d^2 x\right) \tag{9.45}$$

To solve the problem described by Eqs. (9.43) and (9.44) using the method of moments, we assume that the solution is of the form

$$f(x) = x(1-x)$$

The approximation to the potential is therefore

$$V(x) = Ax(1-x)$$

Substitution of this approximate solution in the differential equation in Eq. (9.43) results in a residual $R(x)$

$$R(x) = 2A - \frac{\rho x}{\varepsilon d} \tag{9.46}$$

Now we define a set of weighting functions. As mentioned above, the weighting functions can be defined in different ways. Here we will use the Galerkin's form

$$W(x) = f(x) = x(1-x) \tag{9.47}$$

The weighted average of the residual is calculated and set to zero

$$\int_0^d x(1-x)\left[2A - \frac{\rho x}{\varepsilon d}\right]dx = 0$$

Performing the integration results in:

$$Ad^2 - 2A\frac{d^3}{3} - \frac{\rho d^3}{3\varepsilon d} + \frac{\rho d^4}{4\varepsilon d} = 0$$

or, calculating A we get

$$A = -\frac{\rho}{4\varepsilon}\left(\frac{4-3d}{3-2d}\right)$$

Substitution of this into the approximation for V provides the solution:

$$V(x) = -\frac{\rho}{4\varepsilon}\left(\frac{4-3d}{3-2d}\right)\left(x - x^2\right) \qquad (9.48)$$

This solution is obviously different than the analytical solution in Eq. (9.45). This is not surprising since we expected it to be different. Yet, it is a remarkably good solution as Figure 9.20 shows. In fact, for $x=d/2$, the two solutions happen to be identical.

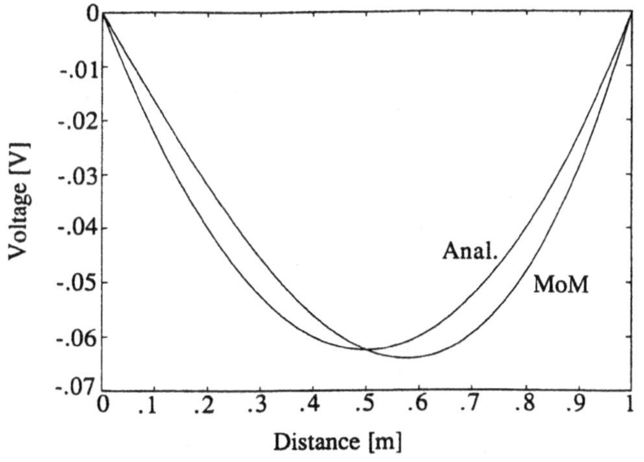

Figure 9.20. Voltage distribution in the parallel plate capacitor as calculated by the method of moments and analytically for $\rho/\varepsilon=1$ and $d=1$ m.

An improved solution can be obtained by increasing the order of the approximation. The result in Eq. (9.48) has a second order term in the solution as the highest order term while the exact solution has a fourth order term. If we include the next higher order term in our approximation, the solution will contain a third order term.

$$V(x) = A_1 x(1-x) + A_2 x(1-x^2)$$

The two weights are now

$$W_1(x) = f_1(x) = x(1-x)$$

$$W_2(x) = f_2(x) = x(1-x^2)$$

The residual is

$$R(x) = 2A_1 + 6A_2 x - \frac{\rho x}{\varepsilon d}$$

and the following two equations are obtained by using the two weights

$$\int_0^d x(1-x)\left[2A_1 + 6A_2 x - \frac{\rho x}{\varepsilon d}\right] dx = 0$$

$$\int_0^d x(1-x^2)\left[2A_1 + 6A_2 x - \frac{\rho x}{\varepsilon d}\right] dx = 0$$

To simplify calculation we will assume that $d=1$ and perform the two integrals above. This gives:

$$\frac{A_1}{3} + \frac{A_2}{2} = \frac{\rho}{12\varepsilon}$$

$$\frac{A_1}{2} + \frac{4A_2}{5} = \frac{\rho}{15\varepsilon}$$

We can write this as a system of two equations in two unknowns as

$$\begin{bmatrix} c_{11} & c_{12} \\ c_{21} & c_{12} \end{bmatrix} \begin{bmatrix} A_1 \\ A_2 \end{bmatrix} = \begin{matrix} q_1 \\ q_2 \end{matrix}$$

or

$$\begin{bmatrix} \frac{1}{3} & \frac{1}{2} \\ \frac{1}{2} & \frac{4}{5} \end{bmatrix} \begin{bmatrix} A_1 \\ A_2 \end{bmatrix} = \begin{bmatrix} \frac{\rho}{12\varepsilon} \\ \frac{\rho}{15\varepsilon} \end{bmatrix}$$

As an extenssion, we can calculate an arbitrary order solution by choosing an approximation of the order

$$V(x) = A_n x(1-x^n) \tag{9.49}$$

Similarly, the weighting functions are

$$W_m(x) = f_m(x) = x(1-x^m) \tag{9.50}$$

The residual now looks as

$$R(x) = A_n n(n+1)x^{n+1} - \frac{\rho x}{\varepsilon d}$$

The weighted residual leads to

$$\int_0^d x(1-x^m)\left[A_n n(n+1)x^{n+1} - \frac{\rho x}{\varepsilon d}\right]dx = 0$$

Performing this integration we get

$$A_n\left[nd^{n+1} - \frac{n(n+1)}{m+n+1}d^{m+n+1}\right] - \frac{\rho x}{\varepsilon d}\left[\frac{d^3}{3} + \frac{d^{m+3}}{m+3}\right] = 0 \tag{9.51}$$

From this we can construct a general order solution by writing

$$[c_{m,n}][A_n] = [q_n] \tag{9.52}$$

This is equivalent to taking the residual for each value of n with each weight and calculating the coefficients of the system of equations.

9.6.2. Subsectional Bases

The approximation in Eq. (9.49) was chosen such as to satisfy the boundary conditions and to approximate the solution over the entire domain of the operator (i.e. solution domain). Because of the nature of the moment of methods, the approximation does not have to be chosen over the whole domain. The weighting or basis functions can be chosen over a subsection of the solution domain. Such an approximation is only valid over the domain chosen but this procedure can simplify the solution considerably. The choice of the approximation over a subdomain still leaves the weighting functions to be chosen independently. Thus, the use of subsectional domains can be combined with the method of point matching or with any other method of choosing the weighting functions.

To be truly effective the functions chosen for approximation over subdomains should be relatively simple. Two functions are used extensively for this purpose. One is the step or pulse function shown in Figure 9.21. The second is the pyramidal ro triangular function shown in Figure 9.22. These functions are often used either for the approximation, weighting function, or both. As an example, the pyramidal function may be used for to approximate the solution while the pulse functions may be used as weighting functions.

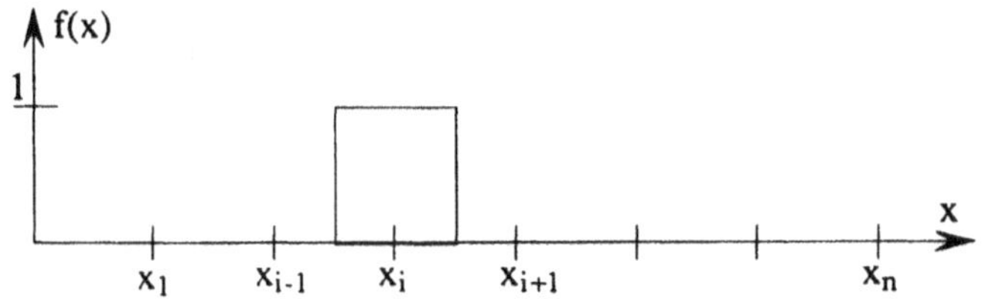

Figure 9.21. The pulse subsectional basis function.

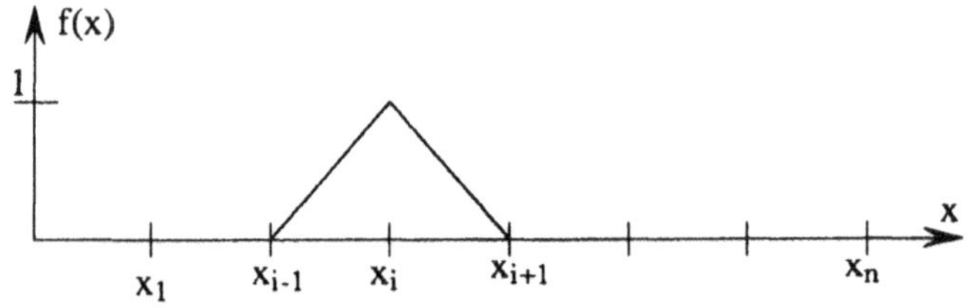

Figure 9.22. The pyramidal subsectional basis function.

9.6.3. The Method of Moments for Integral Operators

The previous sections discussed the method of moments in a general form for a linear operator. This however, is quite similar to, say, the finite element method and, in principle, the finite difference method. The main difference is in the way we construct the approximating functions and weighting functions.

The method of moments is particularly effective for solution of integral equations. The method solves a general integral equation of the form

$$V(z) = \int_{\Omega'} K(z,z')\rho(z')d\Omega' \tag{9.57}$$

where Ω is the domain over which the integration is performed. The function $K(z,z')$ is assumed to be known and is called the kernel of the integral or the Green's function for the problem. The unknown to be determined is $\rho(z')$ and it may be a charge density or a current density on a surface, or a current in a wire, depending on how the problem is set up. The variable z indicates the observation point while z' indicates the source point (i.e. the location of the source). Thus, as was mentioned above, we obtain the equivalent sources on the boundaries of the domain from which the fields may be determined. As an example, assume we need to calculate the magnetic field due to an antenna. Using the method of moments, we will reformulate the field equations into the form above and calculate the currents on the surface of the antenna. These currents may then be used to calculate the magnetic field everywhere in space. Before proceeding with this problem, we will look at a much simpler problem: the calculation of capacitance of a rectangular plate to define the steps required. The reason for the use of this simple problem is that Green's function is easily evaluated.

To solve this problem, we first note that the potential at any point in space due to a charge density over a surface is given by

$$V(x,y,z) = \int_{S'} \frac{1}{4\pi\varepsilon R} \rho(x',y'z')dS' \tag{9.58}$$

where R is the distance between the observation point (x,y,z) to the source point (x',y',z'). This distance is given simply as

$$R = \sqrt{(x-x')^2 + (y-y')^2 + (z-z')^2}$$

Now the kernel is self explanatory: it is the potential at point (x,y,z) due to a (unit) point charge located at point (x',y',z'). This function is known at any point. However ρ_s is not known and must be evaluated. To solve the problem we note the following:

1. If the plate is charged, the potential on the plate changes but the capacitance does not. Thus, we may assume that the plate is at a fixed, given potential V_0.
2. The charge density over the plate is not uniform (larger density around the edges) but we may divide the plate into small rectangles and assume that each rectangle has a charge density ρ_{si}, uniformly distributed on the rectangle. If the rectangles are small, this will cause little error and, in the limit $(n \rightarrow inf)$ the solution will approach the exact solution. Figure 9.23 shows this arrangement. Note that the rectangles do not have to be the same in size or, for that matter in shape although we may well want to make them equal to simplify calculation, again, based on the observation that in the limit, it does not matter what the size and shape of each rectangle is.

3. We can now calculate the potential due to each rectangle but, because we are interested in the capacitance of the plate, we need only calculate the potential on the plate itself. Also, since in each rectangle the potential is constant, we calculate the potential at the center of each rectangle. For simplicity, we assume that the charge of each rectangle is concentrated at the center of that rectangle.

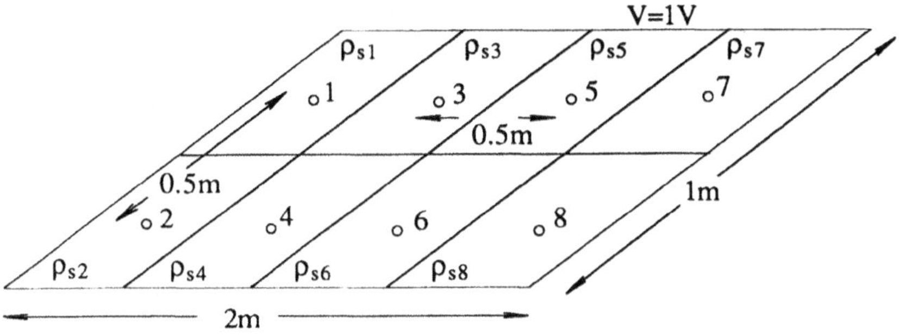

Figure 9.23. Division of a plate into 8 equal parts for the purpose of calculating the charge densities of the sections.

Thus, for example, the potential due to rectangle i at the center of rectangle j is given as

$$V(x_j,y_j,z_j) = \frac{\rho_{si}(x_i',y_i'z_i')\Delta S_i}{4\pi\varepsilon\sqrt{(x_j-x_i')^2+(y_j-y_i')^2+(z_j-z_i')^2}} \tag{9.59}$$

where ΔS_i is the area of rectangle i.

Each rectangle will contribute a potential of this form and the sum of all contributions is equal to V_0.

We note that the potential on rectangle i due to the charge density on itself cannot be calculated from the relation above but can be calculated by integration (using Eq. (9.58) above) and is given as

$$V(x_i',y_i',z_i') = \frac{\rho_{si}(x_i',y_i'z_i')}{4\pi\varepsilon}\left(2a\ ln\frac{b+\sqrt{a^2+b^2}}{a}+2b\ ln\frac{a+\sqrt{a^2+b^2}}{b}\right) \tag{9.60}$$

We will denote this, for convenience as

$$V(x_i',y_i',z_i') = \frac{\rho_{si}(x_i',y_i'z_i')}{4\pi\varepsilon}K$$

where K is the geometric function in brackets in Eq. (9.60).

Now we can calculate the potential at the center of each rectangle as

$$V_1 = \frac{\rho_{s1}}{4\pi\varepsilon}K_1 + \frac{\rho_{s2}\Delta_{s2}}{4\pi\varepsilon R_{21}} + \frac{\rho_{s3}\Delta_{s3}}{4\pi\varepsilon R_{31}} + \dots + \frac{\rho_{s(n-1)}\Delta_{s(n-1)}}{4\pi\varepsilon R_{(n-1)1}} + \frac{\rho_{sn}\Delta_{sn}}{4\pi\varepsilon R_{n1}}$$

$$V_2 = \frac{\rho_{s1}\Delta_{s1}}{4\pi\varepsilon R_{12}} + \frac{\rho_{s2}}{4\pi\varepsilon}K_2 + \frac{\rho_{s3}\Delta_{s3}}{4\pi\varepsilon R_{31}} + \dots + \frac{\rho_{s(n-1)}\Delta_{s(n-1)}}{4\pi\varepsilon R_{(n-1)1}} + \frac{\rho_{sn}\Delta_{sn}}{4\pi\varepsilon R_{n1}}$$

$$\vdots$$

$$V_n = \frac{\rho_{s1}\Delta_{s1}}{4\pi\varepsilon R_{12}} + \frac{\rho_{s2}\Delta_{s2}}{4\pi\varepsilon R_{2n}} + \frac{\rho_{s3}\Delta_{s3}}{4\pi\varepsilon R_{31}} + \dots + \frac{\rho_{s(n-1)}\Delta_{s(n-1)}}{4\pi\varepsilon R_{(n-1)1}} + \frac{\rho_{sn}}{4\pi\varepsilon}K_n \qquad (9.61)$$

Since $V_1, V_2, \dots V_n$, are known (each is equal to V_0), we can write the following matrix equation

$$\begin{bmatrix} \frac{K_1}{4\pi\varepsilon} & \frac{\Delta_{s2}}{4\pi\varepsilon R_{21}} & \frac{\Delta_{s3}}{4\pi\varepsilon R_{31}} & \cdots & \frac{\Delta_{s(n-1)}}{4\pi\varepsilon R_{(n-1)1}} & \frac{\Delta_{sn}}{4\pi\varepsilon R_{n1}} \\ \frac{\rho_{s1}\Delta_{s1}}{4\pi\varepsilon R_{12}} & \frac{K_2}{4\pi\varepsilon} & \frac{\Delta_{s3}}{4\pi\varepsilon R_{32}} & \cdots & \frac{\Delta_{s(n-1)}}{4\pi\varepsilon R_{(n-1)2}} & \frac{\Delta_{sn}}{4\pi\varepsilon R_{n2}} \\ \vdots & \vdots & \vdots & \vdots & \vdots & \vdots \\ \vdots & \vdots & \vdots & \vdots & \vdots & \vdots \\ \frac{\Delta_{s1}}{4\pi\varepsilon R_{1(n-1)}} & \frac{\Delta_{s2}}{4\pi\varepsilon R_{2(n-1)}} & \frac{\Delta_{s3}}{4\pi\varepsilon R_{3(n-1)}} & \cdots & \frac{K_{n-1}}{4\pi\varepsilon} & \frac{\Delta_{sn}}{4\pi\varepsilon R_{(n-1)n}} \\ \frac{\Delta_{s1}}{4\pi\varepsilon R_{1n}} & \frac{\Delta_{s2}}{4\pi\varepsilon R_{2n}} & \frac{\Delta_{s3}}{4\pi\varepsilon R_{3n}} & \cdots & \frac{\Delta_{s(n-1)}}{4\pi\varepsilon R_{n(n-1)}} & \frac{K_n}{4\pi\varepsilon} \end{bmatrix} \begin{bmatrix} \rho_{s1} \\ \rho_{s2} \\ \rho_{s3} \\ \vdots \\ \rho_{s(n-1)} \\ \rho_{sn} \end{bmatrix}$$

$$= \begin{bmatrix} V_0 \\ V_0 \\ V_0 \\ \vdots \\ V_0 \\ V_0 \end{bmatrix} \qquad (9.62)$$

From this, the unknowns ρ_{s1} through ρ_{sn} are calculate by solving the system. These are then the charge densities on the n rectangles. The total charge is calculated as

$$Q = \sum_{i=1}^{n} \rho_{si} \Delta_{si} \tag{9.63}$$

and the capacitance is given by

$$C = \frac{Q}{V_0} \tag{9.64}$$

As a numerical example, consider the plate shown in Figure 9.23. The plate is $1m \times 2m$ in size and is divided into 8 equal rectangles. The coordinates of the centers of the rectangles are given and the potential is given as $1V$. The system of equations is

$$\begin{bmatrix} 62.326 & 4.496 & 4.496 & 3.18 & 2.248 & 1.9 & 1.499 & 1.42 \\ 4.496 & 62.326 & 3.18 & 4.496 & 1.9 & 2.248 & 1.42 & 1.499 \\ 4.496 & 3.18 & 62.326 & 4.496 & 4.496 & 3.18 & 2.248 & 1.9 \\ 3.18 & 4.496 & 3.18 & 62.326 & 3.18 & 4.496 & 1.9 & 2.248 \\ 2.248 & 1.9 & 4.496 & 3.18 & 62.326 & 4.496 & 4.496 & 3.18 \\ 1.9 & 2.248 & 3.18 & 4.496 & 3.18 & 62.326 & 3.18 & 4.496 \\ 1.499 & 1.42 & 2.248 & 1.9 & 4.496 & 3.18 & 62.326 & 4.496 \\ 1.42 & 1.499 & 1.9 & 2.248 & 3.18 & 4.496 & 4.496 & 62.326 \end{bmatrix} \begin{bmatrix} \rho_{s1} \\ \rho_{s1} \\ \rho_{s1} \\ \rho_{s1} \\ \rho_{s1} \\ \rho_{s1} \\ \rho_{s1} \\ \rho_{s1} \end{bmatrix} = \begin{bmatrix} 10^{-9} \\ 1 \\ 10^{-9} \\ 10^{-9} \\ 10^{-9} \\ 10^{-9} \\ 10^{-9} \\ 10^{-9} \end{bmatrix}$$

From which the eight charge densities are

$$\rho_{s1}=\rho_{s2}=\rho_{s7}=\rho_{s8}=0.1240\times10^{-10}\ [Coulomb/m^2]$$
$$\rho_{s3}=\rho_{s4}=\rho_{s5}=\rho_{s6}=0.1146\times10^{-10}\ [Coulomb/m^2]$$

and the total charge on the plate is

$$Q=0.2389\times10^{-10}\ [Coulomb]$$

Thus, the capacitance of the plate is equal to

$$C=24.86\ [microFarad]$$

From these calculations it is obvious that the most difficult part of the solution by moment method (and indeed by any other method) is the formulation of the problem. In the problem we described above, we used the basic solution of the potential of a point charge. However, in a more general sense, this solution is merely a solution to Poisson's equation.

The challenge of integral methods is to find a solution of this type or, in more general terms, to find the Green's function corresponding to the geometry analyzed. This is often a very difficult task.

9.6.4. Method of Moments for Current Distributions

An approach similar to that described above can be used for other applications. One of the most common application of the method of moments is for antenna calculations in which the equivalent currents in the antenna are calculated, from which the radiation pattern or other parameters may be evaluated. Rather than go through the simple approach given above we will now describe the steps involved in the solution in a more formal way. First we must reformulate the problem starting with the field equations. We still require a general integral equation of the form given above except that now we write it as

$$E(z) = \int_{\Omega'} K(z,z')I(z')d\Omega' \qquad (9.65)$$

where z and z' indicate the observation and source points respectively. I is the unknown current and $K(z,z')$ is an appropriate kernel that we must evaluate but which, after it has been defined, is a known function. E is the electric field generated by the equivalent current and this is either known or may be calculated independently. Thus, the only unknown is the current I.

9.6.5. Formulation

We start with the following assumptions:

1. The current is directed in the z direction
2. The wire is thin (not infinitely thin) and perfectly conducting so that all currents are on the surface of the wire.
3. The wire is uniform

Maxwell's equations needed are, (in the frequency domain):

$$\nabla\times\mathbf{E} = j\omega\mu\mathbf{H} \qquad (9.66)$$

$$\nabla\times\mathbf{H} = \mathbf{J} + j\omega\varepsilon\mathbf{E} \qquad (9.67)$$

By defining a magnetic vector potential **A** (see chapter 1)

$$\mathbf{B} = \nabla\times\mathbf{A} \qquad (9.68)$$

The electric field can be directly written as

$$\mathbf{E} = -j\omega\mathbf{A} - \nabla V \tag{9.69}$$

where V is a the electric scalar potential. This can either be verified by substitution in Eq. (9.66) or the formal development in chapter 1 may be followed.
From Eq. (9.67) we can write

$$\nabla\times\mathbf{H} = \frac{1}{\mu}\nabla\times\mathbf{B} = \frac{1}{\mu}\nabla\times(\nabla\times\mathbf{A}) = \mathbf{J} + j\omega\varepsilon\mathbf{E}$$

or, using Eq. (9.68) for the left hand side

$$\nabla\times(\nabla\times\mathbf{A}) = -\nabla^2\mathbf{A} + \nabla(\nabla\cdot\mathbf{A}) = \mu\mathbf{J} + \mu j\omega\varepsilon\mathbf{E}$$

where the relation $\nabla\times(\nabla\times\mathbf{A}) = -\nabla^2\mathbf{A} + \nabla(\nabla\cdot\mathbf{A})$ was used.
Substituting $\mathbf{E}$ from Eq. (9.69)and we get

$$-\nabla^2\mathbf{A} + \nabla(\nabla\cdot\mathbf{A}) = \mu J + \mu j\omega\varepsilon(-j\omega\mathbf{A} - \nabla V) = \mu\mathbf{J} + \omega^2\mu\varepsilon - \nabla(j\omega\mu\varepsilon V)$$

or

$$\nabla^2\mathbf{A} + k^2\mathbf{A} = -\mu\mathbf{J} + \nabla(\nabla\cdot\mathbf{A}) + \nabla(j\omega\mu\varepsilon V)$$

by choosing the divergence as (Lorentz gage)

$$\nabla\cdot\mathbf{A} = -j\omega\mu\varepsilon V \tag{9.70}$$

we get

$$\nabla^2\mathbf{A} + k^2\mathbf{A} = -\mu\mathbf{J} \tag{9.71}$$

and, substituting for $-V$ from Eq. (9.70) we get for the electric field in Eq. (9.69)

$$\mathbf{E} = -j\omega\mathbf{A} + \frac{\nabla(\nabla\cdot\mathbf{A})}{j\omega\mu\varepsilon} \tag{9.72}$$

The two relations in (9.71) and (9.72) are general. We will now assume that the current density in Eq. (9.71) is in the z direction only:

$$\mathbf{J} = \hat{z}J_z$$

Under this condition, the solution to Eq. (9.71) is

$$\mathbf{A} = \hat{z}\frac{\mu}{4\pi}\int_v J_z \frac{e^{-jkR}}{R}\, dv' \tag{9.73}$$

where R is the distance between the source (J) and an arbitrary observation point. This is then the solution to the inhomogeneous wave equation in (9.71).

We can also write for the electric field from Eq. (9.72)

$$j\omega\mu\varepsilon\mathbf{E} = \omega^2\mu\varepsilon\mathbf{A} + \nabla(\nabla\cdot\mathbf{A})$$

and, since **A** is also in the z direction we can write (with $k^2=\omega^2\mu\varepsilon$)

$$k^2 A_z + \nabla^2 A_z = j\omega\mu\varepsilon E_z \tag{9.74}$$

Substituting the solution of the inhomogeneous equation we get

$$\hat{z}\frac{\mu}{4\pi}\int_v \left(\nabla^2 + k^2\right) J_z \frac{e^{-jkR}}{R}\, dv' = j\omega\mu\varepsilon E_z$$

or, in a more convenient form

$$\int_v J_z \left(\frac{\partial^2}{\partial z^2} + k^2\right)\frac{e^{-jkR}}{4\pi R}\, dv' = j\omega\varepsilon E_z \tag{9.75}$$

Now we also note that the electric field on the right hand side of Eq. (9.75) is the field generated by the current density J_z. If instead we want to write this in terms of the incident field on the conductor, (since this is normally known and can be viewed as giving rise to the current density J_z) we note that the incident field is equal to the negative of the scattered field

$$E_z^i = -E_z$$

because the conductor is assumed to be perfectly conducting. Also, for a constant diameter conductor we can write $Jdv=Idl$ and, therefore Eq. (9.75) becomes

$$\int_{-\frac{L}{2}}^{\frac{L}{2}} I_z \left(\frac{\partial^2}{\partial z^2} + k^2\right)\frac{e^{-jkR}}{4\pi R}\, dl' = j\omega\varepsilon E_z = -j\omega\varepsilon E_z^i \tag{9.76}$$

Using the general integral equation we can write the following

$$\int_{-\frac{L}{2}}^{+\frac{L}{2}} I_z(z')\left(\frac{\partial^2}{\partial z^2} + k^2\right)\frac{e^{-jkR}}{4\pi R}\, dz' = j\omega\varepsilon E_z(z) = -j\omega\varepsilon E_z^i(z) \qquad (9.77)$$

This then has exactly the form required, with $I(z')$ as the unknown function and

$$K(z,z') = \left(\frac{\partial^2}{\partial z^2} + k^2\right)\frac{e^{-jkR}}{4\pi R}$$

the kernel. Now the solution proceeds in a manner similar to the previous problem. However, before we do so we note that while R is the distance between a general observation point and the current I, in the case of a wire, we locate the current at the center of the wire and the observation point on the surface of the wire (at which point the electric field is assumed to be known). Figure 9.24 shows this relationship.

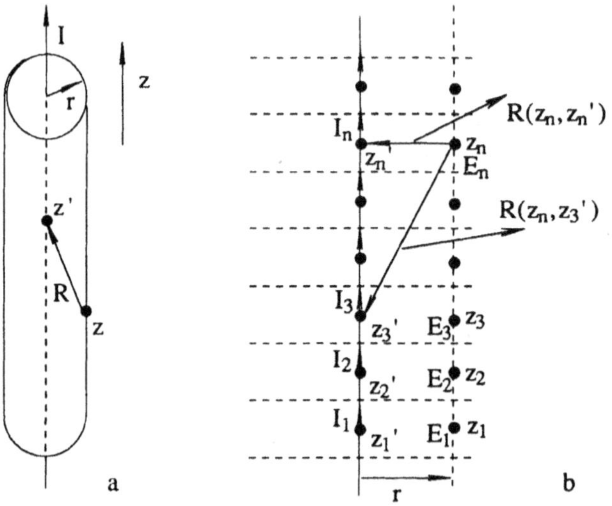

Figure 9.24. A current carrying conductor and its equivalent form as used for method of moments calculation.

To solve the problem we must find an approximation for the unknown values of $I(z)$. One of the simplest forms of approximations is the pulse functions mentioned above. To do so we assume that the approximation to the current I is done by N pulses as

$$I(z') = \sum_{n=1}^{N} I_n f_n(z')$$

where (') indicates that this approximation is on the current itself. The approximating functions f_n are

$$f_n(z') = \begin{cases} 1 \text{ in element } n \\ 0 \text{ in all other elements} \end{cases}$$

Now, the integral equation is approximated as

$$\int_{-\frac{L}{2}}^{+\frac{L}{2}} \sum_{n=1}^{N} I_n f_n(z')K(z,z') \; dz' = -j\omega\varepsilon E_z^i(z) \tag{9.78}$$

Thus, the current element is divided into N segments, and the integral on the left hand side is evaluated at the center of the *ith* element. The center of an element is at a point z_i. Expanding the equation above for an element i gives

$$\int_{-\frac{L}{2}}^{+\frac{L}{2}} \sum_{n=1}^{N} I_n f_n(z')K(z,z') \; dz'' = I_1 \int_{\Delta z_1} K(z_i,z_1')dz' + I_2 \int_{\Delta z_2} K(z_i,z_2')dz' \; + \ldots$$

$$\ldots + I_N \int_{\Delta z_N} K(z_i,z_N')dz' = -j\omega\varepsilon E_z^i(z_i) \tag{9.79}$$

and there are obviously $N-1$ additional equations like this. As with Eq. (9.69) the value on the right hand side must also be known. While we may assume that it is known its evaluation depends on the problem we are trying to solve. If for example we were to analyze a short dipole antenna there are some standard methods that may be used to calculate the electric field in all elements of the solution domain. We will simply assume here that these are known. The matrix equation is therefore

$$\begin{bmatrix} \int_{\Delta z_1} K(z_1,z_1')dz' & \int_{\Delta z_2} K(z_1,z_2')dz' & \cdots & \int_{\Delta z_N} K(z_1,z_N')dz' \\ \int_{\Delta z_1} K(z_2,z_1')dz' & \int_{\Delta z_2} K(z_2,z_2')dz' & \cdots & \int_{\Delta z_N} K(z_2,z_n')dz' \\ \vdots & \vdots & \vdots & \vdots \\ \int_{\Delta z_1} K(z_N,z_1')dz' & \int_{\Delta z_2} K(z_N,z_2')dz' & \cdots & \int_{\Delta z_N} K(z_n,z_n')dz' \end{bmatrix} \begin{bmatrix} I_1 \\ I_2 \\ \vdots \\ I_n \end{bmatrix} = \begin{bmatrix} E_1 \\ E_2 \\ \vdots \\ E_n \end{bmatrix}$$

The method discussed above is only one form of integral method. In fact, this can be generalized in exactly the same way the finite element method can be generalized and it applies equally well to three-dimensional problems.

9.7. Bibliography

The references given here are intended as introductory references. More specific applications and a variety of other formulations will be given in the following two chapters. General aspects of modeling rather than special formulations for the wave equation are stressed here.

Aspects of finite difference modeling are discussed in references [1-4]. The application of finite differences in the time domain (*FDTD*) will be discussed in the following chapter but some aspects of finite difference modeling for the wave equation are given in [5]. Some finite difference formulas, presented in a rather simple way are also given in [5]. Applications of the finite difference method to a number of problems in electromagnetics can be found in [6-10]. The finite element method as first introduced by Oden in 1969 can be found in [11,12]. References [13,14] discuss the finite element method for general engineering aplications. General applications in magnetics can be found in [15-21].

The method of moments has evolved first as a general mathematical tool [22,23]. Its application to electromagnetics started with Harrington's book [24]. A number of applications of the method of moments for scattering, antenna calculations, as well as calculations of capacitance are given in [25-39].

[1] M. Morgan (ed.), "Finite Element and Finite Difference Methods in Electromagnetic Scattering", Elsevier, New York, 1990.

[2] R. D. Richtmyer and K. W. Morton, "Differential Methods for Initial Value Problems," John Wiley, 1967.

[3] G. E. Forsythe and W. R. Wasow, "Finite Difference Methods of Partial Differential Equations", Wiley, 1964.

[4] R.V. Southwell, "Relaxation Methods in Theoretical Physics", Oxford, London, 1946.
[5] M. L. James, G. M. Smith and J. C. Walford, "Applied Numerical Methods for Digital Computation," Harper and Row, Publ., New York, 1977.
[6] E. A. Erdelyi and E. F. Fuchs, "Nonlinear magnetic field analysis of D.C. machines - Part I: theoretical fundamentals," IEEE Transactions on Power Apparatus and Systems, Vol. PAS-89, pp. 1546-1554, 1970.
[7] E. A. Erdelyi and E. F. Fuchs, "Nonlinear magnetic field analysis of D.C. machines - Part II: application of the improved treatment," IEEE Transactions on Power Apparatus and Systems, Vol. PAS-90, pp. 1555-1564, 1970.
[8] J. L. Sieminieniuch and I. Gladwell, "Analysis of explicit difference methods for a diffusion-convection equation," International Journal for Numerical Methods in Engineering, Vol. 12, No. 6, pp. 899-916, 1978.
[9] E. C. Dufort and S. P. Frankel, "Stability conditions in the numerical treatment of parabolic differential equations," in Mathematical Tables and Aids to Computation, Vol. 7, pp. 135-152, 1953.
[10] K.J. Binns and P.J. Lawrenson, Analysis and Computation of Electric and Magnetic Field Problems, Pergamon Press, 1973.
[11] J. T. Oden, "A general theory of finite elements I: Topological considerations," INternal Journal for Numerical Methods in Engineering, Vol. 1, No. 2, 1969, pp. 205-221.
[12] J. T. Oden, "A general theory of finite elements II: Applications," International Journal for Numerical Methods in Engineering, Vol. 1, No. 3, 1969, pp. 247-259.
[13] K. H. Huebner, The Finite Element Method for Engineers, Wiley-Interscience, John Wiley & Sons, Inc., New York, 1975.
[14] O. C. Zienkiewicz, The Finite Element Method in Engineering, third edition, McGraw-Hill Book Co., London, 1977.
[15] C.W. Steele, "Numerical Computation of Electric and Magnetic Fields", Van Nostrand Reinhold Co., New York, 1987.
[16] P.P. Sylvester and R.P. Ferrari, "Finite Elements for Electrical Engineers", Cambridge University Press, Cambridge, 1990.
[17] J.C. Sabonnadierre and J.L. Coulomb, "Finite Element Methods in CAD", Springer Verlag, NY., 1989.
[18] S.R.H. Hoole, "Computer-Aided Analysis and Design of Electromagnetic Devices", Elsevier, NY, 1989.
[19] D.A. Lowther and P.P. Sylvester, "Computer Aider Design in Magnetics", Springer Verlag, NY, 1986.
[20] N. Ida and J. Bastos, "Electromagnetics and Calculation of Fields", Springer Verlag, NY, 1992.
[21] P. L. Arlet et al, "Application of finite elements to the solution of Helmholtz' equation", Proceedings of the IEE, Vol. 115, No. 12, Dec. 1968.
[22] Y. V. Vorobyev, "Method of Moments in Applied Mathematics", Gordon and Breach Science Publishers, New York, 1965.
[23] R.F. Harrington, "Origin and development of the method of moments for field computation", IEEE Antenna and Propagation Society Magazine, Vol. 32, No. 3, 1990, pp. 31-36.
[24] R. F. Harrington, "Field Computation by Moment Method", The Macmillan Company, London, 1967.

[25] J. Moore and R. Pizer, (eds.), "Moment Methods in Electromagnetivs, Techniques and Applications", John Wiley and Sons Inc., New York, 1984.
[26] W. A. Imbriale, "Applications of the method of moments to thin-wire elements and arrays", in Numerical and Assymptotic Techniques in Electromagnetics, R. Mitra, Ed., Springer-Verlag, New York, 1975, pp. 5-88.
[27] A.J. Poggio and E.K. Miller, "Integral equation solution of three-dimensional scattering problems", in R. Mittra, (ed.), Computer Techniques for Electromagnetics, Ch. 4, Pergamon Press, London, 1973.
[28] R. Mitra and C. A. Klein, "Stability and convergence of moment method solutions", in Numerical and Assymptotic Techniques in Electromagnetics, R. Mitra, Ed., Springer-Verlag, New York, 1975, pp. 129-165.
[29] T. K. Sarkar, "A note on the weighting functions in the method of moments", IEEE Transactions on Antennas and Propagation, Vol. AP-33, No. 4, April 1985, pp. 436-441.
[30] D. R. Wilton and S. Govind, "Incorporation of edge conditions in moment method solutions", IEEE Transactions on Antennas and Propagation, Vol. AP-25, No. 6, November 1977, pp. 845-850.
[31] M. M. Ney, "Method of moments as applied to electromagnetic problems", IEEE Transactions on Microwave Theory and Techniques, Vol. MTT-33, No. 10, October 1985, pp. 972-980.
[32] A. Taflove, (ed.), "Special issue on numerical methods for electromagnetic wave interactions", Wave Motion, Vol. 10, No. 6, 1988.
[33] R. F. Harrington, "Matrix methods for field problems", Proceedings of the IEEE, Vol. 55, No. 2, 1967, pp. 136-149.
[34] R.F. Harrington, "The method of moments in electromagnetics", Journal of Electromagnetic Waves and Applications, Vol. 1, No. 3, 1987, pp. 181-200.
[35] E.K. Miller, "A selective survey of computational electromagnetics", IEEE Transactions on Antennas and Propagation, Vol. 36, No. 9, 1988, pp. 1281-1304.
[36] D. K. Reitan and T. H. Higgins, "Accurate determination of the capacitance matrix of a thin rectangular plate," AIEE Transactions, January 1957, pp. 761-766
[37] T. J. Higgins and D. K. Reitan, "Calculation of the Capacitance of a circular annulus by the method of subsurfaces," AIEE Transactions, Vol. 70, 1951, pp. 926-933.
[38] J. E. Storer, "Impedance of thin wire loop antennas, " AIEE Transactions, November 1966, pp. 606-619.
[39] J.C. Toler, "Special issue on electromagnetic wave interactions with biological systems", IEEE Transactions on Microwave Theory and Techniques, Vol. MTT-32, No. 8, Jan. 1984.

Chapter 10. Modeling of the Time-Dependent Wave Equation

10.1. Introduction

The time dependent wave equation is first presented as formulated for solution for waves both in terms of potential functions and fields. The distinction between time-dependent and time-harmonic fields is made to introduce the two most common methods of computation the *FDM* and *FEM*. Finite difference techniques are normally more efficient for time dependent problems, therefore these are introduced here. The general scattering problem is formulated with applications to low and high frequency problems. Emphasis is on efficient solution methods (explicit formulations) and the *FDTD* method. Radiating boundary conditions are used in specific formulations. These include scattering from composites and lossy dielectrics as well as modeling of flaws and inclusions. Finite element methods are much more efficient for time-harmonic fields and will be discussed in detail in the following chapter. Here we merely mention that the method can be used in conjunction with the finite difference method.

The use of finite differences in the time domain and associated aspects of computation and modeling, including radiating boundary conditions and reflections from boundaries, stability criteria and economics of computation are discussed as they relate to testing of composite materials. Examples from microwave testing with lossy and lossless composites are used to emphasize different aspects of modeling.

We first introduce the two-dimensional and axi-symmetric formulations using potential functions and then discuss three-dimensional methods using field variables.

10.2. Formulation of the Time Dependent Wave Equation

10.2.1. The Time Dependent Equations

The formulation of the wave equation in a form suitable for solution in terms of finite differences follows steps similar to those introduced in chapter 1. These represent simple manipulation of Maxwell's equations into a form suitable for solution. We repeat some of those steps here and formulate the equations in two general forms: one is in terms of field variables, the other in terms of potential functions. Only the essentials steps are given since the general considerations including solution of the wave equation and its properties in various media were discusses at length in chapter 1.

The general rule is that potential functions are more convenient in two dimensions while field variables are more appropriate in three dimensions. This however, is not always the case.

The general Maxwell equations are as follows:

$$\nabla\times\mathbf{H} = \mathbf{J} + \frac{\partial\mathbf{D}}{\partial t} \tag{10.1}$$

$$\nabla\cdot\mathbf{B} = 0 \tag{10.2}$$

$$\nabla\times\mathbf{E} = -\frac{\partial\mathbf{B}}{\partial t} \tag{10.3}$$

$$\nabla\cdot\mathbf{D} = \rho \tag{10.4}$$

From these equations, by proper substitution we get wave equations in one of the variables or in terms of an appropriately defined potential function. The resulting equation is called a wave equation because its solution is a propagating wave. We assume here that media are isotropic and linear. Under these conditions Eqs. (10.2) and (10.4) become

$$\nabla\cdot\mathbf{H} = 0$$

$$\nabla\cdot\mathbf{E} = \frac{\rho}{\varepsilon}$$

To obtain a wave equation we use Eq. (10.1) and take the curl on both sides of the equation

$$\nabla\times(\nabla\times\mathbf{H}) = \nabla\times\mathbf{J} + \nabla\times\frac{\partial\mathbf{D}}{\partial t} = \nabla\times\mathbf{J} + \frac{\partial}{\partial t}(\nabla\times\mathbf{D})$$

where the time differentiation and the curl were interchanged. From Eq. (10.3) we have

$$\nabla\times\mathbf{D} = -\varepsilon\frac{\partial\mathbf{B}}{\partial t}$$

Using this and the relation $\nabla\times(\nabla\times\mathbf{H}) = -\nabla^2\mathbf{H} + \nabla(\nabla.\mathbf{H})$ we have

$$\nabla^2\mathbf{H} - \nabla(\nabla\cdot\mathbf{H}) = -\nabla\times\mathbf{J} + \frac{\partial}{\partial t}\left(\varepsilon\frac{\partial\mathbf{B}}{\partial t}\right) \tag{10.5}$$

As for the term $\nabla\times\mathbf{J}$, we use again Eq. (10.3) and the constitutive relation $\mathbf{J}=\sigma\mathbf{E}$

$$\nabla\times\mathbf{J} = \nabla\times\sigma\mathbf{E} = -\,\sigma\frac{\partial\mathbf{B}}{\partial t} = -\,\sigma\mu\frac{\partial\mathbf{H}}{\partial t}$$

Substituting this in Eq. (10.5) and rearranging terms we get

$$\nabla^2\mathbf{H} - \mu\varepsilon\frac{\partial^2\mathbf{H}}{\partial t^2} = \sigma\mu\frac{\partial\mathbf{H}}{\partial t} \qquad (10.6)$$

This is a nonhomogeneous wave equation in terms of the magnetic field intensity. A homogeneous wave equation is obtained by removing the right hand side term in Eq. (10.6) which is due to the current density **J**

$$\nabla^2\mathbf{H} - \mu\varepsilon\frac{\partial^2\mathbf{H}}{\partial t^2} = 0 \qquad (10.7)$$

The distinction between homogeneous (source free) and nonhomogeneous equations indicates the location of the source and the type of solution one expects. With the nonhomogeneous wave equations, the source is in the solution domain and is taken into account in the solution. With the source free equations, the source is outside the solution domain and we expect a solution in terms of the characteristic modes of the problem. This distinction is not merely formal: it is fundamental to the type of solution we choose and the type of results we can obtain.

Instead of starting with Eq. (10.1) we can start with Eq. (10.3) and take the curl on both sides of the equation

$$\nabla\times(\nabla\times\mathbf{E}) = -\,\mu\frac{\partial}{\partial t}(\nabla\times\mathbf{H})$$

where $\mathbf{B} = \mu\mathbf{H}$ was used. Using the relation $(\nabla\times(\nabla\times\mathbf{E}) = -\,\nabla^2\mathbf{E} + \nabla(\nabla\cdot\mathbf{E}))$ and substituting from Eq. (10.1) for $\nabla\times\mathbf{H}$ we have

$$-\,\nabla^2\mathbf{E} + \nabla(\nabla\cdot\mathbf{E}) = -\,\mu\frac{\partial}{\partial t}(\mathbf{J} + \frac{\partial\mathbf{D}}{\partial t}) = -\,\mu\frac{\partial\mathbf{J}}{\partial t} - \mu\frac{\partial^2\mathbf{D}}{\partial t^2}$$

Substituting Eq. (10.4) for $\nabla\cdot\mathbf{E}$, $\sigma\mathbf{E}$ for **J**, and $\varepsilon\mathbf{E}$ for **D**, and rearranging terms we have

$$\nabla^2\mathbf{E} - \mu\varepsilon\frac{\partial^2\mathbf{E}}{\partial t^2} = \nabla\left(\frac{\rho}{\varepsilon}\right) + \mu\sigma\frac{\partial\mathbf{E}}{\partial t} \qquad (10.8)$$

If the sources of the electric field do not exist, (i.e. they are outside the solution domain) we obtain the source free wave equation. In this case, both ρ and **J** must be zero and the two terms on the right hand side of Eq. (10.8) disappear. The source free wave equation in terms of the electric field intensity is

$$\nabla^2\mathbf{E} - \mu\varepsilon\frac{\partial^2\mathbf{E}}{\partial t^2} = 0 \qquad (10.9)$$

Our choice of writing Maxwell's equations in these particular forms was completely arbitrary. We could just as well write them in terms of **B** or **D**. In fact, by multiplying both sides of Eq. (10.6) and (10.7) by μ and using $\mathbf{B}=\mu\mathbf{H}$ we get the homogeneous wave equation and the source free wave equations in terms of **B** rather than **H**

$$\nabla^2\mathbf{B} - \mu\varepsilon\frac{\partial^2\mathbf{B}}{\partial t^2} = \sigma\mu\frac{\partial\mathbf{B}}{\partial t} \qquad (10.10)$$

and

$$\nabla^2\mathbf{B} - \mu\varepsilon\frac{\partial^2\mathbf{B}}{\partial t^2} = 0 \qquad (10.11)$$

Similarly, by multiplying Eqs. (10.8) and (10.9) by ε and using $\mathbf{D}=\varepsilon\mathbf{E}$ we have for the homogeneous wave equation

$$\nabla^2\mathbf{D} - \mu\varepsilon\frac{\partial^2\mathbf{D}}{\partial t^2} = \nabla\rho + \mu\sigma\frac{\partial\mathbf{D}}{\partial t} \qquad (10.12)$$

and for the source free equation

$$\nabla^2\mathbf{D} - \mu\varepsilon\frac{\partial^2\mathbf{D}}{\partial t^2} = 0 \qquad (10.13)$$

In addition to using field variables we can also define potential functions and write the wave equations in terms of these functions. Potential functions may be defined in any way that is convenient and consistent with the field equations. As an example, we can define a magnetic vector potential based on the solenoidal (divergence free) nature of the magnetic field. Because

$$\nabla\cdot\mathbf{B} = 0$$

we can write

$$\nabla\cdot(\nabla\times\mathbf{A}) \equiv 0$$

for any vector **A**. Obviously, we still have to define the divergence of **A**. Thus, a magnetic vector potential can be defined as

$$\mathbf{B}=\nabla\times\mathbf{A}$$

This can now be substituted in Eq. (10.3)

$$\nabla\times\mathbf{E} = -\frac{\partial}{\partial t}(\nabla\times\mathbf{A})$$

In chapter 1 we defined the scalar potential V based on the irrotational nature of the static electric field. We stated that if $\nabla\times\mathbf{E}=0$, then, $\mathbf{E}=-\nabla V$, where V is a scalar potential. Obviously, the time dependent electric field is not irrotational and we cannot define the scalar potential in the same fashion. Instead, we can write the equation above as

$$\nabla\times(\mathbf{E} + \frac{\partial\mathbf{A}}{\partial t}) = 0$$

Now the expression in brackets is irrotational and we can define the time dependent electric scalar potential as

$$\mathbf{E} + \frac{\partial\mathbf{A}}{\partial t} = -\nabla V$$

or, the electric field as

$$\mathbf{E} = -\frac{\partial\mathbf{A}}{\partial t} - \nabla V \qquad (10.14)$$

If the scalar potential is constant in space, or can be neglected, the electric field can be calculated from the magnetic vector potential directly

$$\mathbf{E} = -\frac{\partial\mathbf{A}}{\partial t} \qquad (10.15)$$

This is the case in most *2-D* problems and is often used in *3-D* problems. The definition in Eq. (10.14) is consistent with the static case. The expression above states that the electric field has two sources. One, is due to a charge distribution and exists in the static case as well as the time dependent case $(-\nabla V)$. The second is due to time dependent currents $(\partial\mathbf{A}/\partial t)$. We also note in passing that the vector potential **A** is a vector parallel to **E** and, therefore, to the current density **J**: $(\mathbf{J}=\sigma\mathbf{E})$.

Returning now to Maxwell's equations, we substitute Eq. (10.14) into Eq. (10.1)

$$\nabla\times(\nabla\times\mathbf{A}) = \mu\mathbf{J} + \mu\varepsilon\frac{\partial}{\partial t}\left(-\nabla V - \frac{\partial\mathbf{A}}{\partial t}\right)$$

Using the identity for $\nabla\times(\nabla\times A)$ and rearranging the terms we have

$$\nabla^2\mathbf{A} + \mu\mathbf{J} - \mu\varepsilon\frac{\partial^2\mathbf{A}}{\partial t^2} - \nabla\left(\nabla\cdot A + \mu\varepsilon\frac{\partial V}{\partial t}\right) = 0$$

Since the divergence of **A** has not been defined we can define it such that the last term in the equation above vanishes (Lorentz gage)

$$\nabla\cdot\mathbf{A} = -\mu\varepsilon\frac{\partial V}{\partial t} \qquad (10.16)$$

The nonhomogeneous and the nonhomogeneous (source free) wave equations for the magnetic vector potential are

$$\nabla^2\mathbf{A} - \mu\varepsilon\frac{\partial^2\mathbf{A}}{\partial t^2} = -\mu\mathbf{J} \qquad (10.17)$$

$$\nabla^2\mathbf{A} - \mu\varepsilon\frac{\partial^2\mathbf{A}}{\partial t^2} = 0 \qquad (10.18)$$

To define a wave equation in terms of the electric scalar potential we use the relation

$$\nabla\cdot\mathbf{E} = \frac{\rho}{\varepsilon}$$

Substituting for **E** from Eq. (10.14) we have

$$-\nabla\cdot\left(\frac{\partial\mathbf{A}}{\partial t} + \nabla V\right) = \frac{\rho}{\varepsilon}$$

which, after performing the divergence is

$$\nabla^2 V + \frac{\partial\mathbf{A}}{\partial t}(\nabla\cdot\mathbf{A}) = -\frac{\rho}{\varepsilon}$$

and again using the condition in Eq. (10.16), we get the nonhomogeneous and source free wave equations as

$$\nabla^2 V - \mu\varepsilon\frac{\partial^2 V}{\partial t^2} = -\frac{\rho}{\varepsilon} \qquad (10.19)$$

$$\nabla^2 V - \mu\varepsilon \frac{\partial^2 V}{\partial t^2} = 0 \qquad (10.20)$$

Eq. (10.20) is a scalar wave equation while the others above were vector wave equations.

It is obvious that we can choose any of these variables for representation of the field. The question is: which should we choose for implementation with the finite difference method (or, for that matter, with any other method). Unfortunately, the answer to this question is not obvious and is not unique. In some cases a particular form will suggest itself as the most appropriate form. In others, considerations of solution, boundary conditions, and the type of results needed may dictate the equation to use. Sometimes, when all other aspects are identical or similar, the economics of solution may be dominant.

To introduce the general method of formulation with finite differences consider the general two-dimensional geometry shown in Figure 10.1. The region is made of two parts. The first part is assumed to be free space with parameters $(\varepsilon_a,\mu_a,\sigma_a=0)$. This will be denoted as *domain a*. The source, if there is one, is located in this part of the solution domain. The medium below is the material under testing (*domain m*), typically a lossy dielectric, with parameters $(\varepsilon_m,\mu_a,\sigma_m)$. We require that σ_m be very small compared with the conductivity of a good conductor since we wish to consider high-frequency testing. If high conductivity materials are tested, the equation reduces to a diffusion equation and we may be able to use the low frequency approach (that is: displacement currents may be neglected and the standard eddy current equation is solved). We assume that defects exist in the material, (*domain s*) such as surface variations, inclusions (buried or surface breaking), etc. Since these materials behave like scatterers, we denote their parameters by $(\varepsilon_s,\mu_a,\sigma_s)$. In the notation that follows, the permeability is assumed to be general but the method cannot handle nonlinearities. In most cases of interest in the microwave region it is also possible to assume that the permeability of materials is equal to that of free space μ_0.

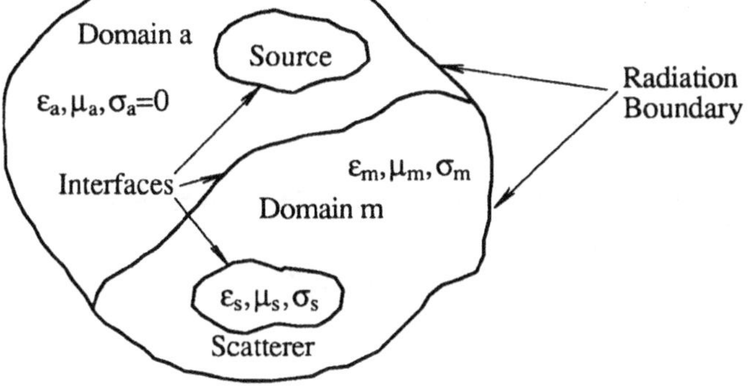

Figure 10.1. A general geometry for modeling. Scatterers are always in *domain m* while sources can only be in located *domain a*.

10.2.2. Alternative Formulation: TE and TM Representation

The formulation may start with any of the wave equations presented above. However, instead of doing this we present here an alternative method, based on the transverse nature of fields. We choose to present the *TM* mode but *TE* modes may be treated similarly. (See section (2.2) for a discussion on *TE* and *TM* waves and corresponding equations).

For *TM* field incidence (that is: the direction of the magnetic field is perpendicular to the direction of propagation of the wave), we have, from Maxwell's equations,

$$\frac{\partial H_x}{\partial x} + \frac{\partial H_y}{\partial y} = 0 \tag{10.21}$$

$$\hat{\mathbf{i}}\frac{\partial E}{\partial y} - \hat{\mathbf{j}}\frac{\partial E}{\partial x} = -\mu_p\left(\hat{\mathbf{i}}\frac{\partial H_x}{\partial t} + \hat{\mathbf{j}}\frac{\partial H_y}{\partial t}\right) \tag{10.22}$$

$$\frac{\partial H_y}{\partial x} - \frac{\partial H_x}{\partial y} = \sigma_p E + \varepsilon_p \frac{\partial E}{\partial t} \tag{10.23}$$

where p denotes three possible materials: free space (will be denoted a), general material (m) and scatterer (normally a conducting material, denoted by s)..

By introducing the magnetic vector potential $\mathbf{A}$ such that $\mu\mathbf{H}=\nabla\times\mathbf{A}$, the wave equation we encountered above is obtained but for *TM* excitation. This means that in this case only the z component of $\mathbf{A}$ exists and we can write the equation as a scalar wave equation in each of the three types of materials mentioned above. The solution consists of solving these equations in their respective domains while matching the solutions on the interfaces between the various materials.

Using the curl relation we get:

$$H_y = -\frac{1}{\mu}\frac{\partial A_z}{\partial x} \quad \text{and} \quad H_x = \frac{1}{\mu}\frac{\partial A_z}{\partial y}$$

Also, from Eq. (10.15) we can write

$$E_z = -\frac{\partial A_z}{\partial t}$$

where we have neglected the scalar potential. Substituting these in Eq. (10.23) we get

$$\frac{\partial}{\partial x}\frac{1}{\mu}\frac{\partial A_z}{\partial x} + \frac{\partial}{\partial y}\frac{1}{\mu}\frac{\partial A_z}{\partial y} = \sigma\frac{\partial A_z}{\partial t} + \varepsilon\frac{\partial^2 A_z}{\partial t^2} \tag{10.24}$$

where μ,ε, and σ are the general material properties for any material in which this equation is used. These can be, in general, anisotropic.

For simplicity in implementation we scale the variables to allow analysis in nondimensional variables. The scaling used here consists of the following:

1. The vector potential A is scaled by A_i (the amplitude of the incident wave). The scaled potential is $\overline{A}$.

$$\overline{A} = \frac{A}{A_i}$$

2. x and y are scaled by a length L. This length is arbitrary but it should be constant and of the order of the scatterer's size (e.g. diameter of a cylindrical scatterer).

$$\overline{x} = \frac{x}{L} \quad and \quad \overline{y} = \frac{y}{L}$$

3. Time is scaled by $1/\omega$:

$$\overline{t} = \omega t$$

where ω is the angular frequency of the incident wave.

Scaling of variables is not an absolute necessity nor is it unique. It is done for purposes of convenience and, in some cases, to simplify either calculation or presentation of results.

With these, and assuming linear, isotropic materials, the three scalar equations in the three different materials are

In free space:

$$\nabla^2\overline{A} = k_a^2\frac{\partial^2\overline{A}}{\partial\overline{t}^2} \tag{10.25}$$

In general media:

$$\nabla^2\overline{A} = k_m^2\frac{\partial^2\overline{A}}{\partial\overline{t}^2} + l_m^2\frac{\partial\overline{A}}{\partial\overline{t}} \tag{10.26}$$

In scatterer:

$$\nabla^2\overline{A} = k_s^2\frac{\partial^2\overline{A}}{\partial\overline{t}^2} + l_s^2\frac{\partial\overline{A}}{\partial\overline{t}} \tag{10.27}$$

where

$$k_a^2=\omega^2L^2\mu_a\varepsilon_a \qquad k_m^2=\omega^2L^2\mu_m\mu_0\varepsilon_m\varepsilon_0 \qquad k_s^2=\omega^2L^2\mu_s\mu_0\varepsilon_s\varepsilon_0$$

$$l_m^2 = \omega L^2 \mu_m \mu_0 \sigma_m \qquad l_s^2 = \omega L^2 \mu_s \mu_0 \sigma_s$$

where μ_a, μ_m, μ_s, ε_a, ε_m, ε_s, are the relative permeabilities and permittivities in the respective materials and σ_m, σ_s are conductivities of the lossy material and scatterer respectively. Interface conditions are obtained from the continuity of the tangential components of electric and magnetic fields,

$$\bar{A}^+ = \bar{A}^- \qquad and \qquad \mu^+\left(\frac{\partial \bar{A}}{\partial n}\right)^+ = \mu^-\left(\frac{\partial \bar{A}}{\partial n}\right)^- \tag{10.28}$$

where the plus and minus signs denote the limits as one approaches the interface. The permeabilities shown are permeabilities on the two sides of the interface.

These equations are easily modified for anisotropic materials by starting with Eq. (10.24) and writing

$$\frac{\partial}{\partial x}\frac{1}{\mu_r}\frac{\partial A_z}{\partial x} + \frac{\partial}{\partial y}\frac{1}{\mu_r}\frac{\partial A_z}{\partial y} = \sigma\frac{\partial A_z}{\partial t} + \varepsilon\frac{\partial^2 A_z}{\partial t^2}$$

With the normalization used above, the following equations are obtained in the three materials

$$\frac{\partial}{\partial \bar{x}}\frac{1}{\mu_a}\left(\frac{\partial \bar{A}}{\partial \bar{x}}\right) + \frac{\partial}{\partial \bar{y}}\frac{1}{\mu_a}\left(\frac{\partial \bar{A}}{\partial \bar{y}}\right) = k_a^2\frac{\partial^2 \bar{A}}{\partial \bar{t}^2} \tag{10.29}$$

$$\frac{\partial}{\partial \bar{x}}\frac{1}{\mu_m}\left(\frac{\partial \bar{A}}{\partial \bar{x}}\right) + \frac{\partial}{\partial \bar{y}}\frac{1}{\mu_m}\left(\frac{\partial \bar{A}}{\partial \bar{y}}\right) = k_m^2\frac{\partial^2 \bar{A}}{\partial \bar{t}^2} + l_m^2\frac{\partial \bar{A}}{\partial \bar{t}} \tag{10.30}$$

$$\frac{\partial}{\partial \bar{x}}\frac{1}{\mu_s}\left(\frac{\partial \bar{A}}{\partial \bar{x}}\right) + \frac{\partial}{\partial \bar{y}}\frac{1}{\mu_s}\left(\frac{\partial \bar{A}}{\partial \bar{y}}\right) = k_s^2\frac{\partial^2 \bar{A}}{\partial \bar{t}^2} + l_s^2\frac{\partial \bar{A}}{\partial \bar{t}} \tag{10.31}$$

with parameters defined as

$$k_a^2 = \omega^2 L^2 \mu_0 \varepsilon_a \qquad k_m^2 = \omega^2 L^2 \mu_0 \varepsilon_m \varepsilon_0 \qquad k_s^2 = \omega^2 L^2 \mu_0 \varepsilon_s \varepsilon_0$$

$$l_m^2 = \omega L^2 \mu_0 \sigma_m \qquad l_s^2 = \omega L^2 \mu_0 \sigma_s$$

Essentially identical forms are obtained for axisymmetric geometries. In most respects the forms are unchanged but we start with the equations in cylindrical coordinates. This is presented next to show how a problem may be formulated and the changes needed.

10.3. The Axi-Symmetric Formulation

The axi-symmetric formulation presented here as an example, is based on the general considerations given before and, for the most part it is similar to the *2-D* formulation. However, there are some differences and these are discussed next. We use a slightly different method here, primarily in the way the various variables are defined. This is primarily to show a different approach to modeling. In terms of implementation of the finite differences and boundary conditions, the problem is identical to the *2-D* case.

The axisymmetric formulation is based on the fact that both **H** and **E** can be obtained from the ϕ component of the magnetic vector potential **A**. Therefore, the whole problem can be modeled using this component of the potential function **A**. Again, the conductivity of air is assumed to be negligible. Using the general geometry in Figure 10.1, the governing equations are:

$$\frac{\partial^2 A}{\partial z^2} + \frac{\partial^2 A}{\partial r^2} + \frac{1}{r}\frac{\partial A}{\partial r} - \frac{A}{r^2} = \mu_a \varepsilon_a \frac{\partial^2 A}{\partial t^2} \quad \textit{in domain a} \tag{10.32}$$

$$\frac{\partial^2 A}{\partial z^2} + \frac{\partial^2 A}{\partial r^2} + \frac{1}{r}\frac{\partial A}{\partial r} - \frac{A}{r^2} = \mu_m \varepsilon_m \frac{\partial A}{\partial t} + \mu_m \varepsilon_m \frac{\partial^2 A}{\partial t^2} \quad \textit{in domain m} \tag{10.33}$$

where *domain a* represents the domain in which the source is located but with the current source excluded. In other words, even though the source is located in this domain, it is assumed to be known and fixed and does not require modeling. The effect of the source is introduced as (Dirichlet) boundary conditions at the interfaces between the source and *domain a*. The interface conditions come from the continuities of tangential electric and magnetic field intensities across interfaces. For the electric field, the continuity condition is

$$\frac{\partial A^-}{\partial t} = \frac{\partial A^+}{\partial t} \tag{10.34}$$

where the minus sign represents the potential in *domain a*, and the plus sign represents that in the lossy medium (*domain m*). The continuity of the tangential components of the magnetic field are treated as follows: From the definition, using cylindrical coordinates,

$$\mathbf{H} = \frac{1}{\mu}\nabla\times A = \frac{1}{\mu}\left[-\frac{\partial A}{\partial z}r + \left(\frac{A}{r} + \frac{\partial A}{\partial r}\right)z\right] \tag{10.35}$$

where z and r are the axial and the radial directions respectively. On the interface boundary, z is the normal direction and r is the tangential direction. The continuity of the tangential magnetic field leads to the following interface condition:

$$\mu_a \frac{\partial A^-}{\partial z} = \mu_m \frac{\partial A^+}{\partial z} \tag{10.36}$$

where μ_a and μ_m are the relative permeability in domains a and m respectively. Scaling is also used here, but a somewhat different scaling technique is adopted. Again, this is a choice that can be made and, in fact, the scaling used in Eqs. (10.25) through (10.31) can also be used here, without modifications. The governing equations after scaling are

$$\frac{\partial^2 A}{\partial z^2} + \frac{\partial^2 A}{\partial r^2} + \frac{1}{r}\frac{\partial A}{\partial r} - \frac{A}{r^2} = \frac{\partial^2 A}{\partial t^2} \tag{10.37}$$

$$\frac{\partial^2 A}{\partial z^2} + \frac{\partial^2 A}{\partial r^2} + \frac{1}{r}\frac{\partial A}{\partial r} - \frac{A}{r^2} = k_m^2 \frac{\partial^2 A}{\partial t^2} + l_m^2 \frac{\partial A}{\partial t} \tag{10.38}$$

where

$$k_m^2 = \frac{\omega^2 L^2 \mu_m \varepsilon_m}{k_a^2} \qquad l_m^2 = \frac{\omega L^2 \mu_m \varepsilon_m}{k_a}$$

and k_a is the same as in Eq. (10.29).

On the boundary between free space and the general material, we have interface conditions:

$$A^- = A^+ \tag{10.39}$$

$$\mu_r \frac{\partial A^-}{\partial z} = \frac{\partial A^+}{\partial z} \tag{10.40}$$

where μ_r is the relative permeability of the lossy dielectric.

Since a time dependent problem is solved, the initial conditions of the problem must also be specified. These depend on the application but, in many cases we can use the following initial conditions:

$$A(r,z,0) = A_i(r,z,0) \tag{10.41}$$

$$\frac{\partial A}{\partial t}(r,z,0) = 0 \qquad \frac{\partial A_i}{\partial t}(r,z,0) = 0 \tag{10.42}$$

where A_i is the incident potential. The boundary conditions of the problem are similar to those for the two-dimensional formulations other than the obvious changes due to the axial symmetry. These are discussed next.

10.4. Radiation Boundary Conditions

In general, the solution domain is infinite. For a practical solution, within the finite difference formalism, one needs to operate within a truncated domain. An artificial boundary condition must be specified on the truncated boundary to simulate the infinite domain. Unlike low frequency applications, where truncation of the solution domain can be used (see Chapter 9), at high frequencies, the reflection of waves from artificial boundaries must be eliminated. This is not merely an inconvenience: artificial boundaries introduce nonphysical waves and therefore incorrect solutions are obtained.

There are some basic methods to account for this type of boundaries. One of these, namely the absorbing or radiating boundary condition is described here. We have also discussed in chapter 9 the use of integral equations which naturally incorporate these conditions in the solution. There are a number of methods of deriving radiation boundary conditions and some of these are pointed to in the bibliography section. Here we give a radiation boundary conditions specifically designed for the formulations above.

In free space the following radiation boundary condition can be used for the *2-D* formulation

$$\left(\frac{\partial}{\partial r} + k_a\frac{\partial}{\partial t} + \frac{1}{2r}\right)A_s = 0 \tag{10.43}$$

Similarly, an absorbing boundary condition for lossy media, has been formulated to allow lossy materials on the boundary.

$$\left(\frac{\partial}{\partial r} + k_m\frac{\partial}{\partial t} + \frac{1}{2r} + \frac{l_m^2}{2k_m}\right)A_s = 0 \tag{10.44}$$

Details of this boundary condition as well as an equivalent form in the frequency domain are given elsewhere (see bibliography). Similar forms for axisymmetric geometries are available. We point however that if the last term in Eq. (10.44) is small (that is: $l_m^2 << k_m$) the radiation condition in (10.44) reduces to the free space radiation condition in Eq. (10.43). This is the case for very low loss dielectrics.

The formulation given here is general and applies equally well to low frequency fields. At low frequencies, the boundary conditions are modified since the boundary conditions in (10.43) and (10.44) are based on a high frequency limit.

For axi-symmetric formulations, both the scattered and incident waves are in the same direction therefore the boundary conditions must be applied to the total or outgoing field. Denoting the outgoing component of the magnetic vector potential by A_o we can write the radiation boundary condition in free space as

$$\frac{\partial A_o}{\partial r} + \frac{\partial A_o}{\partial t} + \frac{1}{r}A_o = O\left(\frac{1}{r^3}\right) \tag{10.45}$$

where the notation $O()$ indicates that the expression is of the order of $(1/r^3)$. In actual calculation this is set to zero by locating the boundary at a reasonably large distance.

Lossy media have a dissipative effect on the propagating waves. The typical wave equation in a lossy medium is shown in Eq. (10.38). Its right side consists of a loss term with the coefficient l_m and a propagating term with the coefficient k_m. The radiation condition in lossy media is

$$\frac{\partial A_o}{\partial r} + k_m \frac{\partial A_o}{\partial t} + A(r) A_o = O\left(\frac{1}{r^3}\right) \tag{10.46}$$

where

$$A(r) = +\frac{l_m^2}{2k_m} + \left(\frac{1}{r^2}\right) \tag{10.47}$$

For $l_m << 2k_m$, the dissipative effect is negligible. In this case, the radiation condition for a dielectric is

$$\frac{\partial A_o}{\partial r} + k_m \frac{\partial A_o}{\partial t} + \frac{1}{r} A_o = O\left(\frac{1}{r^3}\right) \tag{10.48}$$

For $l_m >> 2k_m$, the dissipative term is very large and the scattered waves, which reach the exterior boundary, are negligible. Thus we can use $A_o = 0$ as the boundary condition.

The boundary conditions given above are now incorporated into the finite difference scheme as will be shown next. This guarantees that the artificial boundary has no effect on the field inside the solution domain (no reflections from the artificial boundary).

10.5. Finite Difference Implementation

10.5.1. Two-Dimensional Applications

To obtain the finite difference schemes, the central difference approximation is used to discretize the wave equations in Eqs. (9.29) through (9.31). Central difference schemes are used for the discretization in space and time. The radiation boundary conditions in Eqs. (10.43) and (10.44) are incorporated by projecting $\partial A/\partial r$ to the normal direction of the artificial boundary to obtain $\partial A/\partial n$. The resulting finite difference equations are

$$A_{i,j}^{k+1} = 2A_{i,j}^{k} - A_{i,j}^{k-1} + \left[\frac{\Delta t}{\Delta x}\right]^2 \left[A_{i-1,j}^{k} + A_{i+1,j}^{k} + A_{i,j-1}^{k} + A_{i,j+1}^{k} - 4A_{i,j}^{k}\right] \tag{10.49}$$

for points in *domain a* (Eq. (10.43)). In *domain m* (Eq. (10.44), the finite difference scheme is

$$\left[1+\frac{l_m^2}{2k_m^2}\Delta t\right]A_{i,j}^{k+1} = 2A_{i,j}^{k} + \left[\frac{l_m^2}{2k_m^2}\Delta t - 1\right]A_{i,j}^{k-1} + \frac{1}{k_m^2}\left[\frac{\Delta t}{\Delta x}\right]^2\left[A_{i-1,j}^{k} + A_{i+1,j}^{k} + A_{i,j-1}^{k} + A_{i,j+1}^{k} - 4A_{i,j}^{k}\right] \tag{10.50}$$

for points interior to the mesh. The indices i,j,k, are as shown in the grid in Figure 10.2 and $\Delta x = \Delta y$. On the boundary of the finite difference mesh the total magnetic vector potential is a superposition of the incident and scattered potentials A_i and A_s. The radiation boundary condition only represents the outgoing, or scattered wave therefore, we write for the scattered wave on the boundary:

$$\left[1+a\frac{\Delta t}{k\Delta x}+b\frac{l^2}{k^2}\Delta t\right]A_{i,j}^{(s)k+1} = \left[2 - a\Delta x\left[\frac{1}{R}+c\frac{l^2}{k^2}\right]\cdot\left[\frac{\Delta t}{k\Delta x}\right]^2\right]A_{i,j}^{(s)k} + \left[-1+a\frac{\Delta t}{k\Delta x}+b\frac{l^2}{k^2}\Delta t\right]A_{i,j}^{(s)k-1} + \frac{2}{k^2}\left[\frac{\Delta t}{\Delta x}\right]^2\left[dA_{i-1,j}^{(s)k}+eA_{i+1,j}^{(s)k}+fA_{i,j-1}^{(s)k}+gA_{i,j+1}^{(s)k}-hA_{i,j}^{(s)k}\right] \tag{10.51}$$

where the constants a through h are evaluated from the parameters of the problem and the index (s) indicates the scattered wave. The constants a through h are evaluated to satisfy the radiation boundary conditions. Forward or backward difference schemes are used depending on the location of the boundary. We will not give these here since they are problem dependent. However, an example on how these are evaluated is given in the next section, for the axisymmetric formulation. These equations are then solved directly (explicit schemes, including interface conditions) to obtain the magnetic vector potential. Other quantities, if required are calculated from the magnetic vector potential.

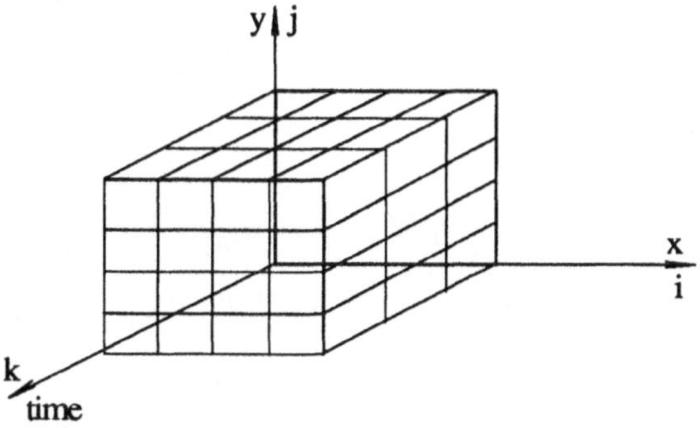

Figure 10.2. Schematic finite difference mesh showing the indices i,j, and k.

As with any finite difference scheme for wave propagation in the time domain, the term $(\Delta t/\Delta x)^2$ appears in the expressions. In order to obtain a stable solution, the discretization in the time domain must be made fine enough to satisfy the Courant stability criterion. (which, in this case is $\Delta t^2/\Delta x^2 \leq 0.5$ because of the normalization used in section 10.2.2). This normally results in a very large number of time steps.

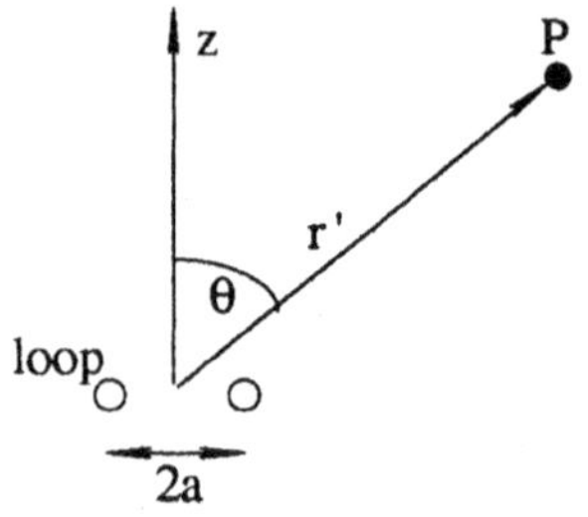

Figure 10.3. Parameters used to calculate the magnetic vector potential of the small loop.

10.5.2. Axi-Symmetric Applications

The problem we consider here has a source in the solution domain. The source generates the incident field component. The source is modeled by by the incident components at a group of points, which are near to, and enclose the source. That is, the equivalent potential values at, or near the source, must be calculated independently, before solution can proceed. To demonstrate this we use the small loop discussed in chapter 9 as an example.

The incident magnetic vector potential of a small loop, carrying spatially uniform current I under time-harmonic excitation, is

$$\bar{A}_i = \frac{Ia}{4\pi}\int_0^{2\pi} f \cos \phi' d\phi' \qquad (10.52)$$

where

$$f = \frac{exp\left(ik_a\left(\sqrt{(r')^2 + a^2 - 2r'asin\theta cos\phi'} - t\right)\right)}{\sqrt{(r')^2 + a^2 - 2r'asinqcosf}} \qquad (10.53)$$

and a is the radius of the loop, r' is the distance of the observation point from the loop's center and θ is the angle which the position vector r' makes with the z-axis, as shown in Figure 10.3. For a small loop $a \rightarrow 0$, and the following approximation can be used:

$$\overline{A}_i \rightarrow \frac{I\pi a^2}{4\pi} e^{ik_a(r'-t)}\left(\frac{ik_a}{r'} + \frac{1}{(r')^2}\right)\sin\theta \tag{10.54}$$

For points on the axis, $\theta=0$. Thus, the incident components of A at these points are zero. Also by axial symmetry, it is easily verified that the scattered components of A at these points cancel each other. Therefore, the condition $A=0$ is imposed on the axis of symmetry.

The finite difference time domain technique is used to discretize the problem both in time and in space. The superscript *'i'* represents the time coordinate and the subscripts *'j'* and *'k'* represent the grid coordinates in the r and z directions respectively, similar to the notation in Figure 10.2. The following difference formulae are used to discretize the wave equations in Eqs. (10.37) and (10.38):

$$A_{j,k}^{i+1} = 2A_{j,k}^{i} - A_{j,k}^{i-1} + \left(\frac{\Delta t}{\Delta z}\right)^2\left(A_{j+1,k}^{i} + \frac{j-1}{j}A_{j-1,k}^{i} + A_{j,k+1}^{i} + A_{j,k-1}^{i} - \frac{4j^2-j+1}{j^2}A_{j,k}^{i}\right) \tag{10.55}$$

in *domain a*, and

$$A_{j,k}^{i+1} = c_1 A_{j,k}^{i} - c_2 A_{j,k}^{i-1} + c_3\left(A_{j+1,k}^{i} + \frac{j-1}{j}A_{j-1,k}^{i} + A_{j,k+1}^{i} + A_{j,k-1}^{i} - \frac{4j^2-j+1}{j^2}A_{j,k}^{i}\right) \tag{10.56}$$

in *domain m* where,

$$c_1 = \frac{2 + \frac{l_m^2}{k_m^2}\Delta t}{1 + \frac{l_m^2}{k_m^2}\Delta t} \qquad c_2 = \frac{1t}{1 + \frac{l_m^2}{k_m^2}\Delta t} \qquad c_3 = \frac{\frac{1}{k_m^2}\left(\frac{\Delta t}{\Delta z}\right)^2}{1 + \frac{l_m^2}{k_m^2}\Delta t}$$

The forward-backward difference schemes discussed in chapter 9 are used to implement the radiation boundary conditions given in Eqs. (10.45) and (10.46). In the present situation, both the incident field and the scattered field are outgoing. Therefore, the radiation conditions are used to take care of the total field. The difference formulae are summarized as follows (refering to Figure 10.4)

$$A_{j,k}^{i+1} = 0 \quad at\ A \tag{10.57}$$

$$A_{j,k}^{i+1} = aA_{j,k}^{i} - b\left[A_{j+1,k}^{i} + A_{j-1,k}^{i}\right] - c\left[A_{j,k}^{i} + A_{j,k-1}^{i}\right] \quad on\ B_4\ to\ B_5 \tag{10.58}$$

$$A_{j,k}^{i+1} = aA_{j,k}^{i} - b\left[A_{j,k}^{i} + A_{j-1,k}^{i}\right] \quad on\ B_3 \tag{10.59}$$

$$A_{j,k}^{i+1} = 0 \qquad at\ A' \tag{10.60}$$

$$A_{j,k}^{i+1} = a'A_{j,k}^{i} - b\left[A_{j,k}^{i} + A_{j-1,k}^{i}\right] - c\left[A_{j,k}^{i} + A_{j,k}^{i}\right] \qquad on\ B_1\ to\ B_2 \tag{10.61}$$

The coefficients a, b, c, and a' are obtained as

$$a = 1 - \frac{1}{r}\frac{\Delta t}{\Delta z} \qquad b = \sin\theta\frac{\Delta t}{\Delta z}$$

$$c = \cos\theta\frac{\Delta t}{\Delta z} \qquad a' = 1 - A(r)\frac{\Delta t}{\Delta z}$$

where $A(r)$ is defined in Eq. (10.47). On the axis of symmetry, $A=0$.

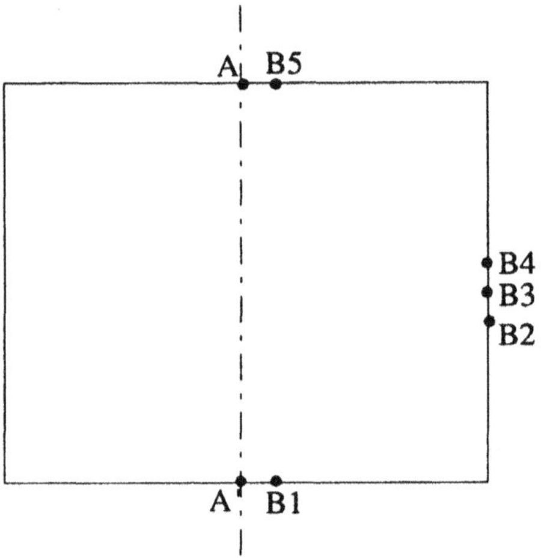

Figure 10.4. General boundary of a geometry used to implement the radiation boundary conditions for the axi-symmetric geometry.

10.6. Examples

10.6.1. Scattering by Embedded Cylinders

A sample *NDT* model is shown in Figure 10.5. In this geometry a wave, at *4 GHz*, incident on the material from free space, is scattered by two cylindrical scatterers embedded in the lossy dielectric material. The medium below the interface is a lossy dielectric

$(\mu_0, \varepsilon_0, \sigma_m = 0.05\ S/m)$ under testing. The scatterers inside the testing material are good conductors $(\mu_0, \varepsilon_0, \sigma_s = 10^7\ S/m)$. The diameter of the cylindrical scatterers is $L=0.2m$. This value is used for normalization as discusses in section 10.6. We use a *101X101* mesh to model the given domain, where one conducting cylinder occupies 9 grid points. Thus, the actual value of Δx is $0.0667\ m$. The normalized value of Δx is $\Delta x/0.2 = 0.3335$. From the requirement for stability criterion, the required Δt is

$$\left[\frac{\Delta t}{\Delta x}\right]^2 \leq 0.5 \qquad \Rightarrow \qquad \Delta t = \sqrt{0.49999 \times (\Delta x)^2} \approx 0.2358$$

This, however, is the normalized value of the time step. The actual time step is obtained by dividing this time step by ω. In this case, the actual time step is 9.38×10^{-12} *sec*. Now it is obvious why the normalization was necessary in the first place. By doing so, we avoided the very large differences in values and therefore the numerical difficulties associated with these differences. On the other hand, it is also obvious that the number of time steps required for solution is large. A single cycle of the wave (*0.25 nanoseconds*) requires almost over *26* time steps for stable solution. The total number of time steps for the solution can be very large.

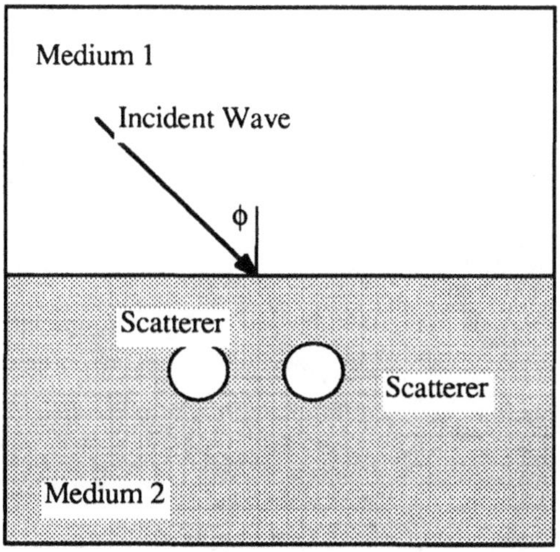

Figure 10.5 Test geometry. Two scatterers embedded in a lossy dielectric are illuminated with an oblique wave at *4 GHz*.

A typical case to which this type of solution applies immediately is testing of composite materials under microwave illumination. We assume that an oblique plane wave is incident from the ambient medium (free space in this case but it can also be a different material). The incident wave penetrates into and reflects from the material surface. Figure 10.6 is the field contour plotting for 45° incidence with respect to the *y*-axis, frequency $f=4\ GHz$ and $\sigma=0.05\ S/m$, without scatterers (or defects) in the material being tested. The field distribution is quite familiar. Now we introduce inclusions into the material. Figure 10.7 shows the wave surface plot for such a case, where the conductivity of the scatterer is $10^7\ S/m$, and the incidence is the same as above. The wave penetrating into the test material is diffracted by the cylinder. The diffracted wave then propagates through free-space, after a refraction at the interface. The field distortion, due to the presence of scatterers is detectable at the surface of the material. Figure 10.8 gives a comparison of field distributions just above the material surface with and without scatterers. If the fields with and without scatterers are subtracted, the field due to scattering by the conducting cylinders is obtained. The corresponding field surface is given in Figure 10.9. The solutions given here are typical of composite materials with conducting fiber reinforcement although the dimensions are larger. Similar results are obtained for testing of reinforced concrete.

Figure 10.6. Contour plotting for oblique incidence.

Figure 10.7. Total wave surface with an embedded cylinder.

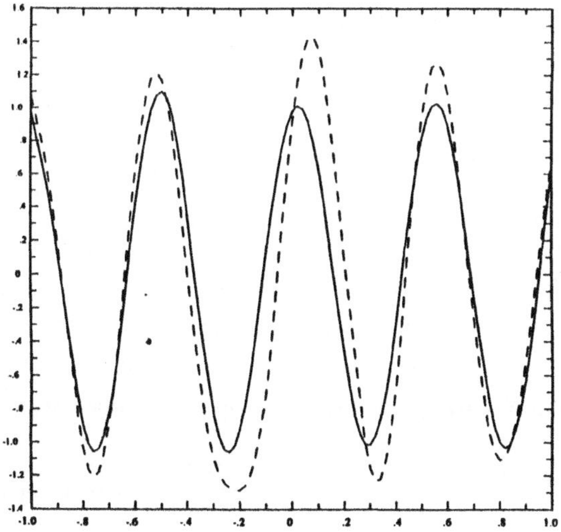

Figure 10.8. Field distribution above the material surface.
- - - - - - with two embedded cylinders
________without scatterers

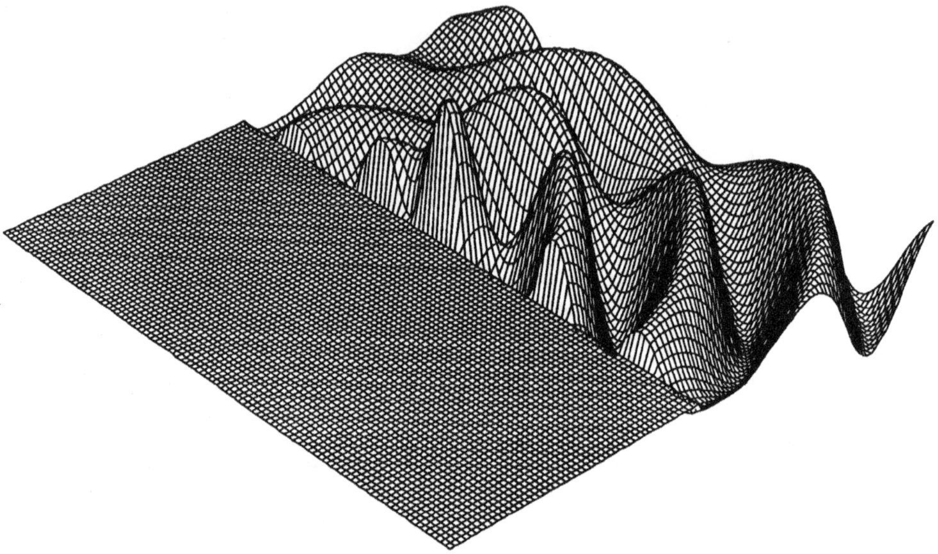

Figure 10.9. Scattered field surface due to the two cylinders.

10.6.2. Waves Due to a Small Loop

A second example to the use of the *FDTD* method is the calculation of waves due to a small loop over a dielectric material as shown in Figure 10.10. The material is assumed to be flat, *12.5 mm* thick and with a relative permittivity of *3*. In this case a lossless dielectric was modeled but lossy dielectrics can be modeled just as easily. However, modeling of scatterers is limited to axi-symmetric scatterers since we use the axisymmetric formulation. The modeling domain is shown in Figure 10.10. The domain is *75 cm* by *75 cm*. The diameter of the loop is *1.6 mm* and the loop is *1 mm* above the dielectric. Calculations are performed at *1 GHz*. A mesh with 121×121 points is used. In this case, the scaling parameter is chosen to be *4.472 mm*. The space step is $\Delta x=6.25$ *mm* and the normalized time step that satisfies the stability criterion is $\Delta t=0.00442$. The actual time step is $\Delta t=7\times10^{-13}$ *sec*. Thus, for *10* cycles of the signal the number of time steps necessary is about *15,000*.

The solution to this problem in the form of the distribution in the near and far field were given in chapter 9 and are also shown in Figures 10.11 and 10.12. The near field solution (given in a 18.5×18.5 *cm* region, or about one half wavelength in each direction) clearly shows the loop above the dielectric and looks very much like the solution of a coil at steady state. However, the far field solution in Figure 10.12 shows the propagating effect, as well as the guiding effect of the dielectric. From the testing point of view this indicates that effects along the slab are easily detected provided they are not too small compared to wavelength.

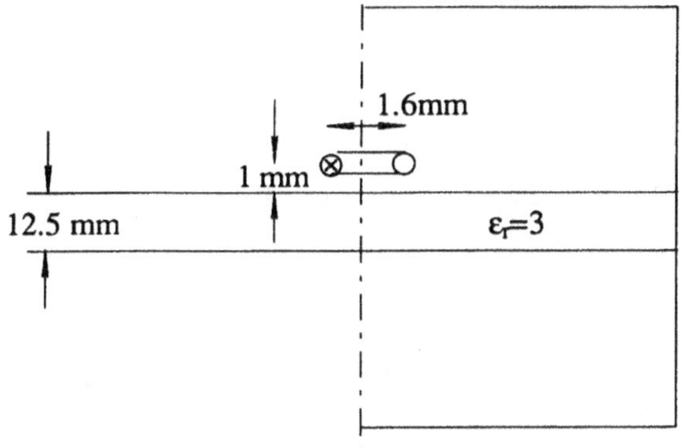

Figure 10.10. A small loop over a dielectric slab. Box on the right shows the solution domain.

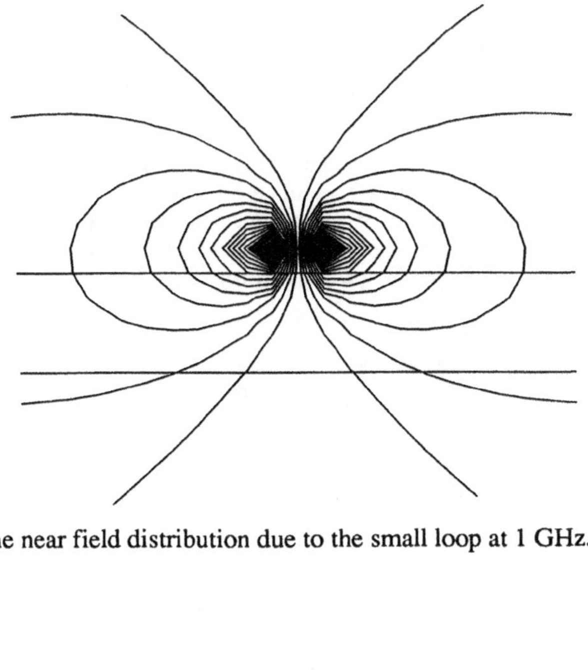

Figure 10.11. The near field distribution due to the small loop at 1 GHz.

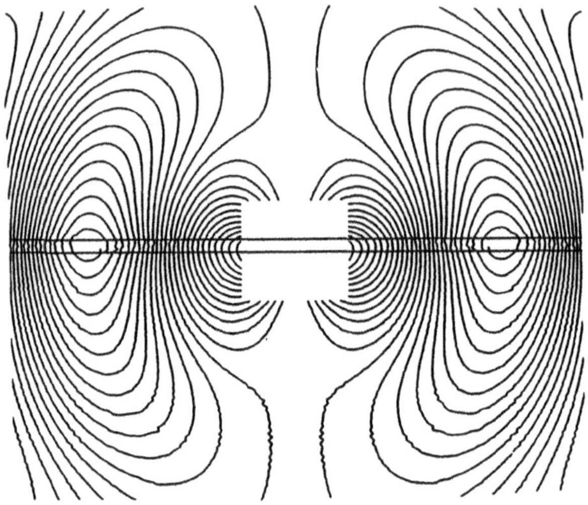

Figure 10.12. The far field distribution due to the small loop at *1 GHz*.

10.7. 3-D Formulation With the Finite Difference Time Domain Method

The formulation presented above in terms of potential functions was particularly useful in two dimensional and axi-symmetric geometries because the magnetic vector potential has a single component under these conditions. Thus, the equations reduced to essentially scalar wave equations.

The *FDTD* method is not restricted to these geometries and, in many ways, its representation in *3-D* geometries is simpler. However, now we must deal with vector functions and therefore there is no advantage in using potential functions.

We show here the formulation in terms of the electric field intensity *E* and the magnetic field intensity *H*. Only the general aspects are given while specific applications are refered to in bibliography since its implementation is similar to that given above for potential functions.

We start with the two curl equations:

$$\nabla\times\mathbf{E} = -\mu\frac{\partial\mathbf{H}}{\partial t}$$

$$\nabla\times\mathbf{H} = \mathbf{J} + \varepsilon\frac{\partial\mathbf{E}}{\partial t} = \sigma\mathbf{E} + \varepsilon\frac{\partial\mathbf{E}}{\partial t}$$

In expanded form these are

$$\hat{\mathbf{x}}\left(\frac{\partial E_z}{\partial y}-\frac{\partial E_y}{\partial z}\right)+\hat{\mathbf{y}}\left(\frac{\partial E_x}{\partial z}-\frac{\partial E_z}{\partial x}\right)+\hat{\mathbf{z}}\left(\frac{\partial E_y}{\partial x}-\frac{\partial E_x}{\partial y}\right)=$$

$$-\hat{\mathbf{x}}\mu\frac{\partial H_x}{\partial t}+\hat{\mathbf{y}}\mu\frac{\partial H_y}{\partial t}+\hat{\mathbf{z}}\mu\frac{\partial H_z}{\partial t}$$

$$\hat{\mathbf{x}}\left(\frac{\partial H_z}{\partial y}-\frac{\partial H_y}{\partial z}\right)+\hat{\mathbf{y}}\left(\frac{\partial H_x}{\partial z}-\frac{\partial H_z}{\partial x}\right)+\hat{\mathbf{z}}\left(\frac{\partial H_y}{\partial x}-\frac{\partial H_x}{\partial y}\right)=$$

$$\hat{\mathbf{x}}\sigma E_x+\hat{\mathbf{y}}\sigma E_y+\hat{\mathbf{z}}\sigma E_z+\hat{\mathbf{x}}\varepsilon\frac{\partial E_x}{\partial t}+\hat{\mathbf{y}}\varepsilon\frac{\partial E_y}{\partial t}+\hat{\mathbf{z}}\varepsilon\frac{\partial E_z}{\partial t}$$

Separating components (and assuming that ε, μ, and σ, are not time dependent) we get the following six equations

$$\frac{\partial H_x}{\partial t}=\frac{1}{\mu}\left(\frac{\partial E_y}{\partial z}-\frac{\partial E_z}{\partial y}\right) \tag{10.62}$$

$$\frac{\partial H_y}{\partial t}=\frac{1}{\mu}\left(\frac{\partial E_z}{\partial x}-\frac{\partial E_x}{\partial z}\right) \tag{10.63}$$

$$\frac{\partial H_z}{\partial t}=\frac{1}{\mu}\left(\frac{\partial E_x}{\partial y}-\frac{\partial E_y}{\partial x}\right) \tag{10.64}$$

$$\frac{\partial E_x}{\partial t}=\frac{1}{\varepsilon}\left(\frac{\partial H_z}{\partial y}-\frac{\partial H_y}{\partial z}-\sigma E_x\right) \tag{10.65}$$

$$\frac{\partial E_y}{\partial t}=\frac{1}{\varepsilon}\left(\frac{\partial H_x}{\partial z}-\frac{\partial H_z}{\partial x}-\sigma E_y\right) \tag{10.66}$$

$$\frac{\partial E_z}{\partial t}=\frac{1}{\varepsilon}\left(\frac{\partial H_y}{\partial x}-\frac{\partial H_x}{\partial y}-\sigma E_z\right) \tag{10.67}$$

Most *FDTD* models use a cell in the finite difference mesh consisting of nine points as shown in Figure 10.13. The electric field and magnetic field components are shown as calculated at the respective points. The spatial increments are assumed to be equal: $\Delta x=\Delta y=\Delta z=h$. The time increment is Δt. Conforming to the most common algorithms (see references), the first order derivatives are calculated as follows:

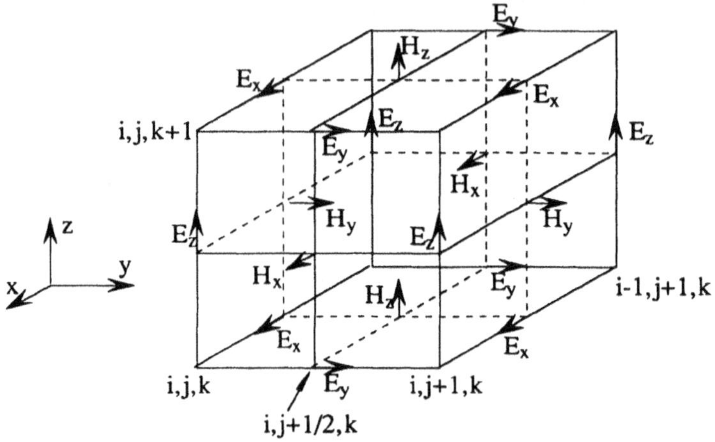

Figure 10.13. The basic cell for finite difference implementation. This cell represents one point in time.

$$\frac{\partial E^m_{i,j,k}}{\partial x} = \frac{E^m_{i+1/2,j,k} - E^m_{i-1/2,j,k}}{h} \tag{10.68}$$

where i, j, k, represent a point on the grid with coordinates x, y, z. m represents a point in time. The difference between this and the central difference representation used in chapter 9 is in the use of half distances.

Similarly, using the half grid distance idea in time, the time derivative is represented as

$$\frac{\partial E^m_{i,j,k}}{\partial t} = \frac{E^{m+1/2}_{i,j,k} - E^{m-1/2}_{i,j,k}}{\Delta t} \tag{10.69}$$

Using these expressions, the six equations are easily and directly represented in the finite difference notation as

$$E_{x_{i+1/2,j,k}}^{m+1/2} = E_{x_{i+1/2,j,k}}^{m} + \frac{\Delta t}{\varepsilon_{i+1/2,j,k} h}\left[H_{z_{i+1/2,j+1/2,k}}^{m+1/2} - H_{z_{i+1/2,j-1/2,k}}^{m+1/2} + H_{y_{i+1/2,j,k-1/2}}^{m+1/2} - H_{y_{i+1/2,j,k+1/2}}^{m+1/2}\right] - \frac{\sigma_{i+1/2,j,k}\Delta t}{\varepsilon_{i+1/2,j,k} h} E_{x_{i+1/2,j,k}}^{m} \tag{10.70}$$

This expression is explicit in time and the material properties (in this case ε) are those at the point at which the electric field is calculated (in this case E_x is calculated and the permittivity entered is that of the point $i+1/2,j,k$). The remaining expressions for **E** are as follows:

$$E_{y_{i,j+1/2,k}}^{m+1/2} = E_{y_{i,j+1/2,k}}^{m} + \frac{\Delta t}{\varepsilon_{i,j+1/2,k}h}\left[H_{x_{i,j+1/2,k+1/2}}^{m+1/2} - H_{x_{i,j+1/2,k-1/2}}^{m+1/2} + H_{z_{i-1/2,j+1/2,k}}^{m+1/2} - H_{z_{i+1/2,j+1/2,k}}^{m+1/2}\right] - \frac{\sigma_{i,j+1/2,k}\Delta t}{\varepsilon_{i,j+1/2,k}h}E_{y_{i,j+1/2,k}}^{m} \tag{10.71}$$

$$E_{z_{i,j,k+1/2}}^{m+1/2} = E_{z_{i,j,k+1/2}}^{m} + \frac{\Delta t}{\varepsilon_{i,j,k+1/2}h}\left[H_{y_{i+1/2,j,k+1/2}}^{m+1/2} - H_{y_{i-1/2,j,k-1/2}}^{m+1/2} + H_{x_{i,j-1/2,k+1/2}}^{m+1/2} - H_{x_{i,j+1/2,k+1/2}}^{m+1/2} - \frac{\sigma_{i,j,k+1/2}\Delta t}{\varepsilon_{i,j,k+1/2}h}E_{z_{i,j,k+1/2}}^{m}\right] \tag{10.72}$$

The finite difference relations for **H** are similar:

$$H_{x_{i,j+1/2,k+1/2}}^{m+1/2} = H_{x_{i,j+1/2,k+1/2}}^{m-1/2} + \frac{\Delta t}{\mu_{i,j+1/2,k+1/2}h}\left[E_{y_{i,j+1/2,k+1}}^{m} - E_{y_{i,j+1/2,k}}^{m} + E_{z_{i,j,k+1/2}}^{m} - E_{z_{i,j+1,k+1/2}}^{m}\right] \tag{10.73}$$

$$H_{y_{i+1/2,j,k+1/2}}^{m+1/2} = H_{y_{i+1/2,j,k+1/2}}^{m-1/2} + \frac{\Delta t}{\mu_{i+1/2,j,k+1/2}h}\left[E_{z_{i+1,j,k+1/2}}^{m} - E_{z_{i,j,k+1/2}}^{m} + E_{x_{i+1/2,j,k}}^{m} - E_{x_{i+1/2,j,k+1}}^{m}\right] \tag{10.74}$$

$$H_{z_{i+1/2,j+1/2,k}}^{m+1/2} = H_{z_{i+1/2,j+1/2,k}}^{m-1/2} + \frac{\Delta t}{\mu_{i+1/2,j+1/2,k}h}\left[E_{x_{i+1/2,j+1,k}}^{m} - E_{x_{i+1/2,j,k}}^{m} + E_{y_{i,j+1/2,k}}^{m} - E_{y_{i+1,j+1/2,k}}^{m}\right] \tag{10.75}$$

While these expressions seem lengthy, they are rather simple. They can be easily extended to anisotropic materials as well as to the solution of the Helmholtz equation (eigenvalue problem).

10.8. Bibliography

Treatment of the wave equation in its various forms can be found in almost any textbook. In particular, treatment as presented here can be found in [1-8]. The various aspects of the finite difference time domain formulation in terms of potential functions is discussed in [9-13]. Various aspects of radiation or absorption boundary conditions are given in [14-19] as well as in [9,13]. Additional applications, including low frequency solutions of the wave equation can be found in [20-22]. The general *FDTD* method in terms of field variables was first introduced by Yee [23]. Numerous variations, applications and additional topics such as radiation boundary conditions in conjunction with the *FDTD* method have been implemented. A sample of these is given in references [24-26].

[1] J. Van Bladel, Electromagnetic Fields, Hemisphere Publishing Co., 1985, pp.376-382.

[2] R.E. Collin, "Electromagnetic Theory of Guided Waves", McGraw-Hill, NY, 1991.

[3] J.A. Kong, "Theory of Electromagnetic Waves," John Wiley & Sons, Inc., New York, 1975.

[4] R.E. Collin, "Foundations for Microwave Engineering", McGraw-Hill, Inc., New York, 1992.

[5] R.F. Harrington, "Time-harmonic Electromagnetic Fields," McGraw-Hill, Inc., New York, 1961.

[6] Chen H. C. , "Theory of Electromagnetic Waves", McGraw-Hill Book Company, New York, 1983.

[7] C.A. Balanis, "Advanced Engineering Electromagnetics", John Wiley & Sons, New York, 1989.

[8] A. Ishimaru, "Electromagnetic Wave Propagation, Radiation, and Scattering", Prentice Hall, Inc., Englewood Cliffs, N.J., 1991.

[9] M. E. Lee, S. I. Hariharan and N. Ida, "Solving general time dependent two dimensional eddy current problems," Journal of Computational Physics, Vol. 89, No. 2, August 1990, pp. 319-348.

[10] M. E. Lee and N. Ida, "Axisymmetric electromagnetic fields coupled to lossy media", in the IEEE Transactions on Magnetics, Vol. 26, No. 2, March 1990, pp. 575-578.

[11] M. E. Lee, S. I Hariharan and N. Ida, "Transient calculations of two-dimensional eddy current problems", in IEEE Transactions on Magnetics, Vol. 25, No. 4, July 1989, pp. 3140-3144.

[12] J.S. Wang, N. Ida and S. I. Hariharan, "Numerical modeling of transient wave propagation for high frequency NDT", in Review of Progress in Quantitative Nondestructive Evaluation, D. O. Thompson and D. E. Chimenti, Eds., Plenum Press, New York, 1989, Vol. 8A, pp. 259-266.

[13] M.E. Lee, "Potential Finite-Difference Time-Domain Methods for Electromagnetic Interface Problems", Ph.D. Dissertation, The University of Akron, May 1989.

[14] B. Engquist and A. Majda, "Absorbing boundary conditions for the numerical simulation of waves", Mathematics of Computation, Vol. 31, 1977, pp. 629-651.

[15] A. Bayliss and E. Turkel, "Radiation boundary conditions for wave-like equations", Communications of Pure and Applied Mathematics, Vol. 33, No. 6, 1980, pp. 707-725.

[16] F.G. Friedlander, "On the radiation field of pulse solutions of the wave equation", Proceedings of the Royal Society, London, Ser. A, Vol. 269, 1962, pp. 53-65.

[17] G. Mur, "Absorbing boundary conditions for finite-difference approximation of the time-domain electromagnetic field equations", IEEE Transactions on Electromagnetic Compatibility, Vol. EMC-24, Nov. 1981, pp. 1073-1077.

[18] F.V. Atkinson, "On Somerfeld's Radiation Condition", Philosophical Magazine, Vol. XL, 1949, pp. 645-651.

[19] C.H. Wilcox, "An expansion theorem for electromagnetic fields", Communications of Pure and Applied Mathematics, Vol. 9, 1956, pp. 115-134.

[20] M.E.Lee and N. Ida, "Hollow Sphere in a Step Field", in COMPEL, Vol. 9, Supplement A, 1990, pp. 278-280.

[21] M.E. Lee and N. Ida, "A Transient Axisymmetric Formulation for Modeling of Composite Materials", in Review of Progress in Quantitative Nondestructive Evaluation, D. O. Thompson and D. E. Chimenti, Eds., Plenum Press, New York, 1990, Vol. 9A, pp. 295-302.

[22] N. Ida, "Microwave NDT", Electrosoft, special issue on NDT Vol. 2, No. 2/3, June/Sept. 1991, pp. 215-237.

[23] K.S. Yee, "Numerical solution of initial boundary value problems involving Maxwell's equations in isotropic media", IEEE Transactions on Antenna and Propagation, Vol. AP-14, No. 3, 1966, pp. 302-307.

[24] A. Taflove, "Application of the finite difference time-domain method to sinusoidal steady state electromagnetic penetration problems", IEEE Transactions on Electromagnetic Compatibility, Vol. EMC-22, No.3, 1980, pp. 192-202.

[25] A. Taflove and K.R. Umashankar, "The finite-difference time-domain method for numerical modeling of electromagnetic wave interactions", Electromagnetics, Vol. 10, 1990, pp. 105-126.

[26] A. Christ and H.L. Hartangel, "Three-dimensional finite-difference method for the analysis of microwave-device embedding", IEEE Transactions on Microwave Theory and Techniques, Vol. MTT-35, No. 8, Aug. 1987, pp. 688-696.

Chapter 11. Modeling of the Time-Harmonic Wave Equation

11.1. Introduction

The finite element method is now introduced for solution of time harmonic fields where the method is most efficient. The various forms of the wave equation in the frequency domain are formulated in terms of the finite element method, using scalar shape functions. The various methods for formulations applicable to high frequency fields are outlined together with the relevant finite elements. Specific formulations in two and three dimensions are given.

For reasons of space, we will restrict detailed discussion to specific methods of modeling but will also point out to extensions and modifications that can be made. The types of problems discussed are those of modeling of materials in waveguides and cavity resonators. There is a very good reason to do so: as with the *FDTD* method, the introduction of an artificial boundary condition creates artificial reflections from boundaries. In waveguides and cavity resonators, the boundary conditions are well defined and there is no propagation beyond the boundaries. This greatly simplifies discussion but also restricts the methods to specific applications.

Three basic formulations are discussed. One is the formulation of the Helmholtz equation and its use in waveguides and cavity resonators. The particular appeal of this equation is its simplicity and therefore will be used to show the formulation in terms of the energy functional and variational techniques. The second is the formulation of the eddy current equation as used for computation of resonant frequencies in waveguides. The *3-D* equation is defined for a linear *3-D* finite element and again, a variational technique is used. While the eddy current equation is general, its use in wave solutions is restricted to calculation of resonant frequencies and relative field quantities. Finally, a more general weak form in conjunction with the Galerkin's technique is used to formulate the more general wave equation in terms of the electric field. This particular form is suitable for scattering problems in an open domain while the first two were restricted to closed boundary applications.

11.2 The Time Harmonic Wave Equations

11.2.1. The Wave Equation

In the previous chapter we introduced various forms of the time dependent wave equation. Particularly useful forms of the wave equation are obtained by assuming sinusoidal fields and using the complex (phasor) notation

$$\mathbf{E} = \mathbf{E}_0 e^{j\omega t} \tag{11.1}$$

where $j = \sqrt{-1}$, $\omega = 2\pi f$ and f is the frequency of the field. Under this condition, the time dependent form of the field is

$$\mathbf{E}(t) = Re\left(\mathbf{E}_0 e^{j\omega t}\right) = \mathbf{E}_0 cos\omega t \tag{11.2}$$

where Re denotes the real part of the phasor.

Sinusoidal variation, as well as phase angles of the field can be introduced in the notation as an angle ϕ

$$\mathbf{E} = \mathbf{E}_0 e^{j\phi} e^{j\omega t} \tag{11.3}$$

Under phasor conditions, the time derivatives of the fields are of the form

$$\frac{\partial \mathbf{E}}{\partial t} = j\omega \mathbf{E} \tag{11.4}$$

and the second order derivative is

$$\frac{\partial^2 \mathbf{E}}{\partial t^2} = -\omega^2 \mathbf{E} \tag{11.5}$$

Now we can introduce these definitions in the wave equations presented in chapters 1 and 10. By doing so we obtain the time harmonic form of the wave equations. Although this can be done for any of the equations, only the equations in terms of **E**, **H**, and **A** are used in the formulations in this chapter therefore only those equations are presented here.

- The nonhomogeneous wave equations in these three variables are

$$\nabla^2 \mathbf{E} - \nabla\left(\frac{\rho}{\varepsilon}\right) - j\omega\mu\sigma\mathbf{E} + \omega^2 \mu\varepsilon\mathbf{E} = 0 \tag{11.6}$$

$$\nabla^2 \mathbf{H} - j\omega\sigma\mu\mathbf{H} + \omega^2 \mu\varepsilon\mathbf{H} = 0 \tag{11.7}$$

$$\nabla^2\mathbf{A} + \mu\mathbf{J} + \omega^2\mu\varepsilon\mathbf{A} = 0 \tag{11.8}$$

In Eq. (11.7) the term $\nabla\times\mathbf{J}_s$ (see Eq. (1.59)), was set to zero. The term $j\omega\sigma\mu\mathbf{H}$ is the eddy current term.

- The equivalent homogeneous or Helmholtz equations are

$$\nabla^2\mathbf{E} + \omega^2\mu\varepsilon\mathbf{E} = 0 \tag{11.9}$$

$$\nabla^2\mathbf{H} + \omega^2\mu\varepsilon\mathbf{H} = 0 \tag{11.10}$$

$$\nabla^2\mathbf{A} + \omega^2\mu\varepsilon\mathbf{A} = 0 \tag{11.11}$$

These are all wave equations and therefore have wave solutions. Depending on the type of formulation we use, one of the wave equations may be more convenient than the other. In most cases, either the equation for **E** or **H** are used. In some cases, as for example, in the two-dimensional applications of the previous chapter the equation for A may be more convenient.

These formulations are discussed next in some detail, outlining their implementation and examples to their solution.

11.2.2. The Generalized Quasi-Static Equation (Eddy Current Equation)

The Helmholtz equations above were obtained by applying the Lorentz gage to the general equation as was shown in chapter 1. This is essentially the same as saying that the equations are obtained from the high frequency point of view. To see this, and to understand the modification made here, we repeat some of the steps involved in defining the wave equation first, and then proceed to modify it.

We start with Maxwell's equations, more specifically with the first two equations using sinusoidal fields (Eqs. (1.11) and (1.12)).

$$\nabla\times\mathbf{H} = \mathbf{J} + j\omega\mathbf{D} \tag{11.12}$$

$$\nabla\times\mathbf{E} = -j\omega\mathbf{B} \tag{11.13}$$

To define the general equation for the time dependent field in terms of the magnetic vector potential, we start with the third of Maxwell's equation

$$\nabla\cdot\mathbf{B} = 0$$

Because this is a divergenceless field we can write

$$\nabla\cdot(\nabla\times\mathbf{A}) \equiv 0$$

for the vector **A** which is a general vector but we identify this as the magnetic vector potential we used previously. We still have to define the divergence of **A**.
Thus, a magnetic vector potential is defined as

$$\mathbf{B}=\nabla\times\mathbf{A}$$

This is now be substituted in Eq. (11.12)

$$\nabla\times\mathbf{E} = -j\omega(\nabla\times\mathbf{A})$$

In chapter 3 we defined the scalar potential V based on the irrotational nature of the static electric field. We stated that if $\nabla\times\mathbf{E}=0$, then, $\mathbf{E}=-\nabla V$, where V is a scalar potential. The time dependent electric field is not irrotational and we cannot define the scalar potential in the same fashion. Instead, we can write the equation above as

$$\nabla\times(\mathbf{E}+j\omega\mathbf{A}) = 0$$

Now the expression in brackets is irrotational and we can define the time dependent electric scalar potential as

$$\mathbf{E}+j\omega\mathbf{A} = -\nabla V$$

or, the electric field as

$$\mathbf{E} = -j\omega\mathbf{A} - \nabla V \qquad (11.14)$$

This definition is consistent with the static case. The expression above states that the electric field has two sources. One, is due to a charge distribution and exists in the static case as well as the time dependent case $(-\nabla V)$. The second is due to time dependent currents $(j\omega\mathbf{A})$. We also note in passing that the vector potential **A** is a vector parallel to **E** and, therefore, to the current density **J**: $(\mathbf{J}=\sigma\mathbf{E})$.

Returning now to Maxwell's equations, we substitute Eq. (11.14) into Eq. (11.11)

$$\nabla\times(\nabla\times\mathbf{A}) = \mu\mathbf{J} + j\omega\mu\varepsilon\left(-\nabla V - j\omega\mathbf{A}\right)$$

Using the identity for $\nabla\times(\nabla\times\mathbf{A})$ and rearranging the terms we have

$$\nabla^2\mathbf{A} + \mu\mathbf{J} + \omega^2\mu\varepsilon\,\mathbf{A} - \nabla\left(\nabla\cdot\mathbf{A} + j\omega\mu\varepsilon V\right) = 0 \qquad (11.15)$$

In the development we followed in chapter 1, we used the Lorentz gage

$$\nabla\cdot\mathbf{A} = -\mu\varepsilon\frac{\partial V}{\partial t}$$

to cancel the last term of the equation. This was permissible since the divergence of the magnetic vector potential has not been defined beforehand. Since we can choose this divergence in any way that is consistent with the field equations, this is not the only gage we can apply. In this case, we use the same condition but in the frequency domain. Lorentz's gage in the frequency domain is

$$\nabla \cdot \mathbf{A} = -j\omega\mu\varepsilon V$$

With this gage, the last term in Eq. (11.15) cancels and we obtain the nonhomogeneous wave equations

$$\nabla^2 \mathbf{A} + \omega^2 \mu\varepsilon\, \mathbf{A} = -\mu \mathbf{J} \qquad (11.16)$$

The current density **J** can be viewed as a general current composed of a source current density $\mathbf{J}_s$ and an induced current density $\mathbf{J}_e$. These induced currents are negligible in the microwave region for many purposes but not for low loss dielectrics. The equation is now

$$\nabla^2 \mathbf{A} + \mu \mathbf{J}_s + \mu\sigma \mathbf{E} + \omega^2 \mu\varepsilon \mathbf{A} = 0 \qquad (11.18)$$

where $\mathbf{J}_e = \sigma\mathbf{E}$ was used. Substitution of the electric field from Eq. (11.14) gives

$$\nabla^2 \mathbf{A} + \mu \mathbf{J}_s - \mu\sigma\nabla V - j\omega\mu\sigma\mathbf{A} + \omega^2 \mu\varepsilon \mathbf{A} = 0$$

If we neglect the electric scalar potential V we get the modified eddy current equation: that is, the eddy current equation with an additional current term due to displacement currents ($\omega^2\mu\varepsilon\mathbf{A}$).

$$\nabla^2 \mathbf{A} + \mu \mathbf{J}_s - j\omega\mu\sigma\mathbf{A} + \omega^2 \mu\varepsilon \mathbf{A} = 0 \qquad (11.19)$$

The term $j\omega\mu\sigma\mathbf{A}$ represents losses due to induced currents. In the high frequency limit these are neglected and we are back to Eq. (11.16). This equation is particularly appropriate for solution in lossy dielectrics where the current source must be taken into account. For example, in a microwave cavity or waveguide, if we need to model the source generating the fields this approach is suitable. The Helmholtz equation cannot handle sources but only modes of the solution. In effect, the solution to the Helmholtz equation are the resonant frequencies of the system (most often in cavities and waveguides but not necessarily so). The eddy current equation requires the frequency be known before hand to produce a solution. Helmholz's equation leads to an eigenvalue problem. The eddy current equation approach is deterministic.

11.2.3. Formulation of the Two-Dimensional Helmholtz Equation

The source free wave equations as given in Eqs. (11.9) through (11.11) lead to characteristic or eigenvalue solutions since, in absence of a source, only a characteristic solution is possible. The formulation presented here is for *TE* and *TM* modes in rectangular waveguides. As is normally done, we assume propagation in the z direction, so that the waveguide cross section is in the *x-y* plane.

The equations we need to solve are (see section 2.2. for detailed derivation of these equations)

$$\frac{\partial^2 E_z}{\partial x^2} + \frac{\partial^2 E_z}{\partial y^2} + (\gamma^2 + k^2)E_z = 0 \tag{11.20}$$

for *TM* modes and

$$\frac{\partial^2 H_z}{\partial x^2} + \frac{\partial^2 H_z}{\partial y^2} + (\gamma^2 + k^2)H_z = 0 \tag{11.21}$$

for *TE* modes. In these equations, E_z and H_z are phasors. A similar equation is obtained for the magnetic vector potential but we will not use it here.

In the first case, the boundary conditions are specified by $E_z=0$ (Dirichlet type) on the walls of the waveguide (perfectly conducting), while in the second, the conditions $\partial H_z/\partial x=0$, $\partial H_z/\partial y=0$ (Neumann type) are used. While the waveguide walls are assumed to be perfectly conducting, the interior of the waveguide can be lossy, as specified by the complex permittivity in the waveguide.

The discussion that follows is for the electric field equation (*TM* modes) but identical steps apply to the magnetic field equation (*TE* modes). The difference in solution is due to the boundary conditions the two waves must satisfy. For convenience, and consistency in notation with Eq. (11.9), we write Eq. (11.21) as

$$\frac{\partial^2 E_z}{\partial x^2} + \frac{\partial^2 E_z}{\partial y^2} + k_c^2 E_z = 0\,, \qquad k_c^2 = \gamma^2 + k^2 \tag{11.22}$$

At cutoff, $\gamma^2=0$ and therefore

$$k_c^2 = \omega^2 \mu\varepsilon$$

With this, we rewrite Eq. (11.22) as

$$\frac{1}{\mu_r}\left(\frac{\partial^2 E_z}{\partial x^2} + \frac{\partial^2 E_z}{\partial y^2}\right) + k_0^2 \varepsilon_r E_z = 0 \tag{11.23}$$

where ε_r and μ_r are the relative permeability and permittivity respectively. Now k_0 is the free space wavenumber and once calculated can be directly interpreted as the cutoff frequency of a particular mode of propagation. We further note that ε_r is in general the complex permittivity, depending on the material.

Following the same steps for *TE* modes, Eq. (11.10) now becomes

$$\frac{1}{\varepsilon_r}\left(\frac{\partial^2 H_z}{\partial x^2}+\frac{\partial^2 H_z}{\partial y^2}\right)+k_0^2\mu_r H_z=0 \tag{11.24}$$

Now either Eq. (11.23) or (11.24) may be formulated in terms of the Galerkin procedure or using an energy functional. We use here the latter method and follow exactly the steps given in chapter 9. An appropriate energy functional for Eq. (11.23) is

$$\int_s \frac{1}{2}\frac{1}{\mu_r}\left[\left(\frac{\partial E}{\partial x}\right)^2+\left(\frac{\partial E}{\partial y}\right)^2\right]ds-\int_s \frac{1}{2}k_0^2\varepsilon_r E\, ds \tag{11.25}$$

and for Eq. (11.24)

$$\int_s \frac{1}{2}\frac{1}{\varepsilon_r}\left[\left(\frac{\partial H}{\partial x}\right)^2+\left(\frac{\partial H}{\partial y}\right)^2\right]ds-\int_s \frac{1}{2}k_0^2\mu_r H\, ds \tag{11.26}$$

In the first case, we solve for the electric field applying the boundary condition

$$E=0 \tag{11.27}$$

while in the second case, the solution is for the magnetic field with boundary condition

$$\frac{\partial H}{\partial n}=0 \tag{11.28}$$

which, as was mentioned in section chapter 9 is implicit in the finite element formulation.

We will follow the steps of the formulation for Eq. (11.25) for the electric field. The finite element we use is the same element as in Figure 9.14 and, for the four nodes of the element, the approximation of the electric field is

$$E_z(x,y)=\sum_{k=1}^{4} E_k N_k$$

Similarly, the approximation for the partial derivatives, the Jacobian and the steps required for integration of the element were given in detail in section 9.5.5. (Eqs. (9.22) through (9.30)).

Therefore, the general term of the elemental matrix due to to the spatial derivatives (Laplacian) of **E** is

$$S_{i,j} = \int_v \left(\frac{\partial N_i}{\partial x} \frac{1}{\mu_r} \frac{\partial N_j}{\partial x} + \frac{\partial N_i}{\partial y} \frac{1}{\mu_r} \frac{\partial N_j}{\partial y} \right) dv \tag{11.29}$$

where the permeability was left in the integral to indicate the fact that this formulation is also applicable to anisotropic materials. After integration, an elemental matrix is obtained. For constant permeability (isotropic, linear case), the elemental matrix can be written as

$$\frac{1}{\mu_r} \begin{bmatrix} S_{11} & S_{12} & S_{13} & S_{14} \\ S_{21} & S_{22} & S_{23} & S_{24} \\ S_{31} & S_{32} & S_{33} & S_{34} \\ S_{41} & S_{42} & S_{43} & S_{44} \end{bmatrix} \begin{bmatrix} E_1 \\ E_2 \\ E_3 \\ E_4 \end{bmatrix} \tag{11.30}$$

Now, E_i are the z components of the electric field intensity at the four nodes of the rectangular element. This is then the elemental contribution (due to the first term of Eq. (11.25)), and is called an elemental stiffness matrix.

After applying the integration to the second term in Eq. (11.25), we get a contribution of the following form.

$$\int_{S_i} k_0^2 \varepsilon_r N^t E dxdy$$

and, upon substitution of the approximation for E and integration, we get the following matrix for the second term of Eq. (11.23)

$$-k_0^2 \varepsilon_r \begin{bmatrix} M_{11} & M_{12} & M_{13} & M_{14} \\ M_{21} & M_{22} & M_{23} & M_{24} \\ M_{31} & M_{32} & M_{33} & M_{34} \\ M_{41} & M_{42} & M_{43} & M_{44} \end{bmatrix} \begin{bmatrix} E_1 \\ E_2 \\ E_3 \\ E_4 \end{bmatrix} \tag{11.31}$$

Now the elemental system of equations is

$$\frac{1}{\mu_r} \begin{bmatrix} S_{11} & S_{12} & S_{13} & S_{14} \\ S_{21} & S_{22} & S_{23} & S_{24} \\ S_{31} & S_{32} & S_{33} & S_{34} \\ S_{41} & S_{42} & S_{43} & S_{44} \end{bmatrix} \begin{bmatrix} E_1 \\ E_2 \\ E_3 \\ E_4 \end{bmatrix} - k_0^2 \varepsilon_r \begin{bmatrix} M_{11} & M_{12} & M_{13} & M_{14} \\ M_{21} & M_{22} & M_{23} & M_{24} \\ M_{31} & M_{32} & M_{33} & M_{34} \\ M_{41} & M_{42} & M_{43} & M_{44} \end{bmatrix} \begin{bmatrix} E_1 \\ E_2 \\ E_3 \\ E_4 \end{bmatrix} = 0 \tag{11.32}$$

Assembly of the elemental equations into a global system of equations gives

$$[SS]\{E\} = k_0^2[M]\{E\} \tag{11.33}$$

A similar system can be written for *TE* modes by replacing μ_r by ε_r (in the first term) and ε_r (in the right term) by μ_r

$$\frac{1}{\varepsilon_r}\begin{bmatrix} S_{11} & S_{12} & S_{13} & S_{14} \\ S_{21} & S_{22} & S_{23} & S_{24} \\ S_{31} & S_{32} & S_{33} & S_{34} \\ S_{41} & S_{42} & S_{43} & S_{44} \end{bmatrix}\begin{bmatrix} H_1 \\ H_2 \\ H_3 \\ H_4 \end{bmatrix} - k_0^2\mu_r\begin{bmatrix} M_{11} & M_{12} & M_{13} & M_{14} \\ M_{21} & M_{22} & M_{23} & M_{24} \\ M_{31} & M_{32} & M_{33} & M_{34} \\ M_{41} & M_{42} & M_{43} & M_{44} \end{bmatrix}\begin{bmatrix} H_1 \\ H_2 \\ H_3 \\ H_4 \end{bmatrix} = 0$$

or

$$[SS']\{H\} = k_0^2[M']\{H\} \tag{11.34}$$

This is an eigenvalue problem in complex variables. In the way this particular system was written in Eq. (11.32) the matrix on the left hand side is real while that on the right hand side is complex (since the complex permittivity must be used). The solution therefore must be in complex variables. The solution proceeds with calculation of the eigenvalues and eigenvectors of either system using any of the standard methods of solution for this type of eigenvalue problem.

The eigenvalues then represent the cutoff frequencies of the various modes and the eigenvectors, the electric field variation in the cross section of the waveguide.

From this, the fields anywhere in the waveguide can then be written using the dispersion relation

$$\gamma^2 = -k^2 + \left(\frac{m\pi}{a}\right)^2 + \left(\frac{n\pi}{b}\right)^2 \tag{11.35}$$

where *m*, *n*, are the mode indices corresponding to the eigenmode. Using the relation

$$\gamma = \alpha + j\beta$$

the electric field in the waveguide, at any point *z* is

$$E_z(x,y,z) = E_0(x,y,z_0)e^{-\alpha z}e^{-j\beta z} \tag{10.36}$$

where E_0 are the fields calculated from Eq. (11.33) at a given point z_0 in the cross section of the waveguide, α is the attenuation constant, β the propagation constant. Normally the electric field is calculated at $z=0$ unless the attenuation effects are needed.

11.2.4. Formulation of the Three-Dimensional Helmholtz Equation

The extension of this formulation to *3-D* applications such as microwave cavities is relatively straight forward. The general 3-D formulation of the Helmholtz equation is not necessary for modeling in cavity resonators. For this purpose, (*TM* or *TE* modes) we only need one component of the electric field or magnetic field. Thus, the formulation essentially involves formulation with a *3-D* finite element, use of the *3-D* Laplacian, and *3-D* (or volume) integration.

We start with the *3-D* equivalent of Eqs. (11.23) or (11.24)

For *TM* fields:

$$\frac{\partial^2 E_z}{\partial x^2}+\frac{\partial^2 E_z}{\partial y^2}+\frac{\partial^2 E_z}{\partial z^2}+(\gamma^2+k^2)E_z=0 \qquad (11.37)$$

while for *TE* fields:

$$\frac{\partial^2 H_z}{\partial x^2}+\frac{\partial^2 H_z}{\partial y^2}+\frac{\partial^2 H_z}{\partial z^2}+(\gamma^2+k^2)H_z=0 \qquad (11.38)$$

These equations are discuses in chapter 2 in conjunction with calculation of fields in resonant cavities.

The energy functionals equivalent to Eqs. (11.37) and (11.38) are

$$\int_s \frac{1}{2}\frac{1}{\mu_r}\left[\left(\frac{\partial E_z}{\partial x}\right)^2+\left(\frac{\partial E_z}{\partial y}\right)^2+\left(\frac{\partial E_z}{\partial z}\right)^2\right]dv-\int_s \frac{1}{2}k_0^2\varepsilon_r E_z\,dv \qquad (11.39)$$

$$\int_s \frac{1}{2}\frac{1}{\varepsilon_r}\left[\left(\frac{\partial H_z}{\partial x}\right)^2+\left(\frac{\partial H_z}{\partial y}\right)^2+\left(\frac{\partial H_z}{\partial z}\right)^2\right]dv-\int_s \frac{1}{2}k_0^2\mu_r H_z\,dv \qquad (11.40)$$

Before discussing the formulation we define a *3-D* element, similar to the way the two dimensional element in section 9.5.5 was defined. A simple three-dimensional element is shown in Figure 11.1. The element has *8* nodes and therefore *8* shape functions. The shape functions are defined as

$$N_i=K(1+\xi\xi_i)(1+\eta\eta_i)(1+\zeta\zeta_i) \qquad (11.41)$$

where ξ_i, η_i, and ξ_i are either *+1* or *−1* and *K* is a constant. The eight shape functions for the element are shown in Table 11.1. for the coordinates in Figure 11.1.

These shape functions are next used to map the global coordinate system using the shape functions themselves for the mapping (isoparametric mapping). The variation of the field variable within the element can now be written in terms of the shape functions and the nodal unknowns as

$$E(\xi,\eta,\zeta) = \sum_{i=1}^{8} N_i(\xi,\eta,\zeta)E_i \tag{11.42}$$

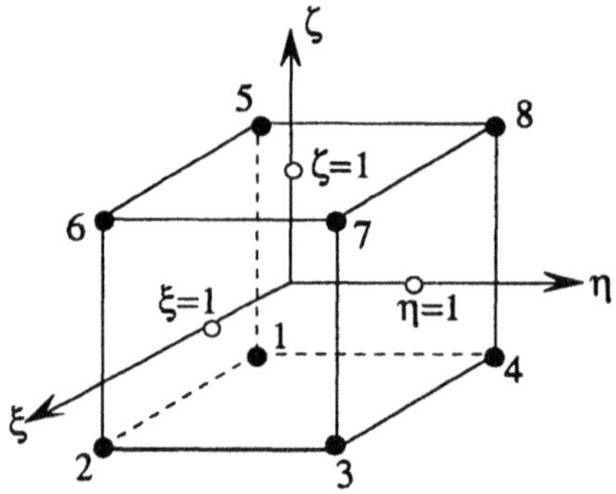

Figure 11.1. Definition of an *8* node three-dimensional finite element in local coordinates.

Table 11.1. Shape functions for the *8* node element in Figure 11.1.

$N_1 = (1/8)(1-\xi)(1-\eta)(1-\zeta)$	$N_5 = (1/8)(1-\xi)(1-\eta)(1+\zeta)$
$N_2 = (1/8)(1+\xi)(1-\eta)(1-\zeta)$	$N_6 = (1/8)(1+\xi)(1-\eta)(1+\zeta)$
$N_3 = (1/8)(1+\xi)(1+\eta)(1-\zeta)$	$N_7 = (1/8)(1+\xi)(1+\eta)(1+\zeta)$
$N_4 = (1/8)(1-\xi)(1+\eta)(1-\zeta)$	$N_8 = (1/8)(1-\xi)(1+\eta)(1+\zeta)$

To evaluate the elemental contribution of the energy functional in Eq. (11.39) (or 11.40) requires in addition the definition of the first derivatives of *E (or H)* with respect to *x* and *y*. These are

$$\frac{\partial E}{\partial x} = \sum_{i=1}^{8} \frac{\partial N_i}{\partial x} E_i \tag{11.43}$$

$$\frac{\partial E}{\partial y} = \sum_{i=1}^{8} \frac{\partial N_i}{\partial y} E_i \tag{11.44}$$

$$\frac{\partial E}{\partial z} = \sum_{i=1}^{8} \frac{\partial N_i}{\partial z} E_i \tag{11.45}$$

The shape functions are derived in local coordinates. Their derivatives are obtained in the local system by using the chain rule of differentiation

$$\frac{\partial N_i}{\partial \xi} = \frac{\partial N_i}{\partial x}\frac{\partial x}{\partial \xi} + \frac{\partial N_i}{\partial y}\frac{\partial y}{\partial \xi} + \frac{\partial N_i}{\partial z}\frac{\partial z}{\partial \xi} \tag{11.46}$$

$$\frac{\partial N_i}{\partial \eta} = \frac{\partial N_i}{\partial x}\frac{\partial x}{\partial \eta} + \frac{\partial N_i}{\partial y}\frac{\partial y}{\partial \eta} + \frac{\partial N_i}{\partial z}\frac{\partial z}{\partial \eta} \tag{11.47}$$

$$\frac{\partial N_i}{\partial \zeta} = \frac{\partial N_i}{\partial x}\frac{\partial x}{\partial \zeta} + \frac{\partial N_i}{\partial y}\frac{\partial y}{\partial \zeta} + \frac{\partial N_i}{\partial z}\frac{\partial z}{\partial \zeta} \tag{11.48}$$

Rewriting these equations in matrix form and inverting the system to obtain the derivatives of the shape functions in the global system the following is obtained

$$\begin{bmatrix} \dfrac{\partial N_i}{\partial x} \\ \dfrac{\partial N_i}{\partial y} \\ \dfrac{\partial N_i}{\partial z} \end{bmatrix} = [J]^{-1} \begin{bmatrix} \dfrac{\partial N_i}{\partial \xi} \\ \dfrac{\partial N_i}{\partial \eta} \\ \dfrac{\partial N_i}{\partial \zeta} \end{bmatrix} \tag{11.49}$$

where $[J]$ is the Jacobian matrix calculated for Eqs. (11.45) through (11.47). Similarly we can write

$$dv = dxdydz = [J]d\xi d\eta d\zeta \tag{11.50}$$

To evaluate the elemental contribution, equations such as (11.43) through (11.45) must be integrated over the element. These integrals are of the form

$$\int_{-1}^{+1}\int_{-1}^{+1}\int_{-1}^{+1} f'(\xi,\eta,\zeta)d\xi d\eta d\zeta \tag{11.51}$$

and, as in the 2-D case, the integration is performed numerically using Gaussian quadrature.

Returning now to the formulation, the approximations in Eqs. (11.43) through (11.50) are inserted in the energy functional. This is then minimized with respect to each unknown value of E, resulting in a typical contribution to the elemental matrix as

$$S_{i,j} = \int_v \left(\frac{\partial N_i}{\partial x}\frac{1}{\mu_r}\frac{\partial N_j}{\partial x} + \frac{\partial N_i}{\partial y}\frac{1}{\mu_r}\frac{\partial N_j}{\partial y} + \frac{\partial N_i}{\partial z}\frac{1}{\mu_r}\frac{\partial N_j}{\partial z} \right) dv \tag{11.52}$$

while the contribution to the second term remains the same. Therefore, the final system of equation to be solved is identical to that in Eq. (11.33) other than the obvious differences in number of nodes and volume integration rather than surface integration in the *2-D* case. This method can then be used to solve for fields in resonant cavities. Now, the eigenvalues obtained are the resonant frequencies of the cavity.

11.2.5. Formulation of the Modified Eddy Current Equation: 2-D Case

The generalized eddy current equation was introduced in Eq. (11.18). We will rewrite it as

$$\frac{1}{\mu}\left[\frac{\partial^2 \mathbf{A}}{\partial x^2} + \frac{\partial^2 \mathbf{A}}{\partial y^2}\right] + \mathbf{J}_s - \sigma\nabla V - j\omega\sigma\mathbf{A} + \omega^2\varepsilon\mathbf{A} = 0 \tag{11.53}$$

For the purpose of this section we will assume that the scalar potential is negligible. Under this condition we get

$$\frac{1}{\mu}\left[\frac{\partial^2 \mathbf{A}}{\partial x^2} + \frac{\partial^2 \mathbf{A}}{\partial y^2}\right] + \mathbf{J}_s - j\omega\sigma\mathbf{A} + \omega^2\varepsilon\mathbf{A} = 0 \tag{11.54}$$

While neglecting the scalar potential seems arbitrary, this in fact is often a very good approximation. However, we might add, that if the complete solution is required, then an additional equation must be added to account for the scalar potential. This is normally introduced through the definition of the divergence of **A** or the divergence of the current density. Also to be noted is the fact that in the two-dimensional case treated here, the magnetic vector potential is a single component vector for which the divergence is zero, and which can be treated as a scalar.

The first three terms of this equation are identical to the eddy current equation discused in section 9.5. Therefore, using the formulation in chapter 9 we can write the following three terms:

For the Laplacian term, the general contribution to the matrix is

$$S_{i,j} = -\frac{1}{\mu}\int_v \left(\frac{\partial N_i}{\partial x}\frac{\partial N_j}{\partial x} + \frac{\partial N_i}{\partial y}\frac{\partial N_j}{\partial y} \right) dv \tag{11.55}$$

The general term accounting for eddy currents (third term in Eq. (11.54) is

$$r_{ij} = \omega\sigma \int_v N_i N_j dv \qquad (11.56)$$

and the input current density is written as

$$q_i = -J_s \int_v N_i dv \qquad (11.57)$$

where the indices i,j vary from *1* to *4* for the quadrilateral two-dimensional element.

The fourth term is due to displacement currents and can be easily evaluated by observing that it is similar to the eddy current term. In fact, if we replace $\omega\sigma$ by $\omega^2\varepsilon$, we get an identical contribution. The displacement current contribution is

$$d_{ij} = \omega^2 \varepsilon \int_v N_i N_j dv \qquad (11.58)$$

As previously, each coefficient above is numerically integrated over the volume of the element using Gaussian quadrature and then summed into a complex elemental matrix of the form

$$\left[[s]_e + [d]_e + j[r]_e\right][a]_e = [q]_e \qquad (11.59)$$

where $[s]+[d]$ is the real part of the elemental matrix, $[r]$ is the imaginary part and $\{a\}$ and $\{q\}$ are the 4×1 unknown and source vectors respectively.

The elemental matrices of all the elements in the solution region are summed into a global system of the form

$$[G][A] = [Q] \qquad (11.60)$$

where there are a total of N equations in N unknowns. N is the total number of nodes in the solution region. This matrix is symmetric and banded, the bandwidth depending on the number of elements, the number of nodes per element and especially on the way the nodes are numbered.

11.2.6. Formulation of the Modified Eddy Current Equation: 3-D Case

The extension of this formulation to three-dimensional geometries is somewhat more complicated that that for the Helmholtz equation because the magnetic vector potential has a single component in the two dimensional case (that is, the magnetic vector potential is always in the direction of the current) while in the *3-D* case, three components must be considered. This complication is, however, easily overcome. With the three components of the magnetic vector potential we write in Cartesian coordinates

$$\left[\frac{1}{\mu_x}\frac{\partial^2 A_x}{\partial x^2}+\frac{1}{\mu_x}\frac{\partial^2 A_y}{\partial y^2}+\frac{1}{\mu_x}\frac{\partial^2 A_z}{\partial z^2}\right]+\left[J_{sx}+J_{sy}+J_{sz}\right]- \\ j\omega\sigma\left[A_x+A_y+A_z\right]+\omega^2\varepsilon\left[A_x+A_y+A_z\right]=0 \tag{11.61}$$

where Coulomb's gage $\nabla\cdot\mathbf{A}=0$ was used explicitly. Now, the process above can be repeated. The energy functional is written as

$$F(A)=\int_v \frac{|\nabla\times\mathbf{A}|^2}{2\mu}dv-\int_v j\omega\sigma\frac{\mathbf{J}\cdot\mathbf{A}}{2}dv+\int_v \frac{\mathbf{A}\cdot\mathbf{A}}{2}dv+\int_v \frac{\varepsilon}{2}\left|\frac{\partial \mathbf{A}}{\partial t}\right|^2 dv \tag{11.62}$$

and the elemental matrices have the following general contributions:

$$s_{3i-2,3j-2}=\int_{v_i}\left[\frac{\partial N_i}{\partial y}\frac{1}{\mu_z}\frac{\partial N_j}{\partial y}+\frac{\partial N_i}{\partial z}\frac{1}{\mu_y}\frac{\partial N_j}{\partial z}\right]dv \tag{11.63}$$

$$s_{3i-1,3j-1}=\int_{v_i}\left[\frac{\partial N_i}{\partial z}\frac{1}{\mu_x}\frac{\partial N_j}{\partial z}+\frac{\partial N_i}{\partial x}\frac{1}{\mu_z}\frac{\partial N_j}{\partial x}\right]dv \tag{11.64}$$

$$s_{3i,3j}=\int_{v_i}\left[\frac{\partial N_i}{\partial y}\frac{1}{\mu_x}\frac{\partial N_j}{\partial y}+\frac{\partial N_i}{\partial z}\frac{1}{\mu_y}\frac{\partial N_j}{\partial z}\right]dv \tag{11.65}$$

$$s_{3i-1,3j-2}=-\int_{v_i}\frac{\partial N_i}{\partial x}\frac{1}{\mu_z}\frac{\partial N_j}{\partial y}dv \tag{11.66}$$

$$s_{3i,3j-1}=-\int_{v_i}\frac{\partial N_i}{\partial y}\frac{1}{\mu_x}\frac{\partial N_j}{\partial z}dv \tag{11.67}$$

$$s_{3i,3j-2}=-\int_{v_i}\frac{\partial N_i}{\partial x}\frac{1}{\mu_y}\frac{\partial N_j}{\partial z}dv \tag{11.68}$$

$$r_{3i-2,3j-2} = r_{3i-1,3j-1} = r_{3i,3j} = \int_{v_i} N_i N_j \, dv \tag{11.69}$$

$$d_{3i-2,3j-2} = d_{3i-1,3j-1} = d_{3i,3j} = \omega^2 \varepsilon \int_{v_i} N_i N_j dv \tag{11.70}$$

$$q_{3i-2} = \int_{v_i} J_x N_i \, dv \tag{11.71}$$

$$q_{3i-1} = \int_{v_i} J_y N_i \, dv \tag{11.72}$$

$$q_{3i} = \int_{v_i} J_z N_i \, dv \tag{11.73}$$

Thus, the *3-D* solution is very similar to the *2-D* solution but with considerably more work. The global matrix now is $3N \times 3N$ in size but in all other respects the solution is the same. Note that the matrix elements described above apply to any finite element since the shape functions used are general.

11.3. The Weak Form Approach

We return now to the Helmholtz equation and show how the same problem may be formulated using the weak form of the equation. In this approach, we use the equation directly rather than looking for an equivalent functional. In this sense, it is more general and applies to any equation. While an energy functional can only be defined for a self adjoint operator. The steps involved in this formulation can be used for any problem, regardless of dimensionality. We will use a generalized weighted residual process and choose the Galerkin method (in which the weights in the weighted residual integral are equal to the derivatives of the approximation function with respect to the unknown). We can start with the Helmholtz equation:

$$\nabla^2 \mathbf{H} + \omega^2 \mu \varepsilon \mathbf{H} = 0 \tag{11.74}$$

or

$$\nabla^2 \mathbf{E} + \omega^2 \mu \varepsilon \mathbf{E} = 0 \tag{11.75}$$

Either of the equations may be used but we choose the electric field as an example. For generality however, we will start with the source free Maxwell equations:

$$\nabla\times\mathbf{E} = -j\omega\mu\mathbf{H} \tag{11.76}$$

$$\nabla\times\mathbf{H} = j\omega\varepsilon\mathbf{E} \tag{11.77}$$

From Eq. (11.76)

$$\mathbf{H} = -\frac{\nabla\times\mathbf{E}}{j\omega\mu}$$

and, substituting this in Eq. (11.77) we get

$$\nabla\times\left(\frac{1}{j\omega\mu}\nabla\times\mathbf{E}\right) = -j\omega\varepsilon\mathbf{E}$$

Rearranging terms

$$\nabla\times\left(\frac{1}{j\omega\mu}\nabla\times\mathbf{E}\right) + j\omega\varepsilon\mathbf{E} = 0$$

By multiplying both sides of the equation by $j\omega\mu_0$ we get

$$\nabla\times\left(\frac{1}{\mu_r}\nabla\times\mathbf{E}\right) - \omega^2\mu_0\varepsilon_0\varepsilon_r\mathbf{E} = 0$$

or, in using the notation $k_0=\omega^2\mu_0\varepsilon_0$ for the free space wavenumber

$$\nabla\times\left(\frac{1}{\mu_r}\nabla\times\mathbf{E}\right) - k_0^2\varepsilon_r\mathbf{E} = 0 \tag{11.78}$$

To obtain the Helmholtz equation in Eq. (11.75) all that is needed is to multiply both sides by μ_r and expand the term $\nabla\times\nabla\times\mathbf{E}$. However, we will not do so here. Rather, we continue with the equation in the form above. Also, we mention that an identical equation (other than the interchange between ε and μ) is obtained if we start with Eq. (11.77)

$$\nabla\times\left(\frac{1}{\varepsilon_r}\nabla\times\mathbf{H}\right) - k_0^2\mu_r\mathbf{H} = 0 \tag{11.79}$$

We note in passing that in Eq. (11.79) the permittivity can be anisotropic while in Eq. (11.78) permeability may be anisotropic. As a rule, we should choose Eq. (11.79) if we expect to analyze anisotropic dielectrics.

Using the relation $\nabla\times(\nabla\times\mathbf{A})=-\nabla^2\mathbf{A}+\nabla(\nabla\cdot\mathbf{A})$, and assuming permeability to be constant, we can write

$$-\left(\frac{1}{\mu_r}\nabla^2\mathbf{E}\right)+k_0^2\varepsilon_r\mathbf{E}=0 \tag{11.80}$$

This equation is now used for finite element implementation. To do so we use the *3-D* finite element defined in Figure 11.1. and the shape functions in Table 11.1. With these, the approximation for **E** is

$$\mathbf{E}=\sum_{k=1}^{8}\mathbf{E}_k N_k$$

A Galerkin process is defined by choosing a set of weighting functions w_m such that

$$\mathbf{E}=\sum_{k=1}^{8}\mathbf{E}_k w_k$$

In other words, the choice for the weighting functions is

$$w_k = N_k$$

where N_k are the shape functions of the element.

Now we can write a weighted residual for Eq. (11.80) by multiplying it by the weights, integrating over the volume of an arbitrary element and setting it to zero:

$$-\int_{V_i}\mathbf{w}.\left(\frac{1}{\mu_r}\nabla^2\mathbf{E}\right)dv+\int_{V_i}k_0^2\varepsilon_r\mathbf{w}.\mathbf{E}=0 \tag{11.81}$$

where the weighting functions were written as vectors. The first integral is equal to

$$-\int_{V_i}\mathbf{w}.\left(\frac{1}{\mu_r}\nabla^2\mathbf{E}\right)dv=-\int_{V_i}\frac{1}{\mu_r}(\nabla\times\mathbf{E}).(\nabla\times\mathbf{w})dv+\int_{V_i}\frac{1}{\mu_r}(\nabla.\mathbf{E})(\nabla.\mathbf{w})dv+$$

$$-\int_{S_i}\mathbf{w}.\left(\frac{1}{\mu_r}\hat{\mathbf{n}}\times(\nabla\times\mathbf{E})\right)ds+\int_{S_i}(\mathbf{w}.\mathbf{n})\frac{1}{\mu_r}(\nabla.\mathbf{E})ds$$

Since the equation is source free, $\nabla \cdot \mathbf{E}=0$ and the second and fourth integrals on the right hand side vanish. Going back to Eq. (11.81) we can write

$$-\int_{V_i} \frac{1}{\mu_r}(\nabla\times\mathbf{E}).(\nabla\times\mathbf{w})dv + \int_{V_i} k_0^2\varepsilon_r \mathbf{w}.\mathbf{E} - \int_{S_i} \mathbf{w}.\left(\frac{1}{\mu_r}\hat{\mathbf{n}}\times(\nabla\times\mathbf{E})\right)ds = 0$$

where the surface is, in general, the surface of the element and $\hat{\mathbf{n}}$ is the unit normal to the surface. We also note from Maxwell's equation in Eq. (11.76) that

$$\hat{\mathbf{n}}\times(\nabla\times\mathbf{E}) = -j\omega\mu(\hat{\mathbf{n}}\times\mathbf{H})$$

Using this, and rearranging terms we get

$$\int_{V_i} \frac{1}{\mu_r}(\nabla\times\mathbf{E}).(\nabla\times\mathbf{w})dv - \int_{V_i} k_0^2\varepsilon_r \mathbf{w}.\mathbf{E} - j\omega\int_{S_i} \mathbf{w}.\left(\hat{\mathbf{n}}\times\mathbf{H}\right)ds = 0 \qquad (11.82)$$

The last term includes an electric surface current density and is a natural boundary condition for the problem. Thus, it is natural to use this formulation with problems for which the boundary are conducting since the current density may be prescribed or, as is often the case, may be assumed to be zero.

Substituting the approximation for $\mathbf{E}$ and $\mathbf{w}$, the first two integrals result in the following terms for the elemental matrices

$$s_{3i-2,3j-2} = \int_{v_i}\left[\frac{\partial N_i}{\partial y}\frac{1}{\mu_z}\frac{\partial N_j}{\partial y} + \frac{\partial N_i}{\partial z}\frac{1}{\mu_y}\frac{\partial N_j}{\partial z}\right]dv \qquad (11.83)$$

$$s_{3i-1,3j-1} = \int_{v_i}\left[\frac{\partial N_i}{\partial z}\frac{1}{\mu_x}\frac{\partial N_j}{\partial z} + \frac{\partial N_i}{\partial x}\frac{1}{\mu_z}\frac{\partial N_j}{\partial x}\right]dv \qquad (11.84)$$

$$s_{3i,3j} = \int_{v_i}\left[\frac{\partial N_i}{\partial y}\frac{1}{\mu_x}\frac{\partial N_j}{\partial y} + \frac{\partial N_i}{\partial z}\frac{1}{\mu_y}\frac{\partial N_j}{\partial z}\right]dv \qquad (11.85)$$

$$s_{3i-1,3j-2} = -\int_{v_i}\frac{\partial N_i}{\partial x}\frac{1}{\mu_z}\frac{\partial N_j}{\partial y}\,dv \qquad (11.86)$$

$$s_{3i,3j-1} = -\int_{v_i} \frac{\partial N_i}{\partial y}\frac{1}{\mu_x}\frac{\partial N_j}{\partial z}\,dv \tag{11.87}$$

$$s_{3i,3j-2} = -\int_{v_i} \frac{\partial N_i}{\partial x}\frac{1}{\mu_y}\frac{\partial N_j}{\partial z}\,dv \tag{11.88}$$

$$r_{3i-2,3j-2} = r_{3i-1,3j-1} = r_{3i,3j} = -k_0^2\varepsilon_r\int_{v_i} N_iN_j\,dv \tag{11.89}$$

These terms are similar to those obtained for the modified eddy current equation. In fact, Eqs. (11.83) through (11.88) are identical to Eqs. (11.63) through (11.68).

As for the current term, we will simply assume here that this term vanishes on the boundary of the geometry (as for example, on the boundary of a microwave cavity). Under this condition the problem reduces to a real eigenvalue problem

$$[S]\{E\} = k_0^2[R]\{E\} \tag{11.90}$$

Although it may seem that this equation does not allow solution for lossy dielectrics, this is not the case. It can be easily modified to take into account losses by using the complex permittivity for ε_r. We will not pursue this any further but the bibliography section of this chapter points to these and other extensions.

11.4. Examples

11.4.1. Resonant Frequency of a Cubic Cavity

To calculate the resonant frequencies of a cavity, we use the modified eddy current equation in section 11.2.4. This formulation requires that the frequency and the source current density be given

At first, this may seem as an impossible task because the resonant frequency is the object of the calculation and also because, in a resonant cavity we normally do not have sources. However, both of these problems are easily overcome. The first problem is solved by assuming the frequency is known and calculating the fields in the cavity. From the fields, the total electric and magnetic energy in the cavity is calculated. A peak in stored energy in the cavity indicates the resonant frequency of the cavity. By repeatedly solving Eq. (11.61) we obtain a scan over a range of frequencies, not unlike adjusting a wavemeter. This can be rather expensive in terms of computational resources if the scan is over a wide range of frequencies. By judicious choice of frequencies this problem can be minimized. For example, an approximate solution may be obtained analytically and the exact resonant frequency found numerically by scanning in a narrow range. The second difficulty, that of a prescribed current density, is overcome by introducing an artificial current source in the cavity. While the current density is artificially introduced, it only affects the modes excited in the cavity, not their frequencies. This is similar to the methods of excitation of cavities discused in chapter 2. Thus, the only important aspect of this modification is to ensure the modes excited are the modes required. Beyond that, the magnitude of the current, location of current, etc., are not important. However, the fields as obtained from this formulation are not physical (in the sense that the excitation current is arbitrary). This again, is not important for the purpose of this calculation since only detection of resonant frequencies is important.

It may seem that this is a very restrictive calculation but this is not the case. From the testing point of view, this is exactly how we need to model the test; the actual fields in the cavity are almost never measured while the shift in resonant frequency, which is very often the only measurement we need, gives no information on the fields in the cavity. Also, and perhaps most important, it is possible to model the correct fields by properly modeling the source. If, for example, the cavity is excited by a current loop (as in Figure 2.21), the loop can be modeled exactly, including currents in the loop and the fields in the cavity obtained based on this model of the current loop.

As an example of the type of calculation this formulation allows, consider the geometry in Figure 11.2a. A cubic cavity, $1m \times 1m \times 1m$ is modeled. The walls are assumed to be perfectly conducting and the interior contains free space. Figure 11.2b shows a simple finite element mesh where only one eighth of the cavity is modeled (because of the three way symmetry in this geometry). Equal sized finite elements are used with a total of *125* elements and *216* nodes.

The element marked with J contains a current density ($J=(1+j1)A/m^2$) in the z direction. Note that even though this current only exists in one element, it is sufficient to generate fields and modes in the cavity. The magnetic vector potential in this case has three components and therefore results in a matrix with *648* unknown potential values. Boundary conditions are rather simple in this case and are marked in Figure 11.2.b. The Neumann boundary conditions on symmetry planes are left unspecified while the external boundary conditions are set to zero on the cavity walls.

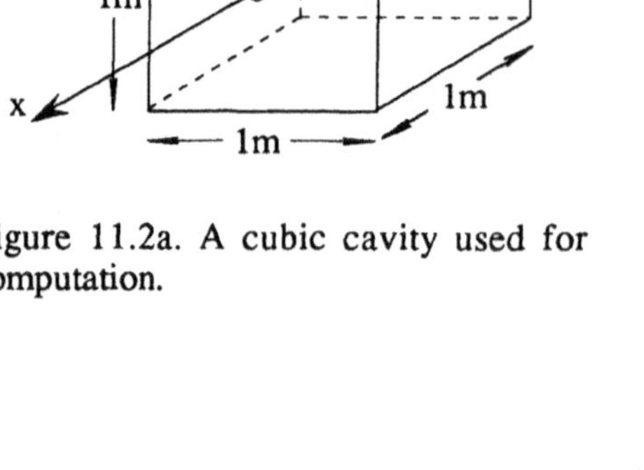

Figure 11.2a. A cubic cavity used for computation.

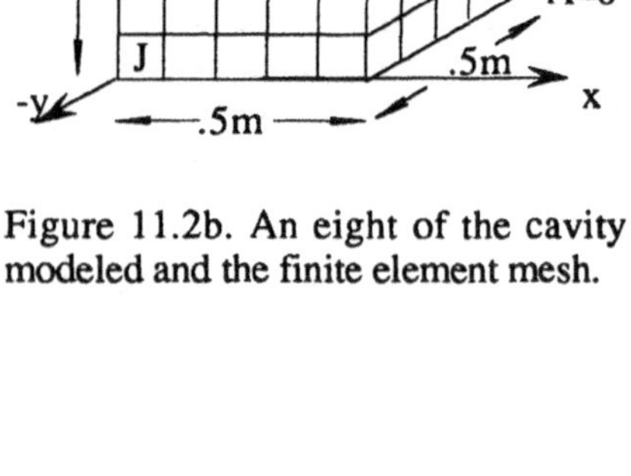

Figure 11.2b. An eight of the cavity as modeled and the finite element mesh.

Solution of Eq. (11.61) proceeds by assuming an initial frequency and calculating the magnetic vector potential based on the nodal contributions given in Eqs. (11.63) through (11.73). Then, the frequency is incremented, and the problem solved again with the new frequency.

From the magnetic vector potential, the energy in the cavity is calculated as follows: The magnetic flux density is calculated from the definition

$$\mathbf{B} = \nabla \times \mathbf{A}$$

The stored magnetic energy in the cavity is calculated as

$$W_H = \int_v \frac{B^2}{2\mu} dv$$

The electric field is calculated as

$$\mathbf{E} = -\frac{\partial \mathbf{A}}{\partial t}$$

and the stored electric energy as

$$W_E = \int_v \frac{\varepsilon E^2}{2} dv$$

where the volume integration is performed over the entire volume of the cavity. The total energy in the cavity is the sum of W_H and W_E.

The results of these calculations are shown in Figure 11.3. The frequency is scanned between *210* and *250 MHz* in increments of *1 MHz*. The resonant frequency is clearly seen at *223 MHz*. To see how this might be used for modeling in *NDT*, consider the result in Figure 11.4. Here the cavity is filled with a dielectric of relative permittivity of *1.1* (e.g. a gas). A frequency scan between *194 MHz* and *223 MHz* shows the resonant frequency to be at *213 MHz*. Thus, the presence of the gas has shifted the resonant frequency by *10 MHz*.

The accuracy of this type of calculation can be demonstrated by calculating the exact resonant frequency from the formula for resonant frequency of a cavity (Eq. (2.122))

$$f_{mnp} = \frac{1}{2\pi\sqrt{\mu\varepsilon}}\sqrt{\left(\frac{m\pi}{a}\right)^2 + \left(\frac{n\pi}{b}\right)^2 + \left(\frac{p\pi}{d}\right)^2}$$

where m,n,p, are the mode indices and in this case, $b=d=a$. For a TM_{110} mode in an empty cavity, this frequency is *213 MHz*. With a full cavity ($\varepsilon_r=1.1$), the resonant frequency is *203 MHz*. Thus, the exact calculation shows resonant frequencies that are about *5%* lower than the finite element calculation but the shift in resonant frequency is almost the same (in this case it is actually exact but in other, more complex cavity loading there is an error in the shift in resonant frequency as well). These errors can be reduced by simply increasing the number of elements in the solution domain.

This comparison also shows another difficulty that is inherent in almost any numerical calculation: while the resonant frequency can be calculated, identification of the modes is not simple. This however can often be found from other information. In this case, only TM_{mn0} modes can be excited because of the direction of current we assumed. In principle, we may scan the frequency range from the lowest expected cutoff frequency in which case the modes will appear in their ascending order.

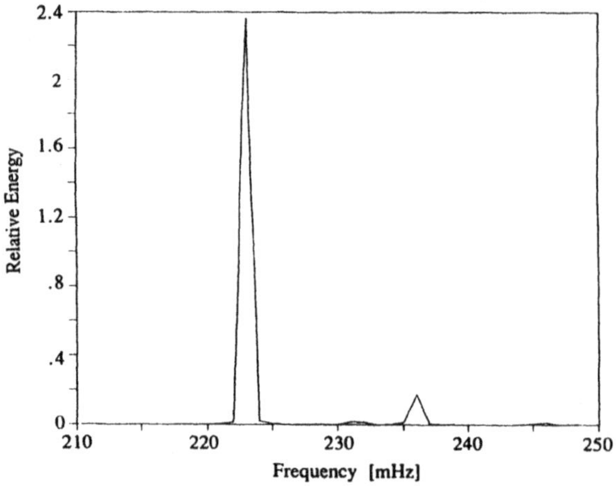

Figure 11.3. A frequency scan for the empty cavity showing resonance at *223 MHz*.

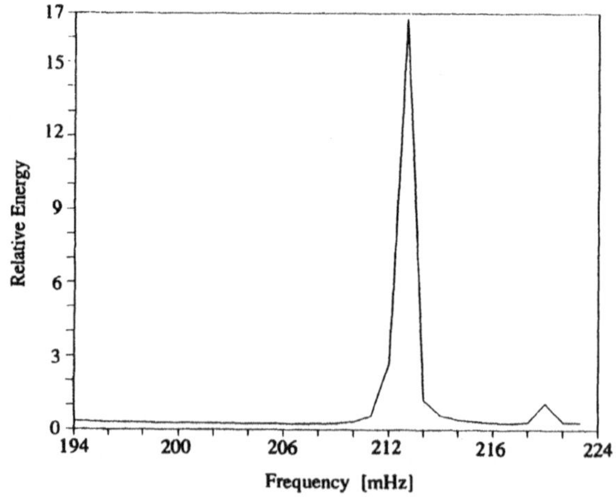

Figure 11.4. A frequency scan of the loaded cavity showing resonance at *213 MHz*.

11.4.2. Modes in Rectangular Waveguides

Modes in rectangular waveguides are calculated using the Helmholtz equation in two dimensions (Eqs. (11.23) or (11.24)). The procedure for solution is similar to the previous example except that now the solution is in the cross section of the waveguide and therefore the two-dimensional formulation is used. A simple waveguide with a schematic mesh and appropriate boundary conditions for *TM* modes is shown in Figure 11.5. The mesh used for calculations is much denser than the one shown in this figure. In this case, *400* elements and *441* nodes were used for solution.

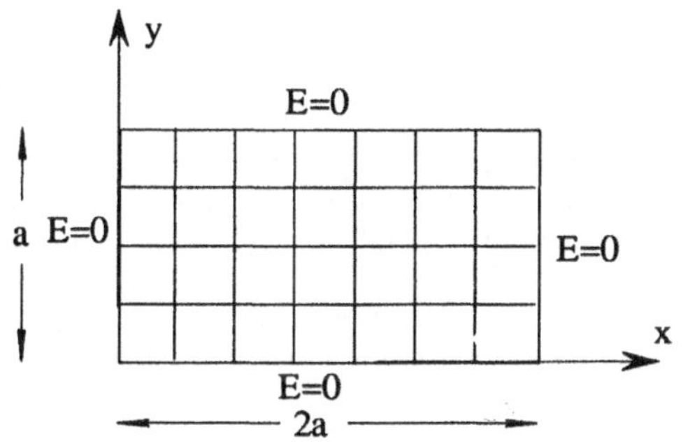

Figure 11.5. Cross-section of a rectangular waveguide with a finite element mesh and boundary conditions for *TM* modes.

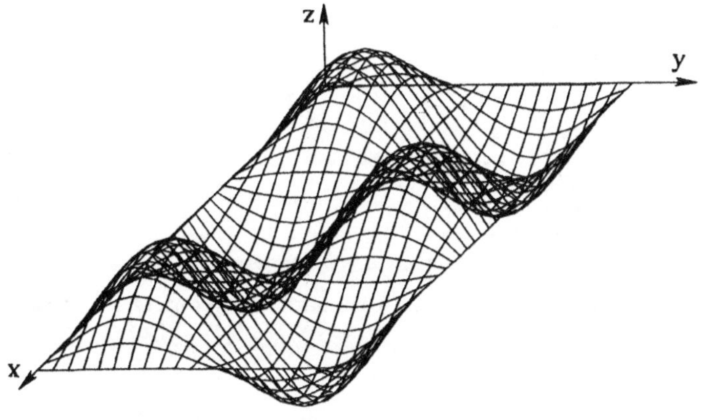

Figure 11.6. Electric field distribution for the TM_{32} mode in the cross section of the waveguide in Figure 11.5.

Solution proceeds with assembly of the matrix to obtain the system in Eq. (11.33). The eigenvalues of this system are calculated using an eigensystem solver to obtain the modes of the waveguide. In this case, the eigenvalues of the system provide k_0 which is the free space cut-off frequencies of the waveguide. For each mode, the electric field eigenvector is obtained as part of the solution. A plot of this eigenvector over the cross section of the waveguide provides the field distribution in the waveguide.

As an example, the field distribution for a TM_{32} mode is shown in Figure 11.6. As with any modal solution, the fields obtained are arbitrary in amplitude. The only way of finding the actual fields is if the value of the field is known at a point (except for boundary points where the electric field is zero) and all points in the solution domain are scaled accordingly. Thus the solution in Figure 11.6. should be viewed as a normalized solution.

11.4.3. Modes in a Loaded Cavity Resonator

A cavity resonator is a *3-D* structure therefore, we must solve the Helmholtz equation in three dimensions. The weak form approach is demonstrated in this example. The solution obtained is for Eq. (11.82) as used to calculate the modes of a loaded cavity. The geometry of the cavity modeled is shown in Figure 11.7. The outer surface of the cavity is a perfect conductor ($E=0$) while the inner surface of the cavity is coated with a good dielectric with relative permittivity of *9*. The inner cavity has a different permittivity depending on the test material that may be inserted.

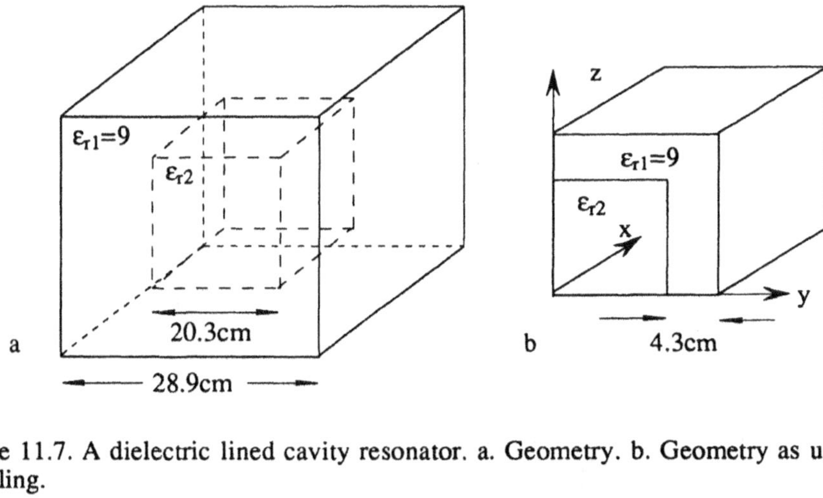

Figure 11.7. A dielectric lined cavity resonator. a. Geometry. b. Geometry as used for modeling.

Finite element discretization is identical to that of Figure (11.2b). One eight of the cavity is modeled for symmetry reasons and the same boundary conditions are used. The resonant frequency for the cavity with liner ($\varepsilon_{r1}=9$) and an empty inner cavity is *465 MHz* (TM_{111} mode). This frequency changes depending on loading. The change in resonant frequency due to changes in the dielectric constant in the inner cavity (ε_{r2}) is shown in Figure 11.8. for the first two modes. As expected, the frequency goes down as the dielectric constant becomes larger.

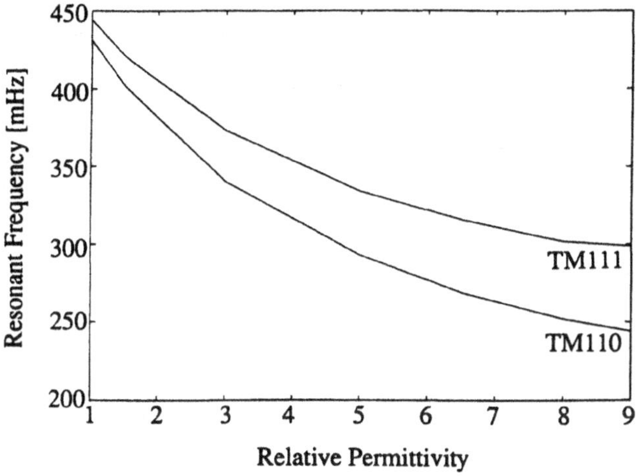

Figure 11.8. Resonant frequency versus dielectric constant in the inner cavity for the TM_{111} and TM_{110} modes.

11.5. Bibliography

Finite element computation of waves and modes in waveguides and cavities started in the mid-1960's and, in fact, preceded many other areas of computation in electromagnetics. The first applications to waveguides and cavities relied primarily on solution of the scalar wave equation in waveguides [1-5]. A large number of applications were treated [6-9] including empty cavities, loaded cavities, and anisotropically loaded waveguides and cavities. The modified eddy current method can be found in references [10-12] while some sources for the weak form formulation are given in [13]. One aspect of calculation of modes using finite elements is the appearance of spurious modes. Although it was not treated in this work, it is a very interesting and important aspect of modeling, especially in cavities. This subject is discused in many articles, some of which are [14-17]. In conjunction with this aspect of calculation, the use of vector finite elements has been found to eliminate the spurious modes. A few representative references are given [18-20]. Additional aspects of modeling in waveguides, cavity resonators and in other microwave applications are given in [21-24].

[1] O.C. Zienkiewicz and Y.K. Cheung, "Finite elements in the solution of field problems", The Engineer, Sep. 24 1965.

[2] P.L. Arlett, A.K. Bahrani and O.C. Zienkiewicz, "Application of finite elements to the solution of Helmholz's equation", Proceedings of the IEE, Vol. 115, No. 12, Dec. 1968, pp. 1762-1766.

[3] Z.J.Csendes and P. Silvester, "Numerical solution of dielectric loaded waveguides: I- Finite Element analysis", IEEE Transactions on Microwave Theory and Techniques, Vol. MTT-18, No. 12, Dec. 1970, pp. 1124-1131.
[4] O.C. Zienkiewicz, A. K. Bahrani and P. L. Arlett, "Solution of three-dimensional field problems by the finite element method", The Engineer, Oct. 27, 1967, pp. 547-550.
[5] A. Wexler, "Computation of electro magnetic fields",IEEE trans on microwave theory and techniques,Vol MTT 17,No.8,Aug 1969.
[6] W.J. English, "Vector variational solutions of inhomogeneously loaded cylindrical waveguide structures", IEEE Transactions on Microwave Theory and Techniques, Vol. MTT-19, No. 1, Jan 1971, pp. 9-18.
[7] Z.J. Csendes and P. Silvester, "Dielectric loaded waveguide analysis program, IEEE Transactions on Microwave Theory and Techniques, Vol. MTT-19, No. 6, June 1971, p789.
[8] M. Otaka, M. Matsuhara and N. Kumagai, "Analysis of guided modes in slab-coupled waveguides using a variational method", IEEE Journal of Quantum Electronics, Vol. WE-12, July 1976, pp. 378-383.
[9] J.B. Davis, F.A. Fernandez and G.Y. Philippou, "Finite element analysis of all modes in cavities with circular symmetry", IEEE Transactions on Microwave Theory and Techniques, Vol. MTT-30, No. 11, Nov. 1982, pp. 1975-1980.
[10] N. Ida and J. Bastos, "Electromagnetics and Calculation of Fields", Springer Verlag, N. Y., 1992.
[11] N. Ida, "Computation of resonant frequencies for nonuniformly loaded cavities using a modified 3D eddy current formulation", in Proceedings of the European TEAM Workshop and International Seminar on Electromagnetic Field Analysis, Y. Krutzen, N.J. Diserens, C.R.I. Emson and D. Rodger, (eds.), 1990, pp. 267-274.
[12] Brauer, J.R., R.H. Valder Heiden and A.B. Bruno, "Finite element modeling of electromagnetic resonators and absorbers", Journal of Applied Physics, Vol. 63, No. 8, April 15, 1988, pp. 3197-3199.
[13] H. Song and N. Ida, "An eddy current constraint formulation for 3D electromagnetic field calculation", in IEEE Transactions on Magnetics, Vol. 27, No. 5, Sept. 1991, pp. 4012-4015.
[14] J.S. Wang and N. Ida, "Eigenvalue analysis in Electromagnetic cavities using divergence free finite elements", in IEEE Transactions on Magnetics, Vol. 27, No. 5, Sept. 1991, pp. 3978-3981.
[15] Bossavit, A., "Solving Maxwell equations in a closed cavity, and the question of spurious modes", IEEE Trans. on Magnetics, 1990.
[16] I.O. Bardi, O. Biro and K. Preis, "Finite element scheme for 3-D cavities without spurious modes", presented at CEFC Toronto.
[17] A. Bossavit, "Simplicial finite elements for scattering problems in electromagnetism", Computer methods in applied mechanics and engineering, Vol. 76, 1989, pp. 299-316.
[18] J.S. Wang and N. Ida, "An Edge Element Solution of Electromagnetic Transmission Through Apertures in Inhomogeneously Loaded Cavities", submitted for presentation at the 1991 AP-S Symposium (will be reviewed for publication in IEEE Transactions on Microwave Theory and Techniques).

[19] J.S. Wang and N. Ida, "Utilization of Geometrical Symmetries in Edge Based Boundary Integral Eddy Current Solutions", submitted for publication to COMPUMAG Conference accepted for publication in the IEEE Transactions on Magnetics.
[20] J.S. Wang and N. Ida, "Eigenvalue Analysis and Anisotropically Loaded EM Cavities using Divergence Free Finite Elements", submitted for publication to COMPUMAG Conference accepted for publication in the IEEE Transactions on Magnetics.
[21] J.S. Wang and N. Ida, "Three-Dimensional FEM-BEM Computation of Electromagnetic Responses of Flaws", presented at the Review of Progress in Nondestructive Evaluation, Bowdoin, Maine, July 29-August 2, 1991.
[22] J.S. Wang and N. Ida, "Simulation of Nondestructive Testing of Materials in Microwave Cavities", presented at the Review of Progress in Nondestructive Evaluation, Bowdoin, Maine, July 29-August 2, 1991.
[23] N. Ida, "Electromagnetic field modeling for nondestructive testing of composite materials", invited paper for ISEM, Sendai, Japan, January 28-30, 1991, in The International Journal of Applied Electromagnetics in Materials, Vol. 2, No.3, Dec. 1991, pp. 253-264.
[24] N. Ida, "Microwave NDT", in Electrosoft, special issue on NDT Vol. 2, No. 2/3, June/Sept. 1991, pp. 215-237.

Chapter 12. Miscellaneous Topics

This short chapter gives some material that is not appropriate for the previous chapters. The tables of material properties are included because these are usually difficult to obtain without extensive search.

12.1. Tables of Material Properties

Table 12.1. Relative permeabilities for some diamagnetic, paramagnetic, and ferromagnetic materials.

Material	*Relative Permeability*
Silver	*0.99998*
Water	*0.999991*
Bismuth	*0.99983*
Copper	*0.999991*
Mercury	*0.999968*
Lead	*0.999983*
Vacuum	*1.0*
Air	*1.00000036*
Aluminum	*1.000021*
Paladium	*1.0008*
Cobalt	*250*
Nickel	*600*
Iron	*6,000*
Supermalloy	*1,000,000*
Steel (0.9%C)	*100*
Silicon Iron (4% Si)	*7,000*
Permalloy (78.5% Ni)	*100,000*
Fe3O4 (Magnetite)	*100*
Ferrites	*5,000*

Table 12.2. Conductivities of selected materials

Material	*Conductivity [S/m]*
Silver	6.1×10^{7}
Copper	5.7×10^{7}
Gold	4.1×10^{7}
Aluminum	3.5×10^{7}
Tungsten	1.8×10^{7}
Brass	1.1×10^{7}
Iron	1.0×10^{7}
Nichrome	1.0×10^{6}
Mercury	1.0×10^{6}
Graphite	1.0×10^{6}
Carbon	3.0×10^{5}
Germanium	2.3
Seawater	4.0
Distiled water	1.0×10^{-4}
Ferrite	1.0×10^{-3}
Water	1.0×10^{-3}
Silicon	3.9×10^{-4}
Bakelite	1.0×10^{-9}
Glass	1.0×10^{-12}
Rubber	1.0×10^{-13}
Mica	1.0×10^{-15}
Quartz	1.0×10^{-17}

Table 12.3. Dielectric constant and attenuation for selected materials.

Material	*RelativeDielectric Constant*	*Attenuation [dB/cm]*
Plexiglas	2.6 (10 GHz)	0.16 (12.4-18 GHz)
Teflon	2.1 (10 GHz)	0.003 (10 GHz)
Polystyrene	2.5 (10 GHz)	0.004 (10 GHz)
Glass epoxy	4.8	0.22 (4 GHz)
Silicon nitrade	7.5 (8-17.5 GHz)	0.2 (13.5 GHz)
Alumina	9-10	
Barium Tutanate	1,200	
Beryllia	6.6	
Epoxy	3.5	

Table 12.3. (continued)

Material	*RelativeDielectric Constant*	*Attenuation [dB/cm]*
Ferrites	*13-16*	
Quartz	*3.8-5*	
GaAs	*13*	
Nylon	*3.1*	
Parafin	*3.2*	
Perspex	*2.6*	
Polystyrene foam	*1.05*	
Polythene	*2.25*	
Teflon	*2.0*	
Vacuum	*1.0*	
Air	*1.0006*	
Rubber	*3.0*	
Paper	*3.0*	
Bakelite	*5.0*	
Glass	*6.0*	
Mica	*6.0*	
Water	*81.0*	
Barium Str. Titanate	*10,000.0*	

Table 12.4. Electromagnetic Units

Symbol	*Quantity*	*Unit*	*Common, Abbr.*
A	*Vector potential*	*Weber/meter*	*wb/m*
B	*Magnetic Flux Density*	*Tesla or Weber/m2*	*T or Wb/m2*
D	*Electric Flux Density*	*Coulomb/meter2*	*C/m2*
E	*Electric Field Intensity*	*Volt/meter*	*V/m*
F	*Force*	*Newton*	*N*
f	*Frequency*	*cycles/second or Hertz*	*c/s or Hz*
H	*Magnetic Field Intensity*	*Ampere/meter*	*A/m*
I	*Current*	*Ampere*	*A*
J	*Current Density*	*Ampere/meter2*	*A/m2*
L	*Inductance*	*Henry*	*H*
M	*Magnetic Moment Density*	*Ampere/meter*	*A/m*
m	*Magnetic Moment*	*Ampere.meter2*	*A.m2*
V	*Voltage*	*Volt*	*V*

Table 12.4. (Continued)

Symbol	*Quantity*	*Unit*	*Common, Abbr.*
v	*Velocity*	*meter/second*	*m/sec*
W	*Work*	*Joule*	*J*
δ	*Skin Depth*	*meter*	*m*
ε	*Permitivity*	*Farad/meter*	*F/m*
μ	*Permeability*	*Henry/meter*	*H/m*
σ	*Conductivity*	*Siemens/meter*	*S/m*
ω	*Angular Frequency*	*Radian/second*	*Rad/s*

Table 12.5. Physical Constants

Constant	*Value*
Permitivity of free space	8.8542×10^{-12} *[F/m]*
Permeability of free space	4×10^{-07} *[H/m]*
Charge of an electron	-1.602x10-19 *[C]*
Speed of light (in free space)	2.9979×10^{08} *[m/Sec]*
Intrinsic impedance of free space	$376.7\ \Omega$

12.2. Hyperbolic and Exponential Functions

Exponential functions are often used to describe hyperbolic functions since these are often easier to evaluate. Both hyperbolic and exponential functions are used extenssively in this work, especially for evaluation of impedances in transmission lines.The following gives some of the more common hyperbolic functions as used in this work.

$$sinh\alpha = \frac{e^{\alpha} - e^{-\alpha}}{2}$$

$$cosh\alpha = \frac{e^{\alpha} + e^{-\alpha}}{2}$$

$$sinh\alpha = -j\ sin\ (j\alpha)$$

$$cosh\alpha = cos\ (j\alpha)$$

$$tanh\alpha = -j\ tan\ (j\alpha)$$

$$e^{x} = coshx + sinhx$$

$$e^{-x} = coshx - sinhx$$

$$e^{j\beta xcos\theta} - e^{-j\beta xcos\theta} = j2sin(\beta xcos\theta)$$

$$e^{j\beta xcos\theta} + e^{-j\beta xcos\theta} = 2cos(\beta xcos\theta)$$

$$sin\beta x = \frac{e^{-j\beta x} - e^{j\beta x}}{2j}$$

$$cos\beta x = \frac{e^{j\beta x} + e^{-j\beta x}}{2}$$

12.3. Euler's Equation

In chapters 9 and 11 we discussed the energy functional for the quasistatic and the time-harmonic wave equation. We pointed out that these functionals are an equivalent formulation of the original equations and as such provide a correct solution to the equation. One question however was not addressed: how do we "find" a functional that is appropriate for the equation?.

The answer is not as simple as one might suspect. Although a functional is no more than a statement of energy ballance, this is somewhat misleading because "energy" terms used are not always apparent. To see this, consider the equation we used in chapter 9:

$$\nabla^2 \mathbf{A} + \mu \mathbf{J}_s - j\omega\mu\sigma\mathbf{A} = 0$$

with the "energy" functional

$$F(A) = \int_v \left[\frac{1}{2\mu} \left(\left| \frac{\partial A}{\partial x} \right|^2 + \left| \frac{\partial A}{\partial y} \right|^2 \right) + \frac{j\omega\sigma}{2} A^2 - J_s A \right] dv$$

That the first term represents energy, is obvious since the spatial derivatives of **A** are part of the curl relation and therefore the square represents magnetic field squared. In this case, the first term is

$$\int_v \frac{1}{2\mu} \left(\left| \frac{\partial A}{\partial x} \right|^2 + \left| \frac{\partial A}{\partial y} \right|^2 \right) dv = \int_v \frac{1}{2\mu} \left(|B_y|^2 + |B_x|^2 \right) dv$$

and this looks as the potential energy of the system. However, since the magnetic vector potential is, in general a complex number, there is also an imaginary term in this relation and, since we cannot properly discuss "imaginary energy", we must call this term an "energy-related" term rather than "energy". This name indicates that the relations above are related to energy but are not energy in the strict definition of the term. Similarly, the other terms are treated as energy-related terms.

To see how the Euler equation is derived, we consider the eddy current equation for axisymmetric geometries:

$$\frac{1}{\mu}\left(\frac{\partial^2 A}{\partial r^2}+\frac{1}{r}\frac{\partial A}{\partial r}+\frac{\partial^2 A}{\partial z^2}-\frac{A}{r^2}\right)=-J_s+j\omega\sigma A \qquad (12.1)$$

The Euler equation to be satisfied for Eq. (12.1) is

$$\frac{\partial}{\partial r}\left(\frac{\partial F(A)}{\partial\left(\frac{\partial A}{\partial r}\right)}\right)+\frac{\partial}{\partial z}\left(\frac{\partial F(A)}{\partial\left(\frac{\partial A}{\partial z}\right)}\right)-\frac{\partial F(A)}{\partial A}=0 \qquad (12.2)$$

Where F(A) is the energy related functional in terms of A. For **A**$=(0,0,A)$ (that is, **A** has only a *z*-directed component), the energy functional is

$$F(A)=\left[\frac{1}{2\mu}\left[\left(\frac{\partial A}{\partial z}\right)^2+\left(\frac{\partial A}{\partial r}+\frac{A}{r}\right)^2\right]-J_sA+\frac{j\omega\sigma}{2}A^2\right]r$$

Then, for the first term in brackets in Eq. (12.2) we get

$$\frac{\partial F(A)}{\partial\left(\frac{\partial A}{\partial r}\right)}=\frac{1}{\mu}\left(\frac{\partial A}{\partial r}+\frac{A}{r}\right)r$$

and the first term in Eq. (12.2) is

$$\frac{\partial}{\partial r}\left(\frac{\partial F(A)}{\partial\left(\frac{\partial A}{\partial r}\right)}\right)=\frac{1}{\mu}\left(\frac{\partial^2 A}{\partial r^2}+\frac{2}{r}\frac{\partial A}{\partial r}\right)r$$

Similarly, for the second term:

$$\frac{\partial F(A)}{\partial\left(\frac{\partial A}{\partial z}\right)} = \frac{1}{\mu}\frac{\partial A}{\partial z} r$$

$$\frac{\partial}{\partial z}\left(\frac{\partial F(A)}{\partial\left(\frac{\partial A}{\partial z}\right)}\right) = \frac{1}{\mu}\frac{\partial^2 A}{\partial z^2} r$$

The third term results in

$$\frac{\partial F(A)}{\partial A} = \left[\frac{1}{\mu}\left(\frac{\partial A}{\partial r} + \frac{A}{r}\right)\frac{1}{r} - J_s + j\omega\sigma A\right] r$$

Collecting terms we have

$$\left[\frac{1}{\mu}\frac{\partial^2 A}{\partial r^2} + \frac{1}{\mu}\frac{\partial^2 A}{\partial z^2} - \frac{1}{\mu}\frac{A}{r^2} + \frac{1}{\mu}\left(\frac{1}{r}\frac{\partial A}{\partial r}\right) + J_s - j\omega\sigma A\right] r = 0$$

and, since the equation inside the square bracket is the original partial differential equation, the functional above is the correct functional for this equation.

Appendix A
Vector Relations

A.1. The Gradient, Divergence, and Curl

A.1.1. The ∇ Operator

The nabla or del (∇) operator is a vector, which, in Cartesian coordinates, can be written in terms of its components as

$$\nabla = \hat{\mathbf{x}}\frac{\partial}{\partial x} + \hat{\mathbf{y}}\frac{\partial}{\partial y} + \hat{\mathbf{z}}\frac{\partial}{\partial z} \tag{A.1}$$

where $\hat{\mathbf{x}}$, $\hat{\mathbf{y}}$, $\hat{\mathbf{z}}$ are the orthogonal unit vectors in the Cartesian system of coordinates. This is merely an operator that applies to other functions. By itself it has no geometrical meaning.

A.1.2. The Gradient

The gradient of a scalar function $U(x,y,z)$ is a vector, perpendicular to a surface on which U is constant and that it points in the direction of increasing U. ∇U points in the direction of maximum rate of increase in U. The function U has the following derivatives in Cartesian coordinates

$$\frac{\partial U}{\partial x}, \quad \frac{\partial U}{\partial y}, \quad \frac{\partial U}{\partial z}$$

The gradient of U is

$$\nabla U = \left(\hat{\mathbf{x}}\frac{\partial}{\partial x} + \hat{\mathbf{y}}\frac{\partial}{\partial y} + \hat{\mathbf{z}}\frac{\partial}{\partial z}\right)U = \hat{\mathbf{x}}\frac{\partial U}{\partial x} + \hat{\mathbf{y}}\frac{\partial U}{\partial y} + \hat{\mathbf{z}}\frac{\partial U}{\partial z} \tag{A.2}$$

A.1.3. The Divergence

The divergence of a vector function **A**, at a point, is defined as the net outward flux of the function **A** per unit volume as the volume tends to zero:

$$\text{div}\mathbf{A} = \nabla.\mathbf{A} = \lim_{\Delta v \to 0} \frac{\oint_S \mathbf{A}.\mathbf{ds}}{\Delta v} \qquad (A.3)$$

We can write this directly using the del operator in Eq. (A.1) as

$$div\ \mathbf{A} = \nabla\cdot\mathbf{A} = \left(\hat{\mathbf{i}}\frac{\partial}{\partial x} + \hat{\mathbf{j}}\frac{\partial}{\partial y} + \hat{\mathbf{k}}\frac{\partial}{\partial z}\right)\cdot\left(\hat{\mathbf{i}}A_x + \hat{\mathbf{j}}A_y + \hat{\mathbf{k}}A_z\right) \qquad (A.4)$$

or

$$\nabla\cdot\mathbf{A} = \frac{\partial A_x}{\partial x} + \frac{\partial A_y}{\partial y} + \frac{\partial A_z}{\partial z} \qquad (A.5)$$

The divergence of a vector is a scalar and it can be viewed as a measure of the source enclosed in the volume Δv. If the volume Δv does not enclose a source, the divergence at that point is zero.

A.1.4. The Curl

The curl of a vector field **A** is defined as a vector whose magnitude is the maximum circulation of the vector **A** per unit area Δs as this area tends to zero. The direction of the curl is normal to the area Δs while the area is oriented such that the circulation is maximum:

$$curl\mathbf{A} = \nabla\times\mathbf{A} = lim_{\Delta s \to 0} \frac{\left[\oint_c \mathbf{A}\cdot\mathbf{dl}\right]_{max}}{\Delta s} \qquad (A.6)$$

Again, using the cross product between ∇ and **A**, we get

$$\nabla\times\mathbf{A} = det\begin{bmatrix} \hat{\mathbf{x}} & \hat{\mathbf{y}} & \hat{\mathbf{z}} \\ \frac{\partial}{\partial x} & \frac{\partial}{\partial y} & \frac{\partial}{\partial z} \\ A_x & A_y & A_z \end{bmatrix} \qquad (A.7)$$

or

$$\nabla\times\mathbf{A} = \hat{\mathbf{x}}\left(\frac{\partial A_z}{\partial y} - \frac{\partial A_y}{\partial z}\right) + \hat{\mathbf{y}}\left(\frac{\partial A_x}{\partial z} - \frac{\partial A_z}{\partial x}\right) + \hat{\mathbf{z}}\left(\frac{\partial A_y}{\partial x} - \frac{\partial A_x}{\partial y}\right) \tag{A.8}$$

A.2. Vector Theorems

A.2.1. The Divergence Theorem

The divergence theorem states that the volume integral of a vector field is equal to the total outward flux of the vector through the closed surface that bounds the volume.

$$\int_V (\nabla\cdot\mathbf{A})dv = \oint_S \mathbf{A}\cdot d\mathbf{s} \tag{A.9}$$

We will not prove this relation here but merely point out that it is a very useful relation that allows conversion of volume to surface integration (and vice versa) and, in a more physical sense, connects the vector on the surface of a volume to the vector inside the volume. This is useful in calculations and measurements such as power relations.

A.2.2. Stokes' Theorem

Stokes' theorem states that the open surface integral of the curl of a vector field is equal to the closed line integral of the vector over the contour bounding the surface.

$$\int_S (\nabla\times\mathbf{A})\cdot d\mathbf{s} = \oint_C \mathbf{A}\cdot d\mathbf{l} \tag{A.10}$$

Similar to the divergence theorem, Stokes' theorem allows transformation of open surface integrals to closed contour integrals. This is useful in a variety of instances such as in calculation of potential.

A.2.3. Helmholtz's Theorem

Helmholtz's theorem states that a vector field is completely determined (to within an added constant) if its curl and divergence are specified everywhere.

Thus, a general vector can always be separated into a curl free (irrotational) field and a divergenceless (solenoidal) field. A solenoidal field has zero divergence but nonzero curl while an irrotational field has zero curl but nonzero divergence. Since a solenoidal field can always be written as the curl of another vector and a solenoidal field can be written as the gradient of a scalar function, a general field E, can be written as

$$\mathbf{E} = -\nabla V + \nabla\times\mathbf{A} \tag{A.11}$$

A.3. Vector Identities

The following are some of the more commonly used vector identities. In the following *U*, *V* are scalars, **A**, **B**, are vectors.

$$\nabla\cdot(\nabla U) = \nabla^2 U \tag{A.12}$$

$$\nabla\times(\nabla U) = 0 \tag{A.13}$$

$$\nabla\cdot(\nabla\times\mathbf{A}) = 0 \tag{A.14}$$

$$\nabla\times(\nabla\times\mathbf{A}) = \nabla(\nabla.\mathbf{A}) - \nabla^2\mathbf{A} \tag{A.15}$$

$$\nabla(UV) = U(\nabla V) + V(\nabla U) \tag{A.16}$$

$$\nabla\cdot(U\mathbf{A}) = U(\nabla\cdot\mathbf{A}) + (\nabla U)\cdot\mathbf{A} \tag{A.17}$$

$$\nabla\cdot(\mathbf{A}\times\mathbf{B}) = -\mathbf{A}\cdot(\nabla\times\mathbf{B}) + (\nabla\times\mathbf{A})\cdot\mathbf{B} \tag{A.18}$$

$$\nabla\times(U\mathbf{A}) = U(\nabla\times\mathbf{A}) + (\nabla U)\times\mathbf{A} \tag{A.19}$$

A.4. The Laplacian

The Laplacian in Cartesian coordinates is defined by the following relation

$$\nabla^2 U = \nabla\cdot(\nabla U) = \left(\hat{\mathbf{x}}\frac{\partial}{\partial x} + \hat{\mathbf{y}}\frac{\partial}{\partial y} + \hat{\mathbf{z}}\frac{\partial}{\partial z}\right)\cdot\left(\hat{\mathbf{x}}\frac{\partial U}{\partial x} + \hat{\mathbf{y}}\frac{\partial U}{\partial y} + \hat{\mathbf{z}}\frac{\partial U}{\partial z}\right) \tag{A.20}$$

or

$$\nabla^2 U = \frac{\partial^2 U}{\partial x^2} + \frac{\partial^2 U}{\partial y^2} + \frac{\partial^2 U}{\partial z^2} \tag{A.21}$$

It is also possible to define the Laplacian of a vector as

$$\nabla^2\mathbf{A} = \nabla(\nabla\cdot\mathbf{A}) - \nabla\times(\nabla\times\mathbf{A}) \tag{A.22}$$

where $\nabla^2\mathbf{A}$ is called the "vector Laplacian" of **A**. This is written as

$$\nabla^2\mathbf{A} = \hat{\mathbf{x}}\nabla^2 A_x + \hat{\mathbf{y}}\nabla^2 A_y + \hat{\mathbf{z}}\nabla^2 A_z \tag{A.23}$$

where, for example, the component in the Ox direction is

$$\nabla^2 A_x = \frac{\partial^2 A_x}{\partial x^2} + \frac{\partial^2 A_x}{\partial y^2} + \frac{\partial^2 A_x}{\partial z^2}$$

A.5. Expressions in Cartesian, Cylindrical and Spherical Coordinates

The expressions below give the gradient, divergence, curl, and scalar Laplacian in Cartesian, cylindrical and spherical coordinates. A vector function **A** and a scalar function U are used.

In Cartesian coordinates x, y, z:

$$\nabla U = \hat{\mathbf{x}}\frac{\partial U}{\partial x} + \hat{\mathbf{y}}\frac{\partial U}{\partial y} + \hat{\mathbf{z}}\frac{\partial U}{\partial z} \tag{A.23}$$

$$\nabla\cdot\mathbf{A} = \frac{\partial A_x}{\partial x} + \frac{\partial A_y}{\partial y} + \frac{\partial A_z}{\partial z} \tag{A.24}$$

$$\nabla\times\mathbf{A} = \hat{\mathbf{x}}\left(\frac{\partial A_z}{\partial y} - \frac{\partial A_y}{\partial z}\right) + \hat{\mathbf{y}}\left(\frac{\partial A_x}{\partial z} - \frac{\partial A_z}{\partial x}\right) + \hat{\mathbf{z}}\left(\frac{\partial A_y}{\partial x} - \frac{\partial A_x}{\partial y}\right) \tag{A.25}$$

$$\nabla^2 U = \frac{\partial^2 U}{\partial x^2} + \frac{\partial^2 U}{\partial y^2} + \frac{\partial^2 U}{\partial z^2} \tag{A.26}$$

In cylindrical coordinates r, ϕ, z:

$$\nabla U = \hat{\mathbf{r}}\frac{\partial U}{\partial r} + \hat{\boldsymbol{\phi}}\frac{1}{r}\frac{\partial U}{\partial \phi} + \hat{\mathbf{z}}\frac{\partial U}{\partial z} \tag{A.27}$$

$$\nabla\cdot\mathbf{A} = \frac{1}{r}\frac{\partial}{\partial r}(rA_r) + \frac{1}{r}\frac{\partial A_\phi}{\partial \phi} + \frac{\partial A_z}{\partial z} \tag{A.28}$$

$$\nabla\times\mathbf{A} = \hat{\mathbf{r}}\left(\frac{1}{r}\frac{\partial A_z}{\partial \phi} - \frac{\partial A_\phi}{\partial z}\right) + \hat{\boldsymbol{\phi}}\left(\frac{\partial A_r}{\partial z} - \frac{\partial A_z}{\partial r}\right) + \hat{\mathbf{z}}\left(\frac{1}{r}\frac{\partial (rA_\phi)}{\partial r} - \frac{1}{r}\frac{\partial A_r}{\partial \phi}\right) \tag{A.29}$$

$$\nabla^2 U = \frac{\partial^2 U}{\partial r^2} + \frac{1}{r}\frac{\partial U}{\partial r} + \frac{1}{r^2}\frac{\partial^2 U}{\partial \phi^2} + \frac{\partial^2 U}{\partial z^2} \tag{A.30}$$

In spherical coordinates R, θ, ϕ:

$$\nabla U = \hat{\mathbf{R}}\frac{\partial U}{\partial R} + \hat{\boldsymbol{\theta}}\frac{1}{R}\frac{\partial U}{\partial \theta} + \hat{\boldsymbol{\phi}}\frac{1}{R\sin\theta}\frac{\partial U}{\partial \phi} \tag{A.31}$$

$$\nabla \cdot \mathbf{A} = \frac{1}{R^2}\frac{\partial}{\partial R}(R^2 A_R) + \frac{1}{R\sin\theta}\frac{\partial}{\partial \theta}(A_\theta \sin\theta) + \frac{1}{R\sin\theta}\frac{\partial A_\phi}{\partial \phi} \tag{A.32}$$

$$\nabla \times \mathbf{A} = \hat{\mathbf{R}}\frac{1}{R\sin\theta}\left(\frac{\partial}{\partial \theta}(A_\phi \sin\theta) - \frac{\partial A_\theta}{\partial \phi}\right) + \hat{\boldsymbol{\theta}}\frac{1}{R}\left(\frac{1}{\sin\theta}\frac{\partial A_R}{\partial \phi} - \frac{\partial}{\partial R}(RA_\phi)\right)$$

$$+\hat{\boldsymbol{\phi}}\frac{1}{R}\left(\frac{\partial(RA_\theta)}{\partial R} - \frac{\partial A_R}{\partial \theta}\right) \tag{A.33}$$

$$\nabla^2 U = \frac{1}{R}\frac{\partial^2 (RU)}{\partial R^2} + \frac{1}{R^2\sin^2\theta}\frac{\partial^2 U}{\partial \phi^2} + \frac{1}{R^2\sin\theta}\frac{\partial}{\partial \theta}\left(\sin\theta\frac{\partial U}{\partial \theta}\right) \tag{A.34}$$

Subject Index

DEVELOPMENTS IN ELECTROMAGNETIC THEORY AND APPLICATIONS

1. J. Caldwell and R. Bradley (eds.): *Industrial Electromagnetics Modelling.* 1983 ISBN 90-247-2889-4
2. H.G. Booker: *Cold Plasma Waves.* 1984 ISBN 90-247-2977-7
3. J. Lekner: *Theory of Reflection of Electromagnetic and Particle Waves.* 1987 ISBN 90-247-3418-5
4. A.E. Lifschitz: *Magnetohydrodynamics and Spectral Theory.* 1989 ISBN 90-247-3713-3
5. H.C.K. Rawer: *Waves in Ionized Media.* (forthcoming) ISBN 0-7923-0775-5
6. *Cancelled*
7. K.I. Hopcraft and P.R. Smith: *An Introduction to Electromagnetic Inverse Scattering.* 1992 ISBN 0-7923-0777-1
8. *Cancelled*
9. C. Altman and K. Suchy: *Reciprocity, Spatial Mapping and Time Reversal in Electromagnetics.* 1991 ISBN 0-7923-1339-9
10. N. Ida: Microwave NDT. 1992 ISBN 0-7923-2007-7

KLUWER ACADEMIC PUBLISHERS – DORDRECHT / BOSTON / LONDON

9 780792 320074

GPSR Compliance

The European Union's (EU) General Product Safety Regulation (GPSR) is a set of rules that requires consumer products to be safe and our obligations to ensure this.

If you have any concerns about our products, you can contact us on ProductSafety@springernature.com

In case Publisher is established outside the EU, the EU authorized representative is:

Springer Nature Customer Service Center GmbH
Europaplatz 3
69115 Heidelberg, Germany

Batch number: 10809439

Printed by Printforce, the Netherlands